The Value of Water: Concepts, Estimates, and Applications for Water Managers

Awwa Research Foundation

About the Awwa Research Foundation

The Awwa Research Foundation (AwwaRF) is a member-supported, international, nonprofit organization that sponsors research to enable water utilities, public health agencies, and other professionals to provide safe and affordable drinking water to consumers.

The Foundation's mission is to advance the science of water to improve the quality of life. To achieve this mission, the Foundation sponsors studies on all aspects of drinking water, including supply and resources, treatment, monitoring and analysis, distribution, management, and health effects. Funding for research is provided primarily by subscription payments from approximately 1,000 utilities, consulting firms, and manufacturers in North America and abroad. Additional funding comes from collaborative partnerships with other national and international organizations, allowing for resources to be leveraged, expertise to be shared, and broad-based knowledge to be developed and disseminated. Government funding serves as a third source of research dollars.

From its headquarters in Denver, Colorado, the Foundation's staff directs and supports the efforts of more than 800 volunteers who serve on the board of trustees and various committees. These volunteers represent many facets of the water industry, and contribute their expertise to select and monitor research studies that benefit the entire drinking water community.

The results of research are disseminated through a number of channels, including reports, the Web site, conferences, and periodicals.

For subscribers, the Foundation serves as a cooperative program in which water suppliers unite to pool their resources. By applying Foundation research findings, these water suppliers can save substantial costs and stay on the leading edge of drinking water science and technology. Since its inception, AwwaRF has supplied the water community with more than $300 million in applied research.

More information about the Foundation and how to become a subscriber is available on the Web at **www.awwarf.org**.

The Value of Water: Concepts, Estimates, and Applications for Water Managers

Prepared by:
Robert S. Raucher, **David Chapman**, **James Henderson**, **Marca L. Hagenstad**, and **John Rice**
Stratus Consulting Inc.
P.O. Box 4059, Boulder, CO 80306-4059

James Goldstein and **Annette Huber-Lee**
Tellus Institute/Stockholm Environmental Institute

William DeOreo and **Peter Mayer**
Aquacraft, Inc.

Brian Hurd (New Mexico State University)

Ron Linsky (National Water Research Institute)

Ed Means (McGuire Environmental Consultants)

and

Mary Renwick (University of Minnesota)

Sponsored by:
Awwa Research Foundation
6666 West Quincy Avenue, Denver, CO 80235-3098

Published by:

Awwa Research Foundation

Advancing the Science of Water®

DISCLAIMER

CONTENTS

TABLES

FIGURES

FOREWORD

The Awwa Research Foundation is a nonprofit corporation that is dedicated to the implementation of a research effort to help utilities respond to regulatory requirements and traditional high-priority concerns of the industry. The research agenda is developed through a process of consultation with subscribers and drinking water professionals. Under the umbrella of a Strategic Research Plan, the Research Advisory Council prioritizes the suggested projects based upon current and future needs, applicability, and past work; the recommendations are forwarded to the Board of Trustees for final selection. The foundation also sponsors research projects through the unsolicited proposal process; the Collaborative Research, Research Applications, and Tailored collaboration programs; and various joint research efforts with organizations such as the U.S. Environmental Protection Agency, the U.S. Bureau of Reclamation, and the Association of California Water Agencies.

This publication is a result of one of these sponsored studies, and it is hoped that its finding will be applied in communities throughout the world. The following report serves not only as a means of communicating the results of the water industry's centralized research program but also as a tool to enlist the further support of the nonmember utilities and individuals.

Projects are managed closely from their inception to the final report by the foundation's staff and large cadre of volunteers who willingly contribute their time and expertise. The foundation serves a planning and management function, and awards contracts to other institutions such as water utilities, universities, and engineering firms. The funding for this research effort comes primarily from the Subscription Program, through which water utilities subscribe to the research program and make an annual payment proportionate to the volume of water they deliver and consultants and manufacturers subscribe based on their annual billings. The program offers a cost-effective and fair method for funding research in the public interest.

A broad spectrum of water supply issues is addressed by the foundation's research agenda: resources, treatment and operations, distribution and storage, water quality and analysis, toxicology, economics, and management. The ultimate purpose of the coordinated effort is to help water suppliers provide the highest possible quality of water economically and reliably. The true benefits are realized when the results are implemented at the utility level. The foundation's trustees are pleased to offer this publication as a contribution toward that end.

This report on the value of water covers a broad range of concepts, empirical measures, and possible applications to water management decisions. The objective is to help water professionals identify and promote water resource strategies or options that provide the greatest amount of total well-being for all members of society combined, applying water resources in a manner that promotes sustainability and the highest and best uses of the resource. The value of water depends to a large extent on several factors, including how the water is used, the quality of the water, the time and location at which it is available, its relative scarcity, and other attributes. For example, water can be extracted from rivers or aquifers for use by the residential and commercial customers of municipal or investor-owned community water systems (CWSs). Alternatively, the same water could be extracted for irrigation by farmers. Or, the water can be left instream to support recreation (e.g., fishing and boating) and ecological purposes (e.g., providing critical habitat for fish and other species, some of which may be designated as threatened or endangered). In any of these possible competing uses, the water would generate values. The challenge for water resource professionals is to understand which uses promote the highest overall values to society. This requires not only an understanding of the relative magnitude of the various possible values associated with the competing use options but also insight on who will realize the benefits and who will bear the costs of a given water management decision.

Walter J. Bishop
Chair, Board of Trustees
Awwa Research Foundation

James F. Manwaring, P.E.
Executive Director
Awwa Research Foundation

ACKNOWLEDGMENTS

This report was funded by the Awwa Research Foundation (AwwaRF). The project was managed and administered by Mr. Frank Blaha. The researchers extend their great appreciation to AwwaRF, and to Frank Blaha for his management, patience, and encouragement. We also wish to thank the project advisory committee (PAC), which provided many very useful comments, guidance, support, and critiques of this project. The PAC helped shape and improve this effort. The PAC members were:

Sandra Archibald (University of Washington)
David Hanson (Pinellas County Utilities)
Bruce Macler (U.S. EPA, Region IX)
Bret Weingart (Water and Wastewater Utilities Department, City of Oklahoma City)

The researchers also wish to acknowledge the useful input and support provided by several individuals who helped write and/or review several portions of this work at various stages of its development, including all the participants at the NGT Workshop, held in Pomona, Calif., on September 23-25, 2003. Thanks are also due to Ron Linsky and his associates at the National Water Research Institute for their excellent work in helping to organize, host, and facilitate the workshop.

In addition, we would like to thank the following participating utilities and organizations, and individuals from these organizations who made key contributions: Cobb County-Marietta Water Authority (Rob Fowler), Denver Water Department (David LaFrance), El Paso Water Utilities (especially Nick Costanzo and colleagues), Lehigh County Water Authority (Aurel Arndt), Massachusetts Water Resources Agency (Steven Estes-Smargiassi), Passaic Valley Water Authority (Joe Bella), Philadelphia Water Department (George Kunkel), City of Phoenix Water Services Department (Steve Rossi), Pinellas County Utilities (Todd Tanberg and Wayne West), San Diego County Water Authority (especially Maria Mariscal), San Francisco Public Utilities Commission (Cheryl Davis and Andrew DeGraca), and Water Replenishment District of Southern California (Jason Weeks and Ted Johnson).

It also is important to credit individuals from the research team for their specific key contributions. The following members of the research team made notable contributions in the following portions of this report: David Chapman (chapters 4 and 5), Bill DeOreo and Peter Mayer (North American section of chapter 3, the commercial, institutional, and industrial (CII) portion of chapter 5, and appendix E), James Goldstein (appendix D), Marca Hagenstad (in-stream portions of chapter 5, and appendix F), Jim Henderson (much of chapters 7 and 8, and appendices H and I), Annette Huber-Lee (global portion of chapter 3 and appendix C), Brian Hurd (the agricultural water values text in chapters 4 and 5, and appendix A), Ed Means (chapter 6), Mary Renwick (residential demand and price elasticity section of chapter 5), John Rice (water rights section of chapter 2, reliability value section of chapter 5), and John Whitcomb (appendix G).

Last, but not least, the researchers wish to thank the many individuals at Stratus Consulting who helped execute the technical tasks and produce this report, especially Diane Callow, Erin Miles, Shiela DeMars, Glenda Mostek, and Christina Thomas.

Thank you all very much.

Robert S. Raucher, PhD (Principal Investigator)
Stratus Consulting Inc.
Boulder, Colo. (December 2004)

EXECUTIVE SUMMARY

This report on the value of water covers a broad range of concepts, empirical measures, and possible applications to water management decisions. Most people understand that water is essential to life and livelihoods and, accordingly, that water has a very high value. However, there is ambiguity and confusion about what the term "value" really means, how it can be measured for water, and why something as valuable and essential to life as water can be purchased at the tap at a price of less than a penny per gallon.

This report attempts to shed light on these and related questions, and to provide insights to water sector professionals on how the concepts and measures of value can be applied to help inform and improve water resource management decisions. The ultimate objective of this report is to help water professionals identify and promote water resource strategies or options that provide the greatest amount of total well-being for all members of society combined. This is what economists refer to as trying to "maximize social welfare." This entails applying water resources in a manner that promotes sustainability and the highest and best uses of the resource.

Part of the challenge in considering or using the value of water is the confusion that arises due to the oft-interchanged use of three related but very distinct terms: cost, price, and value. Each of these terms has a specific and distinct meaning, yet they are often used interchangeably in a manner that fosters unintended confusion. Cost refers to the expense of producing and delivering a unit of water. Price refers to the rate charged to a customer for the unit of water delivered. Usually prices (rates) are based on costs, since utilities strive to recover their capital and operating expenses (such as by developing rates based on cost of service). Utilities incur capital and operating costs, and set and apply rates; therefore these costs and prices are easily identified and measured. Value, on the other hand, is a more ambiguous concept and one that is harder to measure.

Economists define value as the worth of the commodity or service as measured by the maximum willingness to pay (WTP), in terms of money or other goods and services that a person would exchange for an additional unit of the good or service. The WTP concept is what defines the "demand curve" for water (i.e., the demand curve reflects the WTP-based value of water to its users), whereas the supply curve reflects the costs to the utility of providing that water.

The value of water depends to a large extent on several factors, including how the water is used, the quality of the water, the time and location at which it is available, its relative scarcity, and other attributes. To further complicate matters, different individuals hold different values, so that different individuals might choose to have society use its water resources in different ways. Therefore, the issue for water managers is not always limited to trying to understand the magnitudes of the various values, but also is in some cases a matter of deciding whose values will take prominence (i.e., who will be the winners and who will be the losers in a water decision).

For example, water can be extracted from rivers or aquifers for use by the residential and commercial customers of municipal or investor-owned community water systems (CWSs). Alternatively, the same water could be extracted for irrigation by farmers. Or, the water can be left instream to support recreation (e.g., fishing and boating) and ecological purposes (e.g., providing critical habitat for fish and other species, some of which may be designated as threatened or endangered). In any of these possible competing uses, the water would generate values. The challenge for water resource professionals is to understand which use promotes the highest overall values to society. This requires a need to not just understand the magnitude of the various possible values from competing use options, but also insights on who gains and who losses (i.e., who benefits and who bears the costs) from a water allocation or other water-impacting management decision.

There are many different types and sources of value for water. Some values can be deduced from the marketplace because the water contributes to market-traded goods and services. For example, water applied by agricultural irrigators can be valued based on the contribution that the water makes to increased crop yields, and the yields in turn can be valued based on market prices for the foodstuffs grown. Yet even in these applications, the presence of subsidies and other policies distorts the economic value signals provided by the market. Other water values pertain to "nonmarket" goods and services, such as most typically apply to instream uses of water. In such cases, there are various stated and revealed preference methods that economists have used to estimate the value of water. For residential customers of public water systems, there is extensive available published research on the price elasticity of demand for water, but these data have limited relevance for discerning household WTP for water in different quantities, of various qualities, or for various types of residential use.

In the context of public water supply system deliveries to a residential customer, value refers to how much the household would be willing to pay for an additional thousand gallons (or to avoid having the quantity of water they currently use reduced by a thousand gallons). A household's WTP will depend on whether one is considering the water used for essential purposes such as drinking, cooking, and bathing (i.e., uses that people value highly), or if the water is used for more discretionary purposes (e.g., low valued applications such as hosing off the sidewalk). Household WTP for water is not a measure that is readily observed or measured, especially in much of North America where water typically is available at such a low cost that customers tend to use water in some low value applications (e.g., inefficient turf irrigation practices). This means that what we can observe of household values pertains to water purchased at the margin for low-end uses.

Despite the challenges in measuring values for water in different uses, the available empirical research literature does provide some useful insights into the likely general magnitude of water values:

- For example, for water used for agricultural irrigation, empirical investigations suggest a mean value of $21 and $837 per acre foot (AF) for hops and potatoes, respectively (note that these and all other monetary values reported here are in 2003 U.S. dollars, unless stated otherwise). This translates to $0.08 to $2.59 per thousand gallons (an AF equals 325,850 gallons). It is important to observe that there can be considerable variation in these values, based on location, timing, and other factors. For example, the mean value of $837/AF for potatoes is drawn from four available empirical estimates that range from $54 to $1,447/AF.
- In contrast, the value of water for residential customers of municipal systems may be on the order of $1,400 to $2,300/AF, although these estimates are based on limited data and studies, and reflect only values at the margin for current use levels.
- Households also appear to place relatively high values on the reliability of water supplies. WTP estimates to avoid drought-related or other restrictions on water use have been estimated to be $109 to $421/household per year (note that these are not

$/AF estimates, but $/household per year). These estimates suggest reliability values for residential customers may be on the order of $4,000 per AF, or higher

- Limited data exist on the value of water for commercial, institutional, and industrial (CII) purposes, with value estimates in the range of $28 to $804/AF.
- Instream uses of water can provide considerable value. Some recreation-based estimates imply a value of $10 to $770/AF. Other recreational and nonuse values for instream uses of water have been developed, but do not always translate directly into meaningful dollar/AF measures. For example, the whitewater rafting enabled by flows in Gore Creek (Vail, Colo.) generates a value of over $1 million per year to boaters and local economies. In peak use summer periods, this implies a value of over $1,600/AF. Several studies of instream flows needed to preserve threatened or endangered species suggest that households region- or state-wide may have an average WTP of $10 or more per year (implying, for example, that if there are one million households in the relevant region, the total value would be $10 million per year).

In reviewing the empirical estimates of value summarized above (and as presented elsewhere in this report), readers are cautioned that the estimated values may be very sensitive to the specific circumstances (e.g., location, uses, time periods) from which they were estimated. There may also be data and methodological limitations associated with the empirical estimates, such that the values may be unreliable, even for the circumstances from which they were developed. Therefore, using these values in a "transfer" to other circumstances should be done with caution and may introduce considerable error.

In addition to the values described above, there are other sources and types of value that water generates. For example, utility-provision of reliable quantities of high quality and reasonably priced waters can be instrumental to sustaining the economic base of a community or region. Water also is a valuable tool for managing growth in ways that are consistent with the community's broader objectives. There also are important cultural and spiritual values associated with water, and aesthetic values associated with water used to provide green space, fountains, and other amenities. One of the challenges faced by water professionals is to identify and characterize the broad range of values that apply to water, and to do so using a non-technical language that effectively resonates with laypersons, stakeholders, and public officials.

There are various factors that may alter water values in the future. These include potential changes in climate and technology, increased populations, and other factors that may alter the supply or demand for water, or both. Water professionals will need to consider these potential future events and trends when making long-term projections or making long-lived infrastructure and water supply decisions that will affect their communities in the decades ahead.

To help water professions use water values to enlighten and guide their planning processes and decisions, this report also provides a series of templates. These forms break down an economic analysis into several logical and manageable steps, and provide a useful organizational framework to help water professionals conduct the analyses. These templates also provide a valuable way to document the work performed and enhance replicability and transparency. These templates also provide a means to help communicate the analyses and findings to utility managers, public officials, and stakeholders.

Several case studies are provided to reveal various issues associated with the value of water, and to indicate how these values (or value-based concepts) can be used to help inform water management decisions. These case studies cover a broad range of water management issues and locations across the United States. Included is an illustration of the "triple bottom line" approach that can help portray the full range of values a water supply agency provides to the community it serves. Other case studies explore value issues for instream and extractive uses, and reveal how using water value concepts and measures could help promote sustainable management of the water resources involved.

While this report provides a fairly exhaustive review of the conceptual and empirical literature available on the value of water, this research also reveals several important gaps that need to be addressed through future research. Included in the agenda for future research are more and better efforts to estimate the value of water to the residential and CII customers of water supply agencies. As increased water scarcity drives the cost of water provision upward in coming years, it will become increasingly important to understand what water uses are most valued and which values outweigh the anticipated escalation in costs.

CHAPTER 1
INTRODUCTION

"When the well's dry, we know the worth of water." (Benjamin Franklin, *Poor Richard's Almanac*, 1746)

VALUING AN ESSENTIAL RESOURCE

Water is essential to life, the basis for livelihoods, and a source of spiritual renewal and cultural identity. Without water, the earth would be barren and devoid of life (as Mars is, now that water is only a remnant of that planet's distant past). Absent sufficient quantities of water, farms cannot yield adequate food supplies, wildlife species teeter on the brink of extinction, businesses close or relocate, forests become prone to wildfire and disease, and yards and gardens wither and die. The careers of urban water supply managers can be affected by the extent to which they succeed in providing enough water to avert water use restrictions in periods of drought or rapid demand growth.

If water is essential to life, does this mean that water has infinite value? If water is so clearly linked to our economic and social welfare, then how can water be priced so much lower than gold or diamonds? Gems and precious metals, while aesthetically pleasing and of some economic use in various applications, ultimately have no bearing on our ability to sustain life.

This report provides some insights on these questions by examining the value of water. The research team also explores how water values can be used in water resource management contexts to enhance decision-making and improve social welfare.

KEY ATTRIBUTES AFFECT THE VALUE OF WATER

Many people have an intuitive appreciation that water has considerable "value." Most people also recognize that the interaction of timing, location, quality, and end use is integral to assessing the value of water. Thus, water professionals, and the general public, are well aware that not all waters are of equal value. There is broad (though often implicit) recognition that the value of water depends on those key attributes or characteristics that are linked with a given unit of water. For example, the value of water varies considerably across types and locations of use. A cup of freshwater to a person stranded in a desert has more value than a gallon of runoff in a rain-swollen river. The value of water thus depends in part on the relative scarcity of water in a location (defined by supply relative to demand) and on the value of the activities that the water use supports (e.g., the market value of additional crop yields enabled by irrigation, or the added well-being of a community that has enough municipal water to support desired activities such as lawn watering and job creation).

Water quality, timing, and reliability are other attributes that play a large role in determining the value of water. A gallon of high quality water that keeps a human being alive and healthy may be "priceless," whereas the same gallon spread on the sidewalk from a misplaced lawn sprinkler may have no value whatsoever (and a gallon of untreated wastewater may have negative value because of the harm it causes when released into the environment). Likewise, an extra hundred acre-feet of water left in a river at the right time and place may preserve an endangered aquatic species and enable downstream farmers to profitably grow their crops, but if that same water were added to a flood-swollen river it might imperil lives and inflict considerable property damage.

The value of water also can be gleaned where waters are subject to limited market-based trading. But even where water markets exist, they are not fully functioning "perfectly competitive" markets of the type required to provide fully meaningful measures of value. Market-based price signals may not reflect values for water because of many possible factors. The presence of externalities and/or public good aspects implies that there are values beyond the signals reflected in the market; and transaction costs and legal or regulatory systems may preclude or limit transactions. Therefore, it often is difficult to quantify how much value water has in a given application, or how the value derived from one water use compares to the value that would be derived from a potential competing use. It is difficult to ascertain the true value of water in quantitative, monetized terms, and it also can be difficult to fully characterize the value of water in relative or *qualitative* terms.

THE CHALLENGE FACED BY WATER RESOURCE MANAGERS

Three-quarters of the world's surface is water, and water constitutes 90% of the body mass of a typical human. Despite this apparent abundance, high quality freshwater supplies—as suitable for potable and other high end uses—are becoming increasingly scarce. Given the pressures of supply and demand, many water utilities may no longer be able provide ample quantities of exceptional quality water at the relatively low prices to which most North Americans have become accustomed. At the same time, regulatory demands and customer quality expectations are on the increase, and security and infrastructure financing needs are putting additional demands on water system managers. The challenge for water professionals is to meet increased demands within the realities and constraints of regional water supply options, and yet help foster the type of social, ecological, and economic climate a utility's service community needs and desires.

Given the challenges of measuring the value of water, it is hard to ensure that this precious resource is being allocated to competing (and complementary) uses in a manner that provides society the highest possible level of value or well-being (what economists refer to as "maximizing social welfare"). As stewards of this essential resource, utility managers and other leaders in the water services profession face critical challenges. Water utility managers need to deliver a product that protects public health while providing ample supplies at a cost that is affordable to low or fixed income households, and at rates low enough to sustain (or stimulate) their communities' economic prosperity. At the same time, utility managers must be mindful of how their decisions about acquiring and diverting water may affect broader social objectives, including values associated with instream ecological conditions, the economic viability of irrigated family farms in the region, or the long-term sustainability of groundwater extraction.

Accordingly, the objective of this research project is to provide utility managers with information, guidance, and practical tools to assist them as they face the emerging issues associated with providing cost-effective and high quality water service within the context of their communities' broader economic realities, environmental concerns, water resource constraints, and social objectives. Increasingly, water managers will need to integrate water values into their deliberations on supply options and management strategies. The value of water will become increasingly important to managers as they attempt to accommodate increased scarcity and cost into their communities' broader efforts to manage growth and development, promote sustainability, enhance equity, and balance competing demands.

CORE QUESTIONS ASSOCIATED WITH THE VALUE OF WATER

Because this research project is about the value of water, this report addresses the following questions:

1. What does the term "value" mean—how is it defined?
2. What are the different types of value that apply to water?
3. How is water used around the globe—and in North America—and what might this suggest about values?
4. Who "owns" water resources, what mechanisms are used to assign rights and uses for water, and what are the implications for water values (and prices) in the United States?
5. What methods are available for estimating water values?
6. What does the available empirical evidence tell us about the likely or potential values for water in different uses or contexts?
7. How might emerging trends and probable future scenarios affect the value of water?
8. How should water value information be portrayed and communicated to policy-makers, stakeholders, and the general public?
9. How should water values be used to make better informed (and welfare enhancing) water resource management decisions?
10. What do various real world case studies reveal about how one might use value information as a tool to help promote sustainable water resource management?
11. What future research might be most helpful for improving the estimation and use of water values?

WHY VALUES MATTER FOR WATER MANAGEMENT

More fundamental than the list of queries posed above is the question: "Why should water resource professionals care about the value of water?" Or, perhaps more to the point, the core issue might best be motivated by the question: "What does one hope to accomplish by using value information in water resource decision-making?"

Ultimately, the answer to the latter question—and hence the main objective of the materials provided in this report—is to help identify and promote water resource management strategies (or options) that provide for the greatest amount of total well-being for all members of society combined. This is what economists refer to as trying to maximize social welfare. This can be achieved, for example, by comparing benefits to costs and attempting to determine what resource choice or policy provides the greatest net benefit to the community as a whole. And, because this objective includes the important notion of weighing tradeoffs between the future and the present, this is also what water professionals would call promoting sustainable water resource management.

Therefore, this report is intended to provide utility managers and other water professionals with information and tools that can help them perform more informative and complete analyses of the options before them. The guiding principle is to help the decision-makers and other involved parties manage water in a manner that does the most total good possible, across all the parties affected (present and future). This implies a need to take into account all the different types of values that are derived from water, including tradeoffs that inevitably arise when some uses and users are winners and others are losers, with the goal of generating the greatest total value possible.

THE OUTLINE OF THIS RESEARCH REPORT

The questions noted above are addressed through a series of chapters and appendices that constitute the balance of this research report. An overview, or roadmap, of the report is as follows:

Chapter 2 discusses what is meant by the value of water, by providing a broad view of the sources of value that various people and organizations may attribute to water. This is a nontechnical discussion, developed with the intent of helping water resource managers gain a better grasp of how their customers, constituencies, stakeholders, and the public at large may perceive water and its values.

Chapter 3 examines how water is used around the globe, and within North America. The overview provides insights on different levels of water extraction and application by geopolitical region (e.g., by continent) and across economic sectors (e.g., agriculture versus power generation versus municipal and industrial use). The components of residential water use in the United States are also summarized.

Chapter 4 delves into the technical specifics of value—using the concepts, terminology, and methods of the economics profession to define what "value" means, delineate different types and sources of value for water, describe what affects water values, and summarize empirical methods available for estimating value.

Chapter 5 summarizes the empirical literature relating to what types and levels of values have been estimated for water, by type of use or value.

Chapter 6 explores how the value of water may be affected over time, based on emerging trends and potential future scenarios.

Chapter 7 offers a series of templates that water professionals can use to organize and portray analyses of water-related policy options that include different types of water values. The objective is to help practitioners better organize, communicate, and defend the basis and findings of their analyses to decision-makers, stakeholders, and the general public.

Chapter 8 offers summaries of several case studies in which the value of water emerges as a core issue, or has been used to help identify and promote more sustainable or otherwise desirable (social welfare enhancing) water resource management strategies. More detailed versions of the case studies are provided as a series of appendices.

Chapter 9 offers an agenda for suggested future research.

CHAPTER 2
THE CONCEPT OF VALUE

"Children of a culture born in a water-rich environment, we have never learned how important water is to us." (W. Ashworth, *Nor Any Drop to Drink,* 1982)

INTRODUCTION

Because this research report is about the value of water, it is important to focus on what the term "value" means. Value often means different things to different people. The notion of what has value (and why), or how much value something has, may vary considerably across individuals according to their cultures, philosophies, and professional disciplines. Since so many different people value water in so many ways, it is important that those with water resource management responsibilities be willing and able to grasp the different types and levels of values held by the affected public.

Economists have a clear-cut and precise technical definition of value, and an established set of categories for different types or sources of value (e.g., use versus nonuse values, market versus nonmarket values). Chapter 4 provides the economist's view of value in considerable detail, along with discussions about factors that affect value and methods used to estimate values. These are all well-established concepts and tools within the economics profession.

However, in the public policy-making domain in which water resource issues typically are deliberated and decisions made, the economist's view of the world may be too limited to reflect what is really important to appointed and elected officials, stakeholders, and the public at large. In the context of public discourse and decision-making, "value" may have different meanings and be expressed in ways that differ from the economists' technical perspective of water as a commodity. Therefore, this chapter provides a broad perspective of perceptions and beliefs about water values, attempting to embody the types of value that might motivate or resonate with the general public, key stakeholder organizations, and those with water governance responsibilities and authorities.

DEFINITIONS OF VALUE

Webster's New World Dictionary (Webster's 1973) provides several useful definitions of value. One definition is "the worth of a thing in money or goods." This is akin to the economist's notion of using "willingness to pay" as the true indication of value. This definition of value reflects a *quantitative* measure, such as might be expressed in dollar terms, applied to an object or commodity. Chapter 4 provides a detailed discussion of this type of value definition.

Another definition offered by Webster's for value is "the quality of a thing that makes it more or less desirable or useful." This *quality*-based definition steers one away from the strictly quantitative notion embodied in the first definition. Instead, this leads one to think of attributes such as the location, purity, timing, and reliability of water as being its values. For example, one might think of values with respect to concentrations of various compounds in water and whether those measures are consistent with potable use.

Finally, a third definition offered by Webster's is "beliefs or standards." In this context, value refers to core ethical conventions and morals that help govern or reflect a culture and society. For example, when a politician refers to "family values," he or she is not referring to the monetary worth or physical characteristics of families. Instead, the value reference pertains to some higher order principles that transcend the notion of markets or attributes. Value in this context refers to what may be collectively viewed as fundamental "truths" about how we as humans relate to each other and the world/universe around us. An example of a belief-oriented notion of value is that because water is essential for life, safe water should be made available to all people, regardless of their ability (or willingness) to pay. Another example is the use of water in various diverse religious practices such as baptisms and ritual bathing, in which water is a vehicle for transcending or washing away sins or otherwise seeking spiritual purity and renewal. Water also serves as a religious or cultural icon that separates the living from the dead in several mythologies (e.g., crossing the river Hades at one's death).

Given the important differences in how value can be defined and viewed, it clearly can be a loaded term. For the purpose of this report, the research team adheres mostly to the economist's notion of value as reflecting the worth of water as a commodity, where that value will be affected by the various attributes (e.g., quality and location) of the water. This is akin to the first definition offered above, but also embodies elements of the second. However, the approach also attempts to recognize and accommodate the broader notions of value that may be at play for the affected public and stakeholders, and for the appointed or elected officials making key water decisions. The objective is to better ensure that all of the applicable values affected by a water resource decision can at least be recognized and described in some fashion that resonates with the people who care about it (and if reasonable empirical measures of value are also available, so much the better).

BALANCING TRADEOFFS BETWEEN DIFFERENT SOURCES AND TYPES OF VALUE

Water is an essential input for nearly everything that people and society need and care about—health (and life itself), security, pleasure, beauty, foodstuffs, jobs, economic well-being, cleanliness, spirituality, and ecosystem function. In many instances, there are conflicts and tradeoffs regarding the use of water to fulfill these many activities and needs. For example, in some locations, there is heated and heartfelt debate over whether water should be left instream to preserve endangered fish species or extracted to enable farmers to earn another year of irrigation-supplemented incomes. Where instream flows are too low to effectively support both uses, a decision needs to be made that selects which of the competing uses gets the desired water. Water resource managers become part of a process in which tradeoffs are weighed about which use or user will be supported and which will not—i.e., which values will be realized and which will be foregone as a consequence.

Furthermore, what may be a core value to some stakeholders may be a lesser concern (and perhaps even a trivial consideration) to others. For example, some people argue that using water for irrigation that bolsters farm incomes is of little consequence relative to preserving wild fish species, and others will argue that human needs should always take precedence over fish, even if the species involved is close to extinction.

In essence, this reflects a conflict over values. Some people and organizations argue that there is more value to be derived by supporting farm uses of extracted water than there is value lost by decreasing instream flows and adversely affecting endangered fish species. Others will argue the opposite—that the value of the instream uses outweighs the values lost by farmers and their communities. Both sides believe that the benefits of their proposed water allocation outweigh the costs, because each side holds different values. Governance decisions about how these waters are to be allocated thus become issues of comparative values. Governance becomes especially complex in such cases because the values held differ across individuals, and it is difficult to estimate the respective levels of these values in either quantitative or relative terms.

It is no wonder then that even though values are an important consideration for sound water resource management, integrating water values into these deliberations is a significant challenge. Often, the accounting of values is incomplete—or may appear to be incomplete to some parties concerned about the water decision. Thus, it is important to try to identify and account for all types of values—and reflect them as fully and clearly as possible—when water resource management options are evaluated and decisions made.

WORKING TOWARD A COMMON PARLANCE TO DESCRIBE THE TYPE OF VALUES THAT PEOPLE DERIVE FROM OR ASSOCIATE WITH WATER

As noted above, people with different backgrounds, belief systems, and interests are likely to value water in different ways, at different levels, and for different reasons. We need to make sure technicians, decision makers, and the public do not speak past each other when deliberating about water resources and their associated values. Ultimately, making good public policy decisions about water requires that the value terminology be clear and reflect what the public, decision makers, and water professionals all perceive, understand, and accept.

The value taxonomy thus needs to be broad and inclusive, reflecting "values" regardless of whether they are amenable to quantification or monetization. The approach also needs to reflect values to all people, including current and future populations and spanning physical and geopolitical boundaries.

In terms of the breadth of value sources or motives, water managers need to consider the following:

- Cultural attributes. This is intended to include different perspectives of water value across various cultures, the extent to which water is revered or symbolic of greater values (e.g., in baptisms), or integral to maintaining that culture's lifestyle or livelihood (e.g., a farming lifestyle and community).
- Intergenerational equity considerations. This refers to how individuals or societies value preserving resources for future generations (what economists refer to as bequest values), where water uses today may affect the water quantity or quality (or both) available for our descendants.
- Environmental function. This is intended to reflect the value and role of water as essential for maintaining ecosystem integrity, enhancing human habitats, providing critical habitat for sustaining wildlife and biological diversity, preserving species, and providing other ecologic services (these are values or functions manifested, in part, in provisions of the federal Endangered Species Act and Clean Water Act).
- Recreational support. This is intended to reflect the value of water in enabling and improving leisure time activities such as instream and near stream uses (including fishing, boating, swimming, hiking, camping, picnicking, strolling, and wildlife observation).

- Economic development, population growth, and community sustainability. This reflects how water can be used as an engine to promote (or constrain) local community and regional economic development and population growth. Water thus has value to those who stand to benefit from accommodated or spurred levels of economic activity in a service region, and it has value to those that see water as a tool for managing the form, pace, and location of growth.
- Institutional and ownership issues. Values can reflect public perceptions and trust regarding jurisdictional issues, water rights, regulatory compliance, planning authorities, land use, community identity, local control, and what may be perceived as suitable roles for the private and public sectors.
- Water supply reliability. This aspect refers to the values people may place on having water reliably available at desired quantities and qualities across seasons and years, such as may be necessary to avoid periodic water use restrictions during drought periods. These values may apply or be relevant to integrated water resource strategies, customer-defined and understandable reliability targets, public involvement in determining reliability targets, contingency planning, water supply security, and quality reliability.
- Infrastructure reliability. This reflects values that may be held for levels of reliability for delivery of treated source waters, such as reliability related to distribution system maintenance, rehabilitation, and replacement. These values might derive from or be used to evaluate water issues and practices, including capital improvement planning and system infrastructure renewal strategies.
- Water quality. This reflects values associated with attaining the desired water quality to achieve a range of household, local, regional, state, or federal objectives or beneficial uses of water.
- Aesthetic values. This reflects the value of water for sustaining and enhancing aesthetic or cultural enjoyment, such as enjoyment derived from viewing and listening to fountains, looking at surface water bodies (lakes, rivers, reservoirs), admiring green space and associated vegetation and wildlife (e.g., agricultural lands in rural areas, urban parks and gardens), and maintaining open space in urban, suburban or rural areas.

Unless the professional water community can clearly identify and articulate how and why water has value to the many people involved with (and affected by) water resource decisions, it cannot fully consider the full range of benefits or costs of relevant water-management options or decisions. It is essential to recognize the types and sources of values that various people and organizations hold for water. This will aid in public education, and better enable stakeholders to understand and hear one another. In addition, this approach will provide a sense of ownership and participation to the affected public, and help them better understand the issues.

In addition, it is important to recognize that each of the sources or motives for value noted above can be associated with negative values. For example, contaminated water that poses a risk to health would generate negative values (i.e., losses in well-being) for persons that drank it. Thus, improving water quality can generate an increase in value, whereas water quality degradation produces a decrease in value. Likewise, providing additional streamflows can generate many positive values for water-limited downstream communities (including increased opportunities to apply extracted waters to cropland irrigation), but too much water as may arise in flood conditions can impose considerable damage on the impacted communities.

THE ECONOMIC DEFINITION OF VALUE

The discussion above attempts to portray a wide range of possible types or sources of value that may be held by people with a real or perceived stake in water resource decisions. While it is important to try to recognize where and how any or all of these values might be affected by a water decision (or how certain individuals perceive those values may be altered), it also becomes unwieldy to explore values across such a broad and loosely defined set of considerations. Therefore, this report focuses mostly on the economic terminology and measures for value. This is explored in considerable depth in chapter 4, but first there is a need to present a cursory overview of the economist's perspective, and then provide a crosswalk between the lay person value issues noted above and the standard terminology used by economists.

First, readers need to recognize an important distinction between three oft-used and frequently interchangeable terms: value, cost, and price. Our focus is on "value"—meaning the worth of water in terms of how much someone would be willing to pay (in terms of money or other goods and services) to obtain the desired use or outcome. For example,

- "Value" to a residential customer of an urban or suburban water utility may be $300 per acre foot (AF) for reliable delivery at their taps of high quality, safe, potable water (e.g., the customer is willing to pay this amount because tap water with those attributes is "worth" that much to him or her).
- "Cost" refers to the expense associated with producing and delivering a unit of water (e.g., to expand its service area, it may cost a utility $200 per AF at the margin to acquire, treat, and pump the water it delivers to its new customers).
- "Price" is the dollar per unit rate that is charged to customers, and often is based on the utility using average cost as a basis for rates (e.g., charging customers $120 per AF may provide sufficient revenues to cover the utility's overall annualized expenses).

Thus, the same utility-furnished water from the example developed above has a value of $300 per AF to the end user, has a marginal cost to supply it of $200 per AF, and is priced at $120 per AF. The three concepts (value, cost and price) have different meaning and different monetary levels. These three concepts are interrelated by the market, as discussed in chapter 4.

Second, in the relatively sterile view of economists, the diverse range of value types described in the preceding section can all be shoehorned into three benefits categories: human health, economic welfare, and ecosystem services.

- The human health category captures aspects of water policies that can affect the risk of illness or risk of premature fatality, and is often predominant in benefit-cost analyses of drinking water quality standards such as Maximum Contaminant Levels (MCLs). A large literature is available on how to quantify and value changes in various types of health risk (e.g., Raucher et al. 2002). The human health category is not often a large factor in water allocation decisions.

- The economic welfare category embraces all values (apart from those related to human health) that arise from the direct or indirect human use of water resources. This category considers economic welfare broadly defined, and includes additions to farm income due to irrigation, the enjoyment of a day of whitewater rafting or recreational angling enabled by instream flows, the enjoyment of having a green lawn, and so forth. The bulk of water values affected by resource allocation, instream flow, and ambient water quality issues will often fall within this category, including conflicts or tradeoffs when one user gains economic welfare values at the expense of another potential user.
- The ecosystem services category is often also referred to as consisting of nonuse (or passive use) values, since it reflects values apart from any current or anticipated human use. This includes values held by some people to preserve various species or ecosystems even though they do not "use" the resource in question. These may be called existence, stewardship, or bequest values, and are intended to embrace the notion that individuals assign values to other species, or natural states of the environment, in their own right or to preserve them for future generations. These ecologic or nonuse values are often difficult to articulate, and even more challenging to quantify and value. Nonetheless, some nonuse values may be especially important to many people and stakeholders, and therefore highly relevant for many water resource decisions.

Because economists tend to categorize the diverse types of value derived by people into these three broad categories, this can sometimes lead those who are not economists to think that some important values may be omitted (because they do not see their key value types expressed in terms they recognize). Therefore, water professionals need to be careful about fully identifying and qualitatively describing the values they are trying to include in their assessments.

Finally, it is important to consider who realizes the benefits, and who bears the cost of a water resource decision. Economists refer to the distribution of impacts as the "incidence" of a policy. The economic perspective is to try to recognize all the benefits and costs, to whomever they might accrue, and to then ascertain whether the total accumulated benefits outweigh the total costs. For economic efficiency (maximizing social welfare), the notion of winners and losers is less important than whether the total gains outweigh the total losses (economists refer to "Pareto preferred" alternatives, wherein the winners gain enough that they could conceivable compensate the losers for their losses, and still remain better off). However, decisions are rarely made on the basis of economic efficiency alone, and equity considerations are often a key part of policy deliberations. Thus, practitioners need to track not just what values are at stake, but who the likely winners and losers will be.

WATER OWNERSHIP, RIGHTS, AND USES

One cannot interpret the value of any resource or commodity without considering property rights and associated institutions that govern resource ownership and allocation. These legal constructs govern who owns water and who has the rights to water, and this in turn dictates where and how water gets used and values generated. The United States has two main legal frameworks for water ownership and rights: the "prior appropriation doctrine" governs water in

17 relatively arid western states, and the "riparian rights doctrine" governs surface water in the states east of the Mississippi River.

Water Allocation and Use in the Western United States

Surface Water

Western water policies and western water law are not uniform or clear-cut. The institutions governing water have evolved over time and vary from state to state; they are the product of 200 years of court and administrative decisions and legislative compacts. "The opportunities as well as the incentives to use, abuse, or protect water supplies are the result of many fragmented local, state, and federal water institutions" (Fredrick 1995, p. 8). However, the general doctrine governing allocation is the same in the 17 western states, with the state (or federal) government claiming ownership of surface water resources. The government owners then issue water rights for specified quantities and water extraction points to individuals and entities. The quantities, extraction locations, and uses are based on the size and seniority of the initial uses and claims.

The prior appropriation doctrine emerged from an informal set of laws that developed when European settlers first arrived in the American West. Similar to a mining claim, the initial laws established that the first person to make a claim on a surface water source was allowed priority over all later claimants. This system created an environment in which water rights were distinctly defined and owned.

By the early 1900s, a series of court decisions and administrative controls modified prior appropriation law in an attempt to curb third party effects (negative externalities) (Anderson and Leal 1988). In the modified system, water rights now were held by the state or federal governments, and users had only priority based claimant rights on water. In addition, courts introduced several riparian law concepts, including the doctrine that dictated that a water right could be lost if not put to "beneficial use" over time.

Groundwater

Groundwater did not become a critical issue in water provision until the 1950s, when a new generation of centrifugal pumps allowed farmers to tap into aquifers on a large scale (Ashley and Smith 2001). Groundwater quickly became a large source of supply for both farmers and growing cities. Groundwater currently provides an estimated 33% of the fresh water used by western states (Bryner and Purcell 2003). As for surface water, no federal policy was created for the governance of groundwater, and it was left up to each state to develop its own approach. In most western states, groundwater use is governed in a fashion similar to surface waters, with the prior appropriation doctrine (modified with beneficial use and some preferential provisions) applied to determine groundwater rights and uses. However, some states have adopted the legal doctrine of correlative rights or reasonable use to govern groundwater use, under which a party can tap as much groundwater as they can access from their property insofar as the water is put to "beneficial use" and applied in an manner that might be deemed "necessary and reasonable." This doctrine, also known as the American rule, is often governed with an eye toward protecting the rights of other parties by controlling third party impacts (Bryner and Purcell 2003).

Alternatively, some western states use a "right of capture" doctrine for groundwater rights, based on the common law notion of absolute ownership. Under this regime, one can pump as much groundwater as one wants from their land, regardless of impacts on others (Nye 2004). This latter approach has been applied in Texas, with the recent exception of the Edwards Aquifer basin. In the Edwards Aquifer basin, growth led to excessive pumping, which in turn caused significant problems (including impacts on endangered species relying on river habitat flows originating from groundwater-fed springs) (Grubb 1996). The groundwater basin is now managed carefully by the Edwards Aquifer Authority. This is an example of how water scarcity and hydrologic interdependencies have led to unintended impacts and conflicting values. The third party consequences of absolute ownership thus led in the Edwards Aquifer region to the evolution of legal and governance frameworks to ensure water was managed more prudently (and by central authority), for the greater public good.

The Federal Role

At the beginning of the 20th century, the federal government become a major player in western water rights and allocations. The most important institution in western water allocation is the U.S. Department of Interior's Bureau of Reclamation (USBR). The original goal of the agency was to reclaim arid lands for productive use by farmers. The Reclamation Act of 1902, and its 1939 Amendments, gave the Bureau control of significant river flows that it allocated to agriculture to encourage growth in the developing West. Today, the Bureau's responsibilities have expanded to include water conservation, hydroelectric generation, municipal and industrial water supplies, flood control, outdoor recreation, and others. Yet, even with the changing federal role, the Bureau is still the largest supplier of irrigation water in the United States. The agency diverts 40% to 85% of major western river systems, and of that quantity, 85% is allocated to agriculture [Congressional Budget Office (CBO) 1997].

The Bureau allocates water to the agricultural industry through long-term contracts with water or irrigation districts. The districts, in turn, provide the water to individual farmers. The contracts between the agency and the districts specify below-cost (subsidized) prices and fixed water allotments to ensure that water is affordable to farmers. The districts may have water supply sources of their own (including rights), but are more likely to be "contractors" for water developed entirely by the federal government and sold to farmers at a price set to cover only supply costs (Anderson 1983).

Agricultural and Urban System Use

Given the role of the Bureau of Reclamation, it is not surprising that agriculture constitutes the predominant user of western water. In the West as a whole, 76% of surface water withdrawals are for agricultural purposes. In many states, well over 90% of water goes to irrigate croplands. Out of the 17 western states, only in 5 does agriculture account for less than 80% of all surface water withdrawals, and these are all Great Plains states, which have access to large quantities of groundwater (CBO 1997).

While agriculture dominates western water use, municipal water use is becoming increasing important as western cities and suburbs continue to grow. About 8% of total surface water withdrawals and 14% of groundwater withdrawals are allocated to western municipalities (Bryner and Purcell 2003). One of the key issues emerging in the West is the transfer of agricultural water rights to municipal systems. As urban system water demands grow, the value of water for these communities has risen to the point where they are willing to pay farmers enough to bid water rights away from irrigation uses. The legal institutions in the West are gradually becoming more flexible in allowing transfers of water rights so that some water can flow to its highest value uses.

Water Allocation Institutions in the Eastern United States

Surface Water

Surface water allocation east of the Mississippi River generally is based on the riparian rights doctrine. The riparian doctrine reflects the realities of a region that until recently was blessed with large quantities of water that have been ample enough to handle multiple demands. The riparian rights doctrine does not provide a basis for a definitive allocation of water resources, but defines water use in association with ownership of adjacent lands. The doctrine states that landowners with property adjacent to or crossed by a natural body of water have the right to use these waters. The doctrine provides for a pro rata sharing of available water supplies and, taken in the most stringent sense, stipulates that downstream landowners have the right to see the water unchanged in quality or quantity when it reaches them. Clearly, if the doctrine were read literally, no one upstream could use the water because it would reduce quality or quantity for other users downstream. Consequently, provisions have been enacted that amend or supercede the riparian doctrine.

Like western water law, eastern water law varies by state, but the states generally apply riparian rules and use state-issued permits to control levels and types of use. In general, riparian land owners have absolute rights to water (rights may be exercised without regard for other users) applied to domestic uses. Domestic use, also called natural use (including drinking, bathing, cooking, laundry, gardening, livestock, and other uses necessary for life and health), is given priority with no regard of the quantity of water left in the river or lake after the use is satisfied.

Nondomestic use, also called artificial use, includes use of water for irrigation, power production, and manufacturing. Artificial use is constrained by the "responsible use doctrine." The responsible use modification to the riparian doctrine states that some reduction in a watercourse's flow is acceptable, as long as other users are not "unreasonably" harmed. The term unreasonable is not well defined, and is interpreted by state courts case by case.

A lack of clearly defined water rights, coupled with the riparian doctrine's philosophy of maintaining equity between all users, has limited the ability of the riparian approach to address the challenges posed when drought and growth reach the point where source waters are no longer sufficient to meet all demands. Therefore, as pressure on water resources in the East has increased, states have adopted aspects and qualities of the prior appropriation doctrine. All but five eastern states have enacted legislation establishing registration or permitting requirements to substantially regulate water resources (Foran, Beecher, and Wilson 1995). Unlike water rights issued under prior appropriation systems, riparian rights do not confer a right to consume a specified amount of water, nor do they grant ownership to a certain quantity. However, under the permit system, existing riparian users are provided with a permit that designates water use and quantity for a specified time period.

Groundwater

In most eastern states, groundwater ownership and use are governed by a slightly modified version of the principle of absolute ownership found in English law. Under this so-called American rule, a landowner may withdraw percolating groundwater from beneath his or her property for any "natural and ordinary" use without regard for other users (Deason, Schad, and Sherk 2001). The original intent of the American rule doctrine was to encourage economic growth. However, states have begun to modify groundwater law and often are now following the path of surface water law. The majority of eastern states have enacted legislation creating permitting or registration systems for groundwater extraction that may apply statewide or to specific groundwater management areas.

Summary

Governing systems for specifying ownership, rights, and uses of water have varied between the eastern and western regions of the United States. However, in both regions, growing demands and diminishing supplies have increased the scarcity of high quality source waters. This in turn has created pressures that are changing some institutional arrangements under which waters are allocated to different uses and generate values. As the pressures of supply and demand grow in the future, there will be increasing need to consider tradeoffs in water uses, and this in turn means that water resource managers will need to be better prepared to ascertain the values that are affected by the allocation decisions and other policy choices before them.

CONCLUSIONS

Water has or contributes to many different types of values. These values play a large role in how people feel about water resource management decisions that may affect them. Utility managers and other water professionals need to recognize and acknowledge the many and varied types and levels of values held by their customers, public officials, and key stakeholders.

Ultimately, as water becomes increasingly scarce, managers should anticipate greater conflicts regarding the tradeoffs that inevitably will become more dominant. An increasing share of water policies and decisions will result in clear winners and clear losers—there will be those who see their values diminished by water management decisions, and those who will benefit from actions that enhance the water-related values they find important. Ultimately, water resource management institutions and decisions will need to become increasingly attuned to value issues and tradeoffs in values.

Doctrines governing water rights and uses in both the eastern and western United States are already evolving in a manner that reflects increased scarcity. Systems of water ownership, rights, and uses will continue to reflect the need for greater governance authority to oversee and manage how much water gets used, by whom, and for what purposes. Whether or not market-like institutions are allowed a greater role in how water is allocated to different parties and uses in the future, the issue of "values" will become an increasingly important factor that water resource managers must comprehend and balance.

CHAPTER 3
WATER USE PATTERNS

"All the water that is here right now has always been here. There is not going to be any more. The water we drink today has been used before, sometime, somewhere in the world, whether by a dinosaur, George Washington, or Madonna." (Paul Mattausch, 2001)

The value of water is intimately connected to how it is used. This section of the report discusses water use patterns in the world and the United States. Water use is broken down into end uses based on the available data.

GLOBAL WATER USE BY REGION

Global water use is unevenly distributed throughout the continents and, unfortunately, often does not match the water resource availability. Global water withdrawals and consumption by continent are given in Table 3.1, and illustrated in Figures 3.1 and 3.2 (note different scaling is used on these figures). Consumptive use is water that is lost from the stream to evaporation or any other process that causes it to fail to return to the local hydrologic system. The annual global freshwater withdrawn in 1995 is estimated at 3,790 km^3, or 910 m^3 (1 km^3 = 0.240 mi^3). Consumption accounted for 2,070 km^3, or 61% of these withdrawals. Withdrawals are projected to increase to 4,430 km^3 by 2010 (of which consumption will account for 2,400 km^3, or 54%) (Shiklomanov 1999, 2000). A detailed discussion of the method used to determine current and future global water usage (by region and sector) can be found in Shiklomanov (1999, 2000). Broadly speaking, quantitative estimates of water use are based on several factors, including socioeconomic development level, population, physiographic features (including climate), and land area.

In 1995, Asia accounted for about 60% of the global freshwater withdrawal and 67% of the global consumption. This is driven in large part, as discussed below, by the fact that the majority of irrigated land is located in Asia. The North American and European continents account for 30% of the global withdrawals and 20% of the global consumption. A significant fraction of water consumption in Africa is dominated by evaporation from reservoirs.

It is estimated that future annual global water withdrawals will increase from 1995 to 2025 by about 40%, reaching approximately 5,240 km^3. The largest growth in withdrawals will occur in South America (55%), Africa (54%), and Asia (46%), and the smallest growth will occur in Europe (23%) and North America (15%).

GLOBAL WATER USE BY SECTOR AND REGION

Water use varies considerably across economic sectors, as shown in Tables 3.2 and 3.3. Of all economic sectors, agriculture is by far the biggest user of freshwater. According to estimates for the year 1995, agriculture accounted for 66% of the global freshwater withdrawn and 85% of the consumption (Shiklomanov 1999, 2000). Of the remaining sectors, 9% was withdrawn for municipal use, 20% was withdrawn for industrial use, and 5% was withdrawn for reservoir storage.

Table 3.1
Global water withdrawals/consumption by continent: 1990-2025 (km^3)

Continent	1990	1995	2000	2010	2025
Europe					
Water withdrawal	482	455	463	535	559
Water consumption	198	189	197	234	256
North America					
Water withdrawal	653	686	705	744	786
Water consumption	221	237	243	255	269
Africa					
Water withdrawal	203	219	235	275	337
Water consumption	150	160	170	191	220
Asia					
Water withdrawal	2,114	2,231	2,357	2,628	3,254
Water consumption	1,315	1,381	1,458	1,593	1,876
South America					
Water withdrawal	152	167	182	213	260
Water consumption	82	89	96	106	120
Australia and Oceania					
Water withdrawal	29	30	33	36	40
Water consumption	16	18	19	20	22
Global total					
Water withdrawal	3,633	3,788	3973	4,431	5,235
Water consumption	1,982	2,074	2,182	2,399	2,764

Sources: Based on data from Shiklomanov 1999, 2000.
1 km^3 = 2.64154 x 10^5 million gallons (mg) = 8.1066 x 10^5 acre feet.

Global water use across sectors varies across regions. For example, in Africa, agriculture consumed 63%, municipal users consumed 8%, and industry consumed 4% of all water withdrawals in 1995. This is in contrast to Europe, for example, where agriculture consumed 37%, municipal users consumed 15%, and industry consumed 45% of all water withdrawals.

3.3 reveals how sectoral shares of water use vary across countries of different levels of economic development, with industrial use dominant in high income countries, and agricultural use dominant in lower income nations (UNESCO 2003).

Agriculture Use

The amount of water involved in the production of food is significant. For example, 1 to 3 m^3 are required to yield 1 kg of rice (approximately 240 gal/lb of rice). Moreover, it is estimated that while only 15% of the total global cultivated lands are irrigated for food crops, these account for almost half of the value of global crop production (Shiklomanov 1999, 2000). UNESCO (2003) calculated that for the average person's calorie intake per day, approximately 1000 m^3 (26,400 gal) per year are required—highlighting the global importance of water for food security.

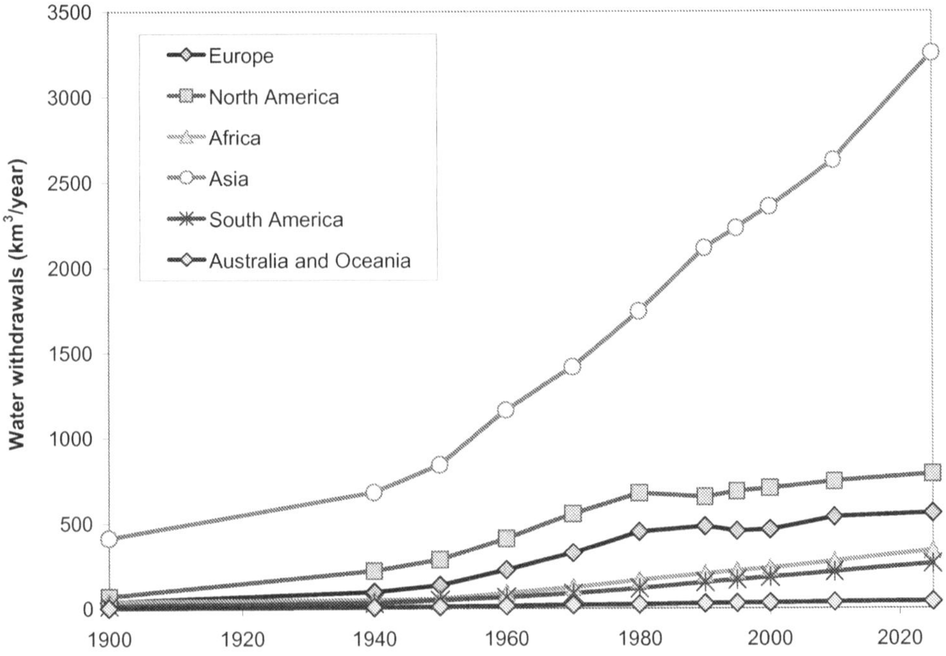

Sources: Based on data from Shiklomanov 1999, 2000.

Figure 3.1 Global water withdrawals by continent from 1900 to 2025

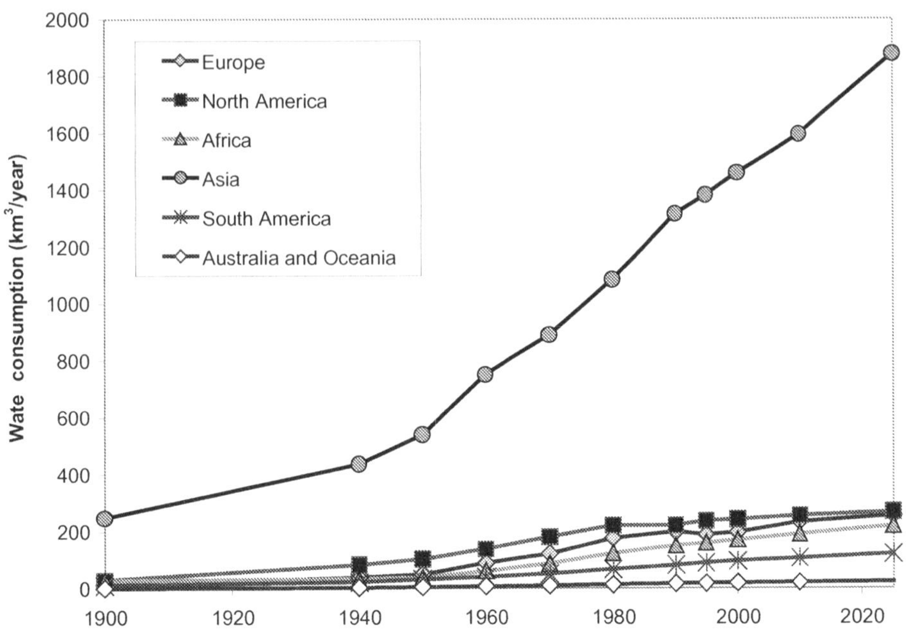

Sources: Based on data from Shiklomanov 1999, 2000.

Figure 3.2 Global water consumption by continent from 1900 to 2025

Table 3.2
Global water withdrawals/consumption by sector: 1990-2025 (km^3)

Sector	1990	1995	2000	2010	2025
Population (million)	5,285	5,735	6,181	7,113	7,877
Irrigated land area (million ha)	243	253	264	288	329
Agricultural use					
Water withdrawal	2,425	2,504	2,605	2,817	3,189
Water consumption	1,691	1,753	1,834	1,987	2,252
Municipal use					
Water withdrawal	305	344	384	472	607
Water consumption	45	50	53	61	74
Industrial use					
Water withdrawal	735	752	776	908	1,170
Water consumption	79	83	88	117	169
Reservoirs	167	188	208	235	269
Global total					
Water withdrawal	3,633	3,788	3,973	4,431	5,235
Water consumption	1,982	2,074	2,182	2,399	2,764

Sources: Based on data from Shiklomanov 1999, 2000.

Table 3.3
The ratio of water withdrawal by sectors of economic activity to total water withdrawal (%) by continents for 1995 and 2025

	1995				2025			
Continent	Agri-culture	Munic-ipal	Indus-try	Reser-voir	Agri-culture	Munic-ipal	Indus-try	Reser-voir
Europe	37.4	14.7	44.8	3.2	37.2	14.0	45.8	3.1
North America	43.5	10.7	41.5	4.4	41.4	12.3	41.3	4.8
Africa	63.0	8.1	4.4	24.7	53.1	18.0	6.0	22.8
Asia	80.0	6.9	9.9	3.2	72.0	9.5	15.2	3.3
South America	58.6	17.2	15.4	8.7	44.2	22.7	23.8	9.2
Australia and Oceania	51.0	10.9	23.5	14.8	46.8	11.3	26.1	15.7
Global total	66.1	9.1	19.9	5.0	60.9	11.6	22.3	5.1

Sources: Based on data from Shiklomanov 1999, 2000.

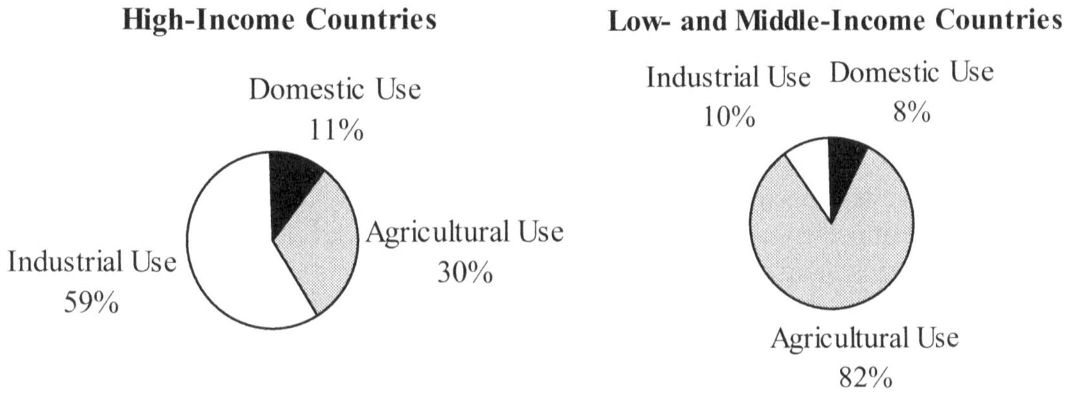

Source: Data from UNESCO 2003.

Figure 3.3 Global water withdrawals by sector

The global irrigated area in 1995 was 253 million hectares (ha) and is expected to grow by nearly one-third to 330 million ha by 2025. Although agriculture water use will decline in terms of the percentage of total global usage, agriculture withdrawals are expected to increase by about 27% to 3,189 km^3 by 2025. Most of this expected increase will occur in the developing world, most notably Asia, which occupies over half of the total net irrigated area (Gleick 1993).

Agriculture water withdrawals are smallest in the north. For instance, in northern Europe withdrawals are between 300 and 5,000 m^3 per ha, while in southern and eastern European countries withdrawals are between 7,000 and 11,000 m^3 per ha. In the United States, where water withdrawals for irrigation accounted for 40% of the total freshwater withdrawals in 1995 (Solley, Pierce, and Perlman 1998), average water withdrawal for irrigation is estimated by different authors to be between 8,000 and 10,000 m^3/ha (approximately 3 AF/acre). In the countries of Asia, Africa, Central and South America, where there is a wide variety of climatic conditions, crops, and irrigation techniques, the values for specific water withdrawal range from 5,000-6,000 m^3/ha to 15,000-17,000 m^3/ha and in individual regions of Africa, from 20,000 to 25,000 m^3/ha. In many places, these withdrawals are primarily from groundwater sources that currently are being extracted at unsustainable rates.

Industrial Use

Water for industrial use includes water for cooling, for transportation, as a solvent, for hydropower, and for nuclear power production. The biggest share of freshwater for industrial use is stored in reservoirs and dams for electrical power generation (57 to 69% is used for hydropower and nuclear power generation, 30 to 40% for industrial processes, and 0.5 to 3% for thermal power generation). However, the volume of water evaporated from reservoirs is estimated to exceed the combined freshwater needs of industry and domestic consumption, which greatly contributes to water losses around the world, especially in hot tropical regions

(Shiklomanov 1999, 2000). Industrial use is expected to increase by about 56% from 752 km^3 in 1995 to 1,170 km^3 in 2025. The percentage of industrial water use is expected to increase significantly in Asia, Africa, and South America. However, the industrial fraction of water use is not expected to be more than 20% in South America, 13% in Asia, and 6% in Africa.

Municipal Use

Municipal use is expected to increase by about 76% from 344 km^3 in 1995 to 607 km^3 in 2025, in part reflecting the growth in population. People in developed countries on average consume about 10 times more water daily than those in developing countries. It is estimated that the average person in the developed countries of Europe and North America uses 500 to 1,000 L/day, compared to 50 to 100 L/day in developing countries of Asia, Africa, and Latin America (Shiklomanov 1999, 2000).

Water consumption for municipal systems (typically about 10 to 15% of water withdrawn) consists of transmission and evaporative water losses, urban irrigation, recreation demands, and, to a lesser extent, water consumed for household needs. For modern systems with a centralized water supply and an efficient distribution system, domestic consumption does not usually represent more than 5 to 10% of the total water withdrawal. For smaller systems, consumption can be as high as 40 to 60% of the total water intake. Consumption will also depend on climate, with higher consumption rates in the relatively warmer regions.

U.S. WATER USE BY REGION

Annually, the United States receives enough precipitation to cover its entire area of 3.5 million mi^2 to a depth of 30 inches. However, distribution of annual precipitation varies widely from region to region: it can be as low as 5 inches in arid areas of the Western region while areas in the Southeast region may receive over 100 inches. Additionally, the amount of water available to recharge surface and groundwater supplies is a small percentage of the annual precipitation and is dependent on the annual precipitation for that region. The four regions defined by the U.S. Geological Survey (USGS) are the Northeast, the Southeast, Central, and Western:

- The Northeast region includes Maine, New Hampshire, Vermont, Massachusetts, Rhode Island, Connecticut, New Jersey, Delaware, New York, Maryland, Pennsylvania, Ohio, Michigan, Wisconsin, Illinois, Indiana, Kentucky, West Virginia, Virginia, and the District of Columbia.
- The Southeast region includes Tennessee, North Carolina, South Carolina, Georgia, Alabama, Mississippi, Florida, the Virgin Islands, and Puerto Rico.
- The Central region includes Montana, North Dakota, South Dakota, Minnesota, Wyoming, Nebraska, Iowa, Missouri, Arkansas, Louisiana, Texas, Colorado, Kansas, Oklahoma, Utah, and New Mexico.
- The Western region includes Washington, Oregon, Idaho, California, Nevada, Arizona, Alaska, and Hawaii.

Consumption of freshwater as a percentage of local average annual precipitation is compiled by United States Department of Agriculture (USDA) and accounts for consumption from all sources including residential, industrial and agriculture. As can be seen in Figure 3.4, many areas in the United States consume more water than they receive in annual precipitation. Most of these areas lie in the Central and Western regions where precipitation tends to be lower. Corresponding groundwater withdrawals are high in these regions as shown in Table 3.4.

However, the largest total withdrawals of water, including groundwater and surface water, occur in the Northeast region, reaching 130,700 million gallons per day (mgd), of which 101,500 mgd (77.7%) is fresh water and 29,100 mgd is saline, as shown in Figure 3.5. This reflects the dense population of this region, which results in a high demand for thermoelectric power (including steam power generation and cooling water use, but excluding hydropower). The water withdrawal for thermoelectric power in Illinois alone exceeds 17,000 mgd. Fresh water withdrawal in the Central and Western regions accounts for more than half of the total fresh water withdrawals in the United States, most of which is used for livestock and irrigation.

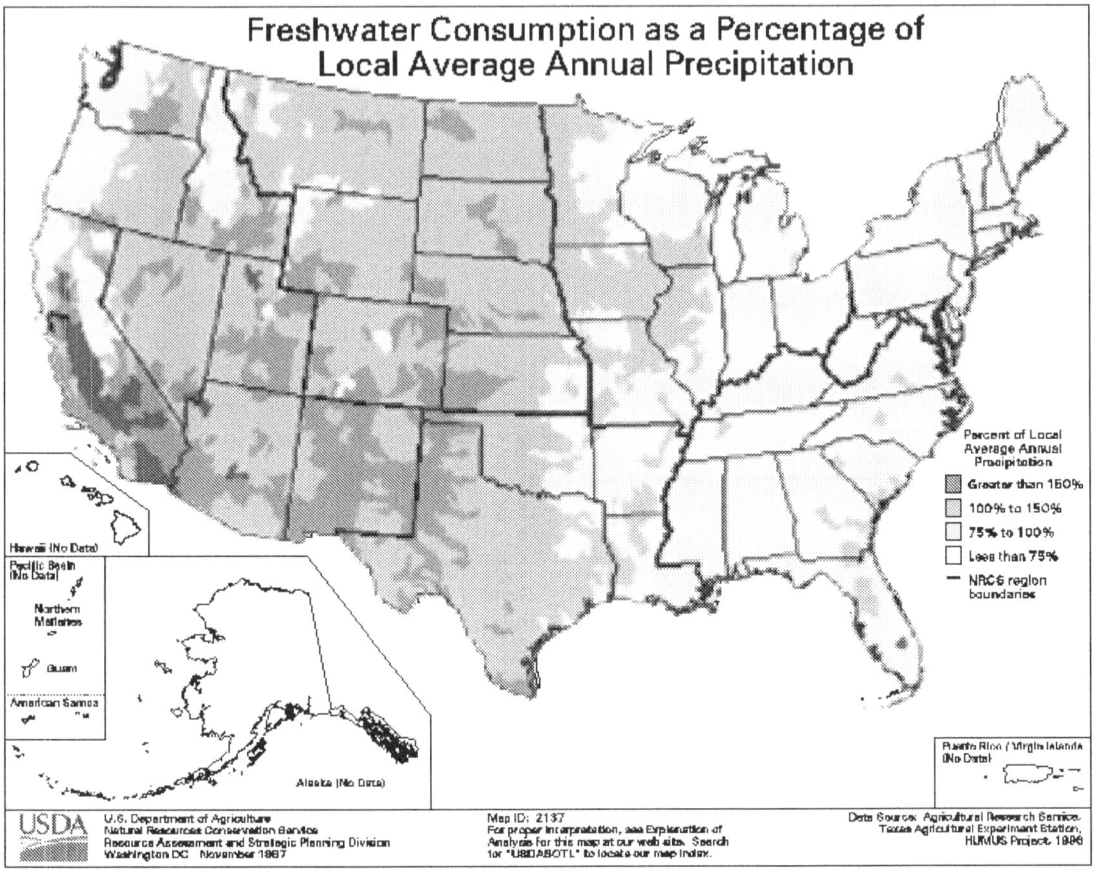

Source: USDA 2003a.

Figure 3.4 USDA map showing freshwater consumption as a percentage of local average precipitation

Table 3.4
Water withdrawals by water-use category and water-resources region, 1995 (mgd)
(figures may not add to totals because of rounding)

Water use category	Northeast withdrawal	Southeast withdrawal	Central withdrawal	Western withdrawal	Total withdrawal
Groundwater — fresh	8,451	10,004	32,772	25,174	76,401
Groundwater — saline	31	16	723	344	1,114
Surface water — fresh	93,056	37,745	76,065	57,581	264,447
Surface water — saline	29,100	15,176	4,860	10,582	59,718
Total — fresh	101,540	47,758	108,670	82,731	340,699
Total — saline	29,130	15,195	5,592	10,930	60,847
Total — fresh and saline	130,670	62,953	114,262	93,661	401,546

Source: Based on USGS 2003a.

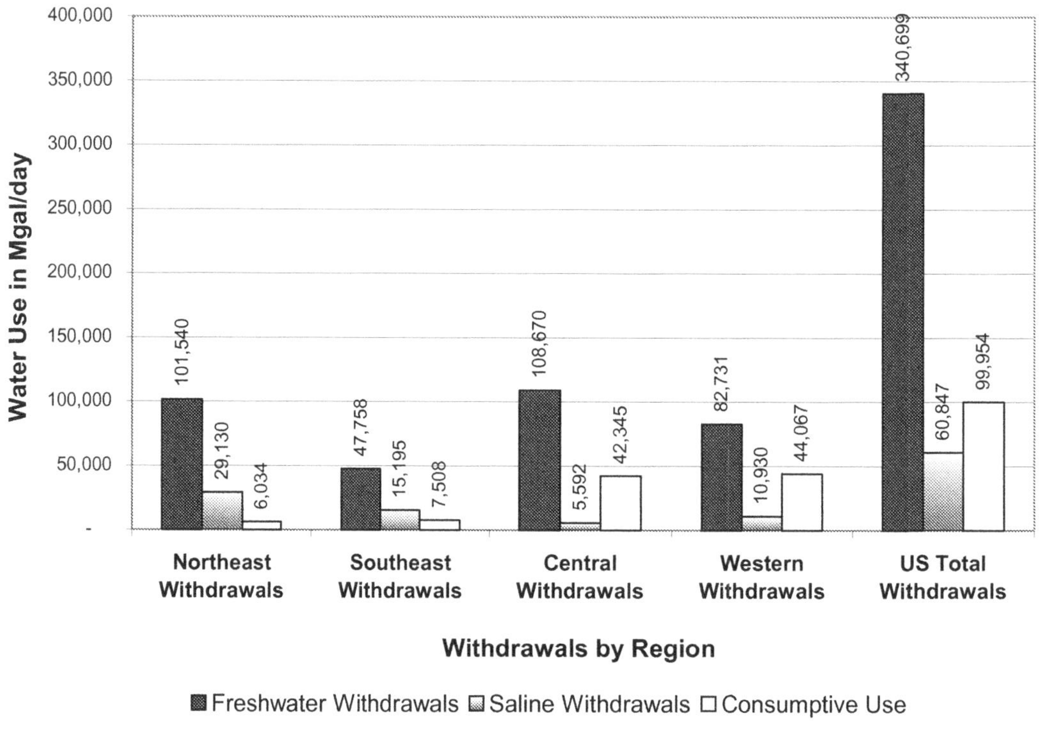

Source: Based on USGS 2003a.

Figure 3.5 Consumptive water use vs. total withdrawal of saline and freshwater

U.S. WATER USE BY SECTOR

In 1995, the United States water withdrawals were estimated to be 402,000 mgd, of which nearly 85% was fresh water. The USGS has not updated their national water use data since 1995. To monitor water use trends and provide information on water use trends, the USGS has defined the major water use sectors as shown in Table 3.5.

Agricultural (irrigation and livestock) and thermoelectric power make up the highest water use categories and account for nearly 81% of water withdrawals in the United States. The thermoelectric power sector requires the largest withdrawal of water, 190,000 mgd (see Figure 3.6), of which 132,000 mgd is freshwater. However, the largest withdrawal of freshwater is made by the irrigation sector (134,000 mgd) and comprises 34% of withdrawals when combined with the livestock sector. Public supply and industrial/mining water use combined account for approximately 17% of withdrawals. The remaining 2% of withdrawals are used for the domestic and commercial sectors. A graph of the water withdrawals by sector is shown in Figure 3.7.

Water withdrawals for thermoelectric power and irrigation decreased from 1990 to 1995 by 3% and 2%, respectively. This decrease occurred despite an increase in population and an increase in the irrigated acreage. The increase in population during this time resulted in an increase in water withdrawn for public supply and domestic use; however, daily per capita use showed its first decline since 1950, largely as a result of conservation. Decreases for water withdrawal also occurred in the mining and industrial sectors (USGS 2003b).

During this same time period, the livestock and commercial sectors showed marked increases (as much as 22%) in water withdrawals. Some of the increase can be attributed to differences in reporting methods and estimates based on population increases rather than actual measurements (USGS 2003b).

Table 3.6 shows the water withdrawal for 1995 in each sector. Quantities of fresh and salt water usage, measured in million gallons per day, are shown. Fresh water includes surface and groundwater withdrawals.

Table 3.5
Water use sectors as defined by USGS

Sector	Water use
Public supply	Water provided for 25 or more users by public or private suppliers. May be used by the domestic, commercial, industrial, or thermoelectric sectors.
Domestic	Water used for household purposes, including drinking, cooking, bathing, clothes washing, and lawn and garden watering.
Commercial	Water used for motels, commercial facilities, civilian institutions, and public supply to golf courses.
Irrigation/livestock	Water applied to crops and golf courses, and water used for livestock and livestock products such as milk, eggs, and wool. Includes captive fish production (except fish hatcheries).
Industrial/mining	Water used for manufacturing processes, including cooling and washing for industries such as petroleum refining and paper products. Also includes water used in the extraction of minerals during the mining process.
Thermoelectric	Water used for the generation of electric power and associated processes such as steam generation and cooling.

Source: Based on USGS 2003b.

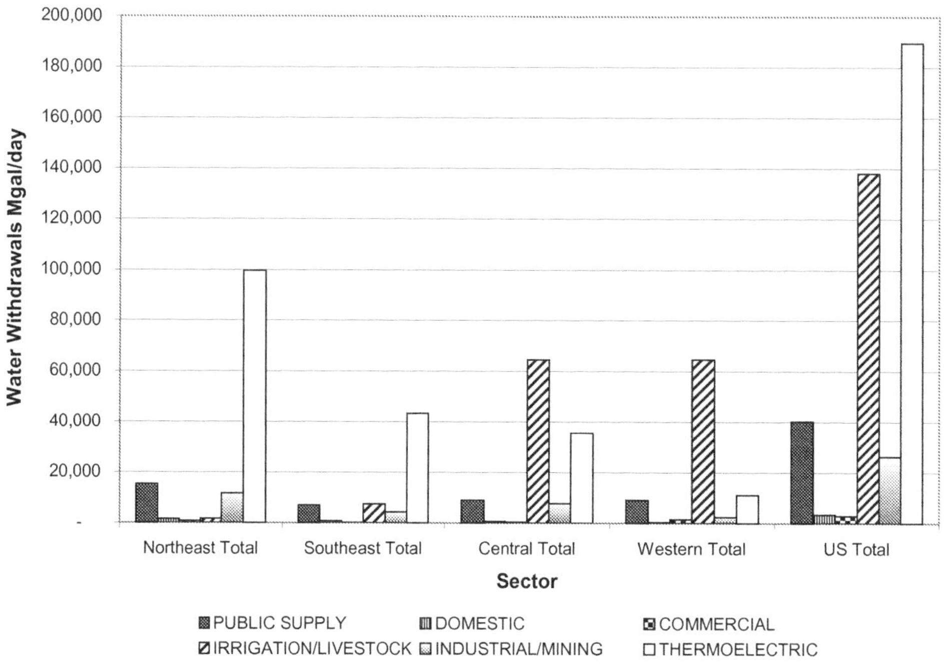

Source: Based on USGS 2003b.

Figure 3.6 Regional water use in the United States by sector

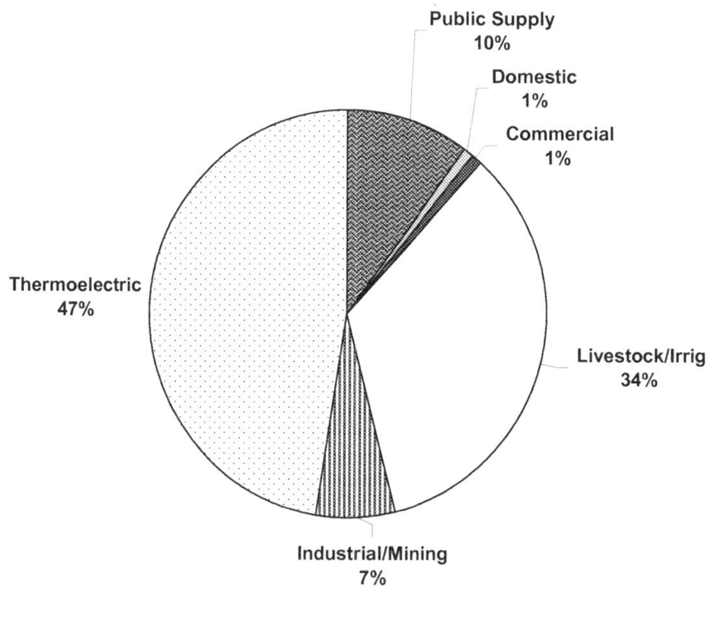

Source: Based on USGS 2003b.

Figure 3.7 Percentage of water use in the United States by sector

Table 3.6
Total U.S. water withdrawals by sector from 1995 USGS data (totals may not sum due to rounding)

Category	Water source	Withdrawals (mgd)	% (of freshwater)
Public supply	Fresh	40,200	11.8
Domestic	Fresh	3,390	1.0
Commercial	Fresh	2,890	0.8
Irrigation	Fresh	134,000	39.3
Livestock	Fresh	5,490	1.6
Industrial	Fresh	20,700	6.1
	Saline	1,660	—
Mining	Fresh	2,560	0.8
	Saline	1,210	—
Thermoelectric	Fresh	132,000	38.7
	Saline	57,900	—
Total	Fresh	341,230	100.0
	Saline	60,770	—
	Total	402,000	—

Source: Based on USGS 2003a.
402,000 mgd = 555 km^3/yr.

END USES OF WATER BY NORTH AMERICAN MUNICIPAL USERS

Of all of the water use sectors, the municipal sector places one of the highest values on water since municipal customers are willing to pay a much higher unit marginal price for water than are other sectors. Municipal customers, however, use water for a wide variety of purposes, and each of these may have a widely varying value to the customer. The AWWA Research Foundation (AwwaRF) conducted a study of end uses of water by municipal customers. The results of this study by Mayer et al. (1999) provide interesting information on where municipal customers use water, as shown in Figure 3.8. This Residential End Uses of Water Study (REUWS) was conducted on homes that were generally not using best available water conserving fixtures and appliances based on 2003 standards. Indoor uses averaged nearly 70 gpcd. Studies conducted in Seattle (DeOreo et al. 2001), East Bay MUD (Mayer et al. 2002), and Tampa have shown that use of high efficiency fixtures and appliances can reduce indoor uses to 40 gallons per capita per day (gpcd) or less without causing any reduction in levels of satisfaction on the part of the customer. Outdoor uses in this data set included customers who did little or no irrigation and those that devoted over 300 gallons per day (gpd) to irrigation. This use is highly dependent on climate, land use, and economic status of the user.

COMPARISON OF U.S. AND GLOBAL WATER USE

It is evident from the preceding text and figures that water use patterns vary considerably between regions of the world. Key differences can be found according to (1) what economic sectors are dominant in each continent (e.g., irrigation in Asia, leading to that continent having by far the highest total water use) and (2) what level of economic development has been realized in each region (with agricultural use dominant in low and middle income countries, and industrial use more prevalent in higher income nations).

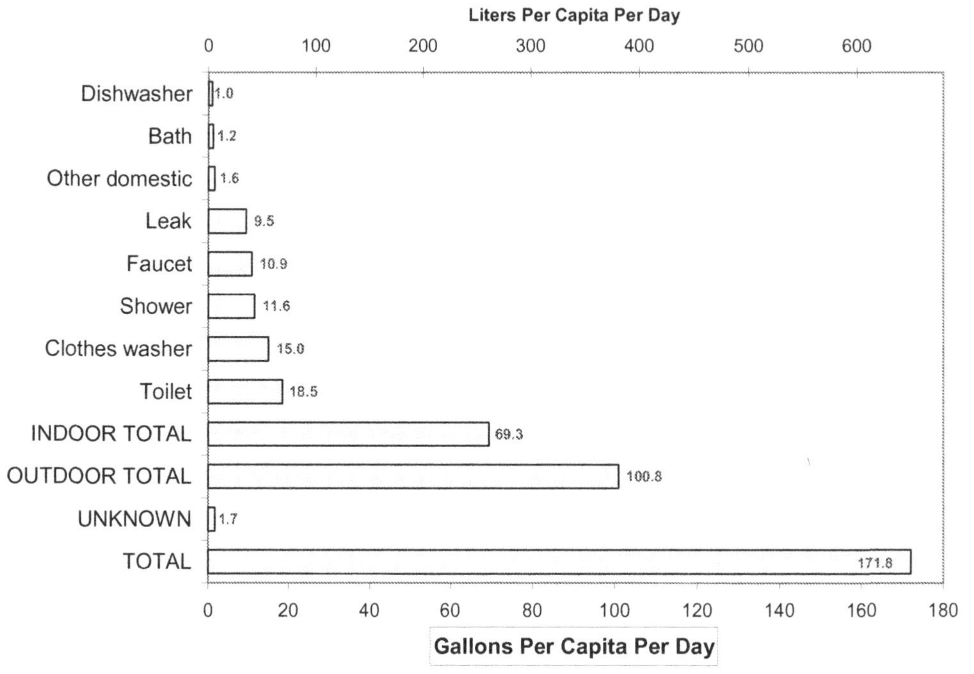

Source: Based on data from Mayer et al. 1999.

Figure 3.8 Breakdown of water use by municipal customers in North America

Within North America, total water use is dominated by thermoelectric power generation and agricultural uses (livestock and irrigation), with public water supply accounting for 10% of total water use. Over the past decade, population growth has led to an increasing share devoted to public water supply and correspondingly lower shares to energy and agriculture. In the residential sector, outdoor use (e.g., lawn watering) typically exceeds indoor uses, especially in summer months. Indoor residential use is largely for sanitation (toilet flushing being the largest residential indoor use, on average), followed by laundry uses and showering.

One useful metric for comparing water use across regions of the globe is to examine levels of water consumption on a per capita basis, as shown in Table 3.7. Per capita water consumption is displayed in graphical form in Figure 3.9. On a per capita basis, North America is by far the largest user of water, with total per capita water consumption of 774 m^3/yr (204 kgal), more than double the global average of 360 m^3/yr (95 kgal). Also of note is that water consumption per capita in North America is nearly triple the level observed in Europe, indicating that water use is not always clearly defined by levels of economic development and cultural norms (features that North America and Europe share, in general). Differences in municipal per capita water use also are striking, with North American levels at 83 m^3/yr (22 kgal) at 250% of the global average of 33 m^3/yr (8.7 kgal), and nearly five times the level in Africa of 17 m^3/yr (6.2 kgal).

Table 3.7
Per capita water use in 2000 (m^3/yr/capita)

	Withdrawals	Consumption	Municipal use*
Europe	637	271	40
North America	2,245	774	83
Africa	296	214	17
Asia	642	397	27
South America	351	185	32
Australia and Oceania	1,048	603	66
Global average	656	360	33

Sources: Based on water use data from Shiklomanov 2000; population data from UN World Urbanization Prospects report (UNEP 2002).
*Municipal use estimated based on sectoral shares portrayed in Table 3.3.

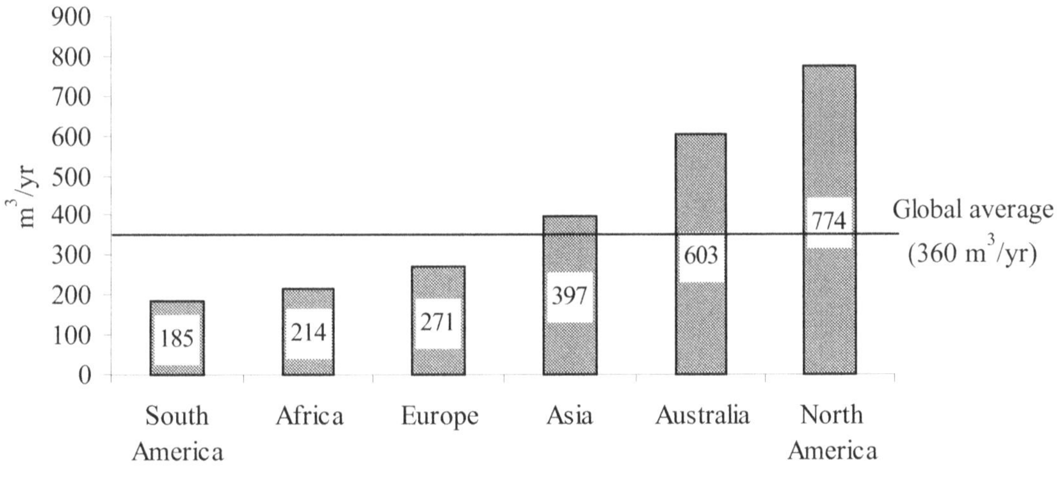

Figure 3.9 Per capita water consumption by region (in 2000, in m^3/yr)

SUMMARY

This chapter has provided a general overview of levels and patterns of water use by global region and economic sector, and compared uses in North America to the rest of the world. This information on patterns and levels of use is provided here to offer some context and perspective for the discussion of water "values" that follows in this report. Of special note, the information that follows will reveal that much of today's largest volume of water use is not aligned to where the highest values are derived. For example, much of water use today is directed toward agriculture, whereas the value added by water in agricultural uses may be quite low relative to water-generated values in other applications.

Likewise, residential water uses in North America are dominated by outdoor applications such as lawn watering, and the largest indoor use is for toilet flushing—uses for which values are modest and the need for high quality water minimal. In contrast, typically all of residential water is supplied at premium (highly treated and monitored) quality as needed to provide safe tap water for the highest value residential uses of meeting essential human needs for drinking and cooking. These high value essential water demands require only a small fraction of total residential water use (i.e., about 1%).

Therefore, the evidence presented here on water use patterns reveals an evident disconnect between the levels of water use and the value added (and quality required) to meet other water demands. The quality and value of water needed to meet essential household and economic applications of water point to a need to better align water qualities and prices with the values of the various types of water uses. A review of the literature on the range of values associated with various instream and extractive water uses is provided in the next two chapters.

CHAPTER 4
OVERVIEW OF THE ECONOMIC CONCEPT AND MEASUREMENT OF VALUE

"Some people know the price of everything and the value of nothing." (Oscar Wilde, attributed)

INTRODUCTION: PROVIDING AN ECONOMIC FOUNDATION FOR VALUE

Chapter 2 of this report introduced the concept of value and discussed what the term "value" might mean to people with respect to water resources and water quality. That chapter discussed the concept and types of value in general and nontechnical terms, to reflect how a broad spectrum of the general populace (including water professionals and stakeholders) may conceive of and use the term value with respect to their attitudes about how water resources should be used and managed. Chapter 2 was broad and all-encompassing, to provide a crosswalk between lay concepts of value and the manner in which value-relevant terms are conceived of and used within the economics profession.

In contrast, this chapter focuses more narrowly on the economics-based definitions and measures of value. This economics-oriented approach is adopted here because the technical concept and empirical measurement of "value" reside squarely within the domain of the economics discipline. Therefore, in this chapter, the research team provides an introductory and simplified overview of the economic concepts associated with valuing water resources. Although the economics profession does not provide the only way to conceive of, define, and measure value, this report needs to provide readers within the water community with a suitable grounding in the sound technical basis that economics provides on this topic. The objective here is not to dwell on economic theory per se. Rather, this chapter is designed to provide a relatively basic discussion of the conceptual basis that applies to water valuation, as well as clear definitions for applicable economic terms that might otherwise be used or interchanged inappropriately by noneconomists. For example, this chapter describes important distinctions between price, cost, and value to minimize the often confusing misapplication of these terms.

The conceptual background is provided here so that empirical discussions in later chapters can be more readily interpreted. In specific, the objective is to reduce potential for confusion related to semantics, and to help readers gain a better understanding of the merits and limitations of the alternative economic valuation methods applied (and the value estimates thereby derived), as reported and used in the following portions of this report.

The overall discussion is aimed at providing information to utility managers to enable effective decision-making on the efficient use and allocation of scarce water resources. To that end, the discussion below uses the framework of water provision by a utility to its customers. The term "consumer" refers to the residential, commercial, and other customers that use utility-furnished water, and the term "supplier" or "producer" refers to the water utility. Clearly, both the consumers and suppliers in this framework are each consumers and suppliers at different levels in the spectrum of water use. The fundamentals discussed below provide the foundation for all discussions on the suitable valuation, pricing, and efficient allocation of scarce resources, including water.

PAYING ECONOMIC RENTS FOR A NATURAL RESOURCE: A BASIC PREMISE

A fundamental principle on which efficient market economies operate is that consumers of goods and services bear the full costs of the resources they consume. By bearing the full costs, consumers use the relevant resources efficiently; i.e., resources are directed by the market only to those uses where the "value" to the consumer exceeds the "full cost" of diverting the resources to that use. The cost of the resource is thus intended to reflect the full "opportunity cost" of using the resource in a given application, based on bidding the resource away from its most highly valued competing use. The market uses "prices" to allocate scarce resources between buyers and sellers. Prices are used to balance supply (which is driven by the cost of providing the good or service) with demand (which is based in the value of the good or service to the consumer).

Gasoline Prices Reflect the Value and Cost of Crude Oil

The notion of paying a price that reflects the full cost of a resource can be illustrated by considering the market for motor fuel. For example, when refueling their cars with gasoline, one component of the price consumers pay for gasoline is for the actual crude oil resources extracted from the ground, in addition to pumping, refining, transportation, and marketing costs. The crude oil resources have a value independent of the other associated costs of the refined gasoline at the pump because these resources are privately owned and in limited supply. The payment for the natural resource itself (in this example, for the crude oil) has a component called "economic rent." The economic rent is the payment received by a resource owner for the resource itself (i.e., the payment received over and above the cost needed to produce the good). Consumer prices for gasoline also reflect the costs of refining and then delivering the final petroleum product to the pump. In other words, consumers typically pay for the raw resources extracted (crude oil) as part of the price of the final product, plus the cost of treatment and distribution. (For petroleum products, there also are market distortions and other factors—including government subsidies, taxes, and international price fixing cartels—that are reflected in the "at the pump" prices faced by consumers, but these factors are ignored here to help simplify the example.) The economic rent paid for the crude oil is accrued by the owners of that resource, because there is clearly defined ownership and control of petroleum and mineral resources under our legal systems. The level of economic rent paid for the crude oil is driven by the market-based forces of supply and demand, and the market encompasses many suppliers and demanders of crude oil. The interaction of supply and demand results in a "price" that dictates how much of the finite natural resource (crude oil) is extracted and sold in any time period, and the market-determined price also determines who will buy and sell (and how much).

It should be noted that while the pump price of gasoline includes many components of the total provision costs, there are many market distortions in the market for crude oil and gasoline that affect the final pump price. Thus while there are active markets for gasoline, these markets are not necessarily efficient because of the number of price distortions associated with the Organization of Petroleum Exporting Countries (OPEC), government subsidies and controls, and other associated regulations.

Tap Water Prices Rarely Reflect the Value (Opportunity Cost) of Raw Source Water

For water to be allocated and used efficiently, the same economic principles seen in the crude oil example must apply—water prices for residential and commercial water consumers need to reflect the full social costs of the potable (or other) water they consume. These costs include (1) the cost of the of the raw water resource itself and (2) the costs of treating, pumping, and conveying the finished product to their taps. Ideally, the prices should also reflect any other social costs of extracting the water for consumer use. For example, if there are adverse impacts on fisheries downstream because flows become too low to support the relevant ecosystems, then such external costs should be reflected in the allocation and pricing as well (external costs are discussed in greater detail later in this chapter).

Water utilities have generally been adept at using their prices (rates) for capturing their out-of-pocket expenses for pumping and treating water (i.e., recapturing the costs of delivering suitable quantities to their customers at the desired locations and qualities). However, consumer water pricing (or in general water resource planning) rarely reflects the value of the raw water resource itself.

There are some exceptions under which source waters are paid for, especially in the western United States, where water rights are based on the property rights legal doctrine of prior appropriations. In such settings, some suppliers need to acquire rights to extract additional waters from surface supplies, and thus they may pay some "price" (in the form of a water rights purchase) for the raw water itself. Yet these surface water rights purchases often do not reflect a suitable economic rent or value for the water (for reasons discussed at greater length below). Moreover, throughout the American West, many water rights are long-standing endowments that were not paid for by the utility, or were purchased many years past and are no longer a borne cost reflected in pricing, or are significantly undervalued, or both. In addition, western utilities with groundwater supply options typically do not have to pay for the rights to those subsurface waters.

OPPORTUNITY COST AS A KEY BASIS FOR WATER RENTS AND PRICING

In many water-abundant regions of the country, the predominant components of the full cost of providing potable water are the expenses associated with treatment and distribution. Because the volume of available water in these settings is large relative to overall demand, typically there is a very low "opportunity cost" for the extracted water resource itself. Opportunity cost is the cost associated with using the unit of water in its next best competing use. It is likely that the opportunity cost of using an additional acre-foot of water is low in water abundant regions because after the utility extracts what it needs, an abundant supply remains instream to meet the demands of competing uses of the resource.

In such cases of water supply abundance, cost recovery by the public water supply utility can be focused on debt service and operating and maintenance expenses associated with their treatment plants, distribution networks, and other fixed assets. Because the opportunity cost of the water extracted by the utility is low, the economic rent associated with use of the water resource itself is relatively minor. The rent for the source water typically is uncollected by the water utility (i.e., the source water itself is obtained for free by the utility, such that its customers also pay no rent for the raw water). But in this type of water-abundant setting, the water is not significantly underpriced: the price generally reflects the full opportunity costs of the source water.

In contrast, in many regions water is in relatively short supply. As a consequence, every acre-foot of water used by water utility customers is an acre-foot of water taken away from an alternative use. The competing uses can include instream uses for recreation or ecological preservation, or extractive uses such as for cropland irrigation. Where water is relatively scarce, the opportunity cost of water allocated to one type use, such as agriculture, can be quite high. The opportunity cost of extracting large volumes of water for municipal and industrial uses can include millions of dollars in foregone cropland yields, and perhaps the loss of local agricultural communities, lifestyles, and associated support sectors. The opportunity cost might also include the loss of important fish species and the associated value of ecological consequences or foregone recreational and commercial opportunities (or both).

The opportunity costs of water extraction for public water supply purposes can also reflect values in competing uses by other residential and commercial utility customers across space or time. For example, nonsustainable mining of groundwater supplies is, in effect, taking water away from other current users (spatially) as well as future users (temporally, leaving higher costs and diminishing water yields to our children and other future residents).

Many utilities are looking to extract more water to meet growing demands from expanding populations and increased economic activity. In the American West, this often manifests as the conversion of surface water rights from farmers to cities. This means that each additional acre-foot consumed by utility customers is an acre-foot that otherwise would have been applied to agricultural lands or other productive uses. Hence, rights to surface water, or access to groundwater, must be obtained at a price. Whether a utility should pay that price depends on how much its customers value the additional water compared to how much value is foregone by altering the water use allocation. Accordingly, a water agency needs to understand both the cost and the value of the water resources it may need to acquire and deliver.

WATER AS NATURAL CAPITAL

The concept of "natural capital" is sometimes used to describe the values associated with water and other natural resources. This section provides a brief discussion of the concept of natural capital and the manner in which it applies to water resource values.

Water, and most natural resources, can be considered capital in the same way that physical buildings, machinery, treatment facilities, and other materials are considered capital in the production of goods and services. As such, water capital is an asset used in the production of a flow of goods and services. The concept of "natural capital" was first attributed to William Vogt's (1948) *Road to Survival.*

The major categories of natural capital are renewable natural capital (living and active) and nonrenewable natural capital (passive). Ecosystems consist largely of renewable natural capital. Renewable natural capital can possibly yield an essentially unlimited steady flow of useful goods and services as long as it remains intact. The volume of the flow, however, is finite. Overuse of renewable natural capital can impair or destroy its ability to regenerate itself and sustain the flow of goods and services. Nonrenewable natural capital is associated with stocks of nonrenewable natural resources such as oil and minerals. This natural capital can be extracted and consumed for a finite period of time. While the total stock of water may be fixed, similar to oil or minerals, the water cycle generates a flow of water much more similar to a renewable natural capital asset.

Service flows from natural capital have economic values that are derived from the contributions that these service flows make to human welfare. The economic value of a natural capital stock is the sum of the discounted present values of the flows of all the services it provides to people (Freeman 1997). This is the same measure of economic value that is applied to any asset or commodity.

Some service flows from natural capital assets are easily valued (e.g., market values of trees harvested), while others are not bought and sold in markets and therefore do not have market prices (e.g., clean water). In these cases, the total economic value of an asset and all of its service flows may not be reflected by its market value (if it has a market value at all). For example, the price of an acre of wetland might reflect its value for potential commercial or residential development, which might be quite different from the value of its ecosystem services as wildlife habitat and as a means of filtering nutrients, controlling floods, and recharging groundwater aquifers.

Unfortunately, the connection between some types of ecosystem functions and an economically valuable service flow can be indirect and subtle. For example, Freeman (1997) discusses how water, as the basis for aquatic ecosystems, is an input to the provision denitrification services. Denitrification has no market value. However, this service can enhance submerged aquatic vegetation, which can enhance commercial and recreational fisheries. Since fish harvest is a service flow that can be valued, the relationship between nitrate concentrations and fish populations could provide a basis for valuing the denitrification function in terms of its contribution to the economic value of fish harvest. If denitrification also supports other economically valuable service flows, the economic value of fish provides only a partial measure of the value of this ecosystem function. The total value of the denitrification function would be the sum of its contributions to the values of all the affected service flows.

The concept of natural capital as applied to water is fully consistent with how the concepts and measurement of value are developed in this report. The key is to recognize that source waters are a valuable asset, and need to be valued and managed in the same way that any other asset or commodity would be.

VALUE, COST, AND PRICE: DISTINCT BUT RELATED CONCEPTS

Three terms—value, cost, and price—are often used interchangeably. However, in economics, these terms have very distinct meanings that in turn have important implications for how available "valuation" data can and should be interpreted. These terms arise in the context of their roles in the "market," and therefore this section clarifies the distinctions between the terms cost, value, and price and then provides a brief overview of how markets function.

Value

Value is a fundamental concept in the field of economics and, while related, is distinct from the concept of price. In economic parlance, the worth, or value, of a commodity is defined in terms of the quantity of other commodities, or money, one is willing to give up to acquire the good of interest. In modern economies, money is typically the unit of exchange. Value depends on the desirability and need of the good, while price also depends on its relative scarcity.

The concept of value can be expressed by two common terms of welfare economics: willingness to pay (WTP) and willingness to accept (WTA). WTP is defined as the amount of money, or other goods, an individual would be willing to pay to acquire the good (in this case, a specific quantity and quality of water). WTA is the quantity of money, or goods, that an individual would be willing to accept to give up the good. WTP and WTA can be driven by any number of characteristics of the good, such as quality, timing, and location. These attributes are all important factors in the value of water. For example, all other things being equal, one would likely be willing to pay more for high quality potable water than for the same quantity of lower quality water. It should be noted that neither WTP nor WTA measures take into account the overall quantity of the good available in the market.

Whether WTP or WTA is the correct measure of value depends on the issue of property rights. If one currently holds rights to the good, such as a water rights, then the amount that one is willing to accept to give up those rights is the correct measure of value. Whereas when one is trying to acquire the rights, WTP is considered the correct measure of value. Therefore, while value may be driven by the quantity and quality of a good, it also is related to the rights of the individual seeking to either acquire or relinquish the good.

To more fully understand the concept of value, and its relationship to price, one needs to grasp the idea of "consumers" and "producers," or suppliers, of water. From the consumer's perspective, the "demand" for a water is simply a tracing of an individual's willingness to pay (i.e., value) for different quantities of water, holding all else equal. People place the highest value on the first unit of a good they seek to consume. For water, the highest value presumably is assigned by consumers to the quantity required to meet their family's drinking and related needs for basic survival.

Typically, the next marginal unit of a water provides less value to individuals than the previous unit. For each customer, the value of each next unit of water tends to decrease as more water is consumed, because after the highest value applications are satisfied (e.g., for basic needs such as human consumption and bathing), subsequent quantities of water go to less valued applications (e.g., lawn irrigation). Hence, the "demand curve" for water—as for most goods and services—slopes downward in a typical "quantity versus value" diagram (where value or price is on the vertical axis and quantity is on the horizontal axis).

Demand curves can be for an individual, a group of people, or society. A demand curve for an individual shows what he or she is willing to pay for various quantities. An aggregate demand curve represents what a group is willing to pay for various quantities; it is the summation of demand across the group. Thus, the area-wide residential demand curve for water faced by a given water utility is an aggregate demand; it can be represented by the horizontal summation of individual household water demand curves whose slope and position vary, in part, due to differences in household characteristics.

In turn, individual household water demand equals the horizontal summation of household components of demand, categorized here as basic necessity indoor demands, indoor discretionary demands, and outdoor discretionary demands (hereafter, landscape irrigation demands). Figure 4.1 shows the components of household water demand and how they combine to portray a household's overall or aggregate demand curve for water.

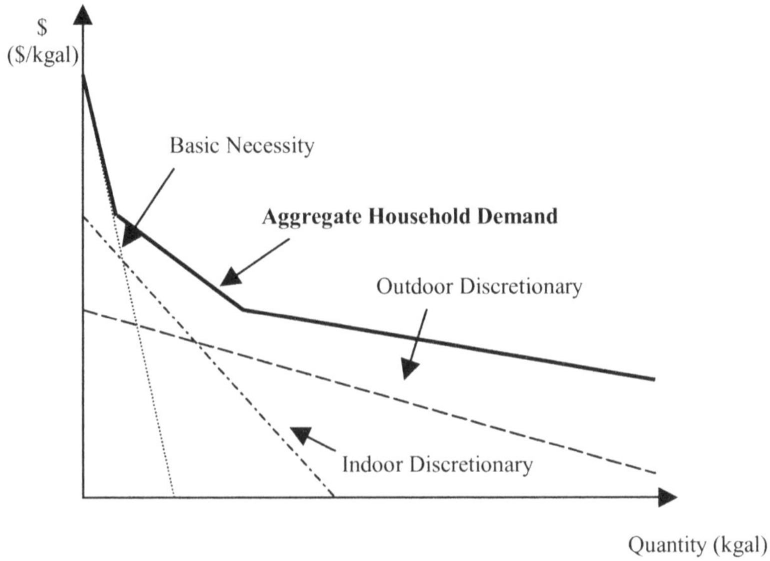

Figure 4.1 Components of household water demand

The nature, relative size, and WTP for each of these components of demand, which varies from household to household, influence the position and slope of the household demand curve. The position of the household demand curve is determined by the summation of water use dedicated to basic necessity and discretionary uses, and the slope of the demand curve depends on the price responsiveness of each demand component and the relative proportion of each component to total household demand, all other things held constant.

Since household characteristics influence the slope and position of household demand curves, they are expected to influence the willingness to pay for water. As a result, the composition of aggregate household demand is an important determinant of WTP. In general, consumers are expected to be willing to pay more for basic necessary indoor uses (such as drinking, cooking, bathing, and sanitation), since there exist no close substitutes for these uses, than they would be willing to pay for discretionary activities. Also, if aggregate demand is decomposed, basic necessary uses would probably constitute a smaller share of total demand than discretionary uses (chapter 3 provides data on water use levels for various indoor and outdoor components of residential water demand).

Cost

From the utility manager's viewpoint, a number of factors contribute to the cost of providing water to customers. This cost often becomes the basis for the price of the water (i.e., the rates) charged to customers. The total cost of water faced by the utility can be divided into two main factors: (1) variable costs associated with the expense of operation and maintenance of utility facilities, and (2) fixed costs for capital outlays in infrastructure (e.g., pipes and treatment plants) and water rights (where applicable). Fixed costs do not vary with the quantity of water provided; variable costs, e.g., chemical, labor, or pumping cost, do vary with the quantity of the water provided.

The supply curve for water reflects the *marginal cost* of providing the resource to the potential buyers. Analogous to the demand curve, the supply curve is a mapping of the supplier's willingness to sell different quantities of water. The quantity of water that utilities are willing to provide historically has depended on the direct costs of providing water. The higher the costs, the less a utility would be willing to provide for a given price.

As a practical matter, most utilities are natural monopolies (the only supplier within a given area), and charge prices to their water users at rates that are intended simply to cover the out-of-pocket financial expense of their operations (including debt service on capital costs and annual operating expenses). Rates are often governed by local governmental entities that own public utilities, or by rate commissions that review and approve rates for investor-owned entities. In either case, utility rates are often kept at relatively low levels that recover costs (plus a modest return or profit for economically regulated private systems). If water rights acquisition or other out-of-pocket expenses are incurred to access and tap source waters, then these rents for the resource are included in the rates charged to customers. However, as a general rule, utilities do not face the expense of source waters (at least not in a manner that reflects full rents and opportunity costs), and thus customers in general do not pay prices that reflect the full or true value of the water they are using.

Water utilities also typically seek to minimize the costs to their customers, and so provide water in a manner that takes advantage of their least cost supply options. A utility will therefore first tap and deliver sources that require relatively little pumping or treatment. If demand grows to exceed the lowest cost supply option, then the utility will add its next lowest cost option to meet the next increment of demand. This results in a de facto water utility supply curve that might resemble Figure 4.2, in which additional quantities are provided in fixed increments or blocks, depending on the quantity of water available under each supply option. The sequencing of the supply options depends on the cost of providing that block of supply. As increasing quantities are needed to meet the demands of a growing community, utilities often find that they face increasing costs to supply each new increment of demand.

As noted above, the costs utilities typically incur to supply water to their customers entail financial expenses paid out of pocket by the utility. These financial costs include expenses such as those made for capital outlays (i.e., the debt service on these capital outlays), operating expenses such as those for labor and fuel, and, to the extent applicable, the costs incurred (if any) to acquire source water. These internal costs often do not reflect the full cost to society of providing waters to utility customers, because the full social costs need to reflect the opportunity costs (i.e., the returns or value) that would otherwise be obtained from applying those waters to their best alternative uses (e.g., cropland irrigation). These opportunity costs increase as more water is diverted from other uses (as water is diverted from uses that have increasing values of their own, such as from low value crops to higher valued returns on irrigation). This is another reason why the supply curve is upward sloping.

In addition to opportunity costs for the resource itself, full social costs of providing water to water utility customers should include any *externalities.* Externalities are third party impacts not reflected in or compensated by the market, such as environmental damage that may occur because of excessive reductions in in-stream flows, disposal of brine waste, habitat loss due to transmission lines and reservoir development, and so forth. The issue of externalities is discussed in greater detail later in this chapter.

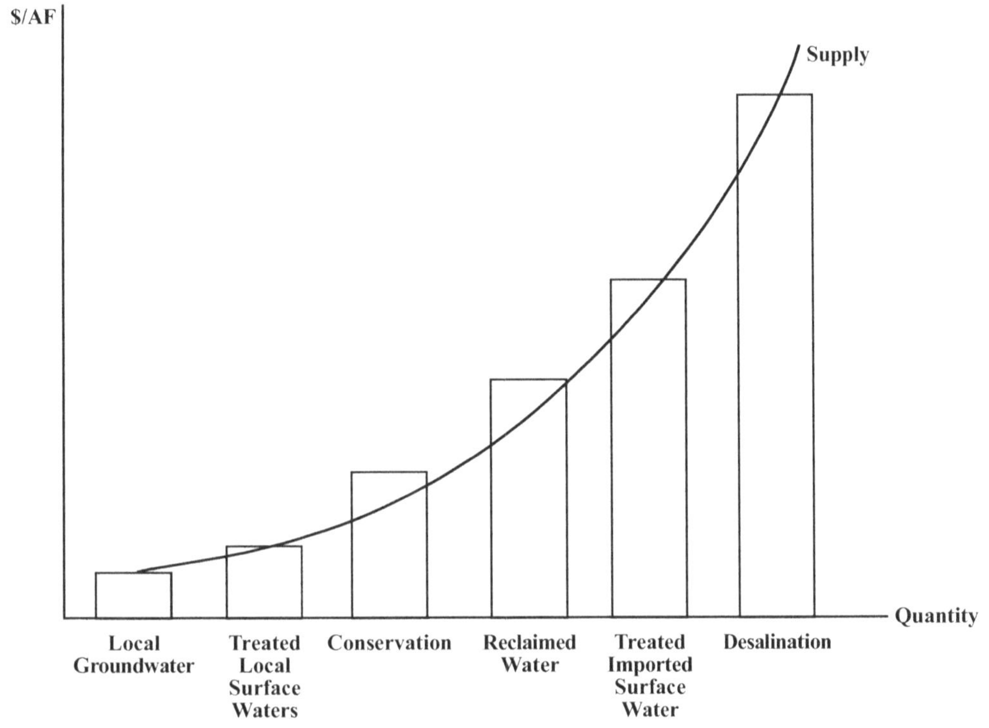

Figure 4.2 The cost of options for delivering potable water shapes a utility's supply curve

Price

In a market, a *price* is determined by where the supply and demand curves intersect. At this unique market-clearing point, *price, cost,* and *value* are all equal numerically. However, the terms still represent different concepts, and at all other quantities the cost and value (WTP) have different dollar values.

Figure 4.3 shows a generalized supply and demand graph, in which the demand curve reflects value to the customer and is downward sloping. The supply curve reflects increasing (upward sloping) costs to the utility to deliver water to its customer base.

In this depiction of an economic market, P* is the price that would clear the market (i.e., the quantity demanded would equal the quantity supplied at that price). The market clearing quantity is depicted as Q*. It is important to note the value of the commodity depicted in this market is greater than either its cost or its price for all but the last unit traded. That is, to the left of Q*, the values of water (embedded in the demand curve) are all higher than the costs (depicted in the supply curve), and value also exceeds P* to the left of Q*. For quantities beyond Q*, however, value is less than cost, and application of the resource for this market use is not efficient beyond that level.

In markets where there are many suppliers and consumers, price is often considered to be a good estimate of the marginal value of the good to both consumers and producers. It is thus considered economically efficient to allow the market process (i.e., prices determined in competitive market) to dictate the allocation of resources.

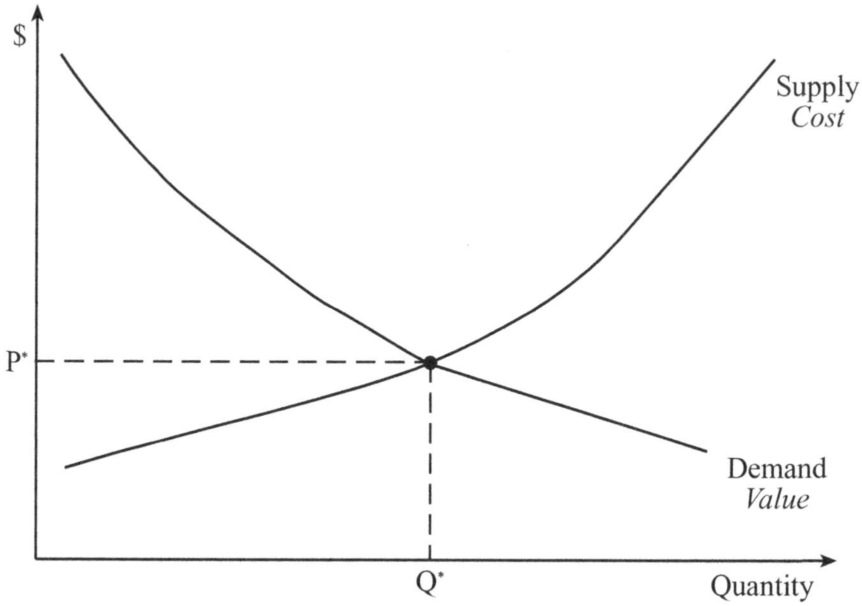

Figure 4.3 Supply and demand: Price relative to value and cost

In the water arena, however, the market typically is not fully functioning, and thus the prices are not always an efficient allocator of water resources across various uses and users. Insofar as the prices charged by utilities for water typically embody only out-of-pocket expenses (and omit or under-reflect opportunity costs for water in alternative uses, or externalities), the market in which utilities supply water to residential and commercial customers is likely to under-price water. This implies too much water is consumed in these markets (consumer values for the water uses exceed the price charged, but not the full social cost of the resource).

OTHER FACTORS THAT INFLUENCE THE VALUE OF WATER

The value of water for a public water supply utility (i.e., municipal and industrial supply) is influenced by several additional factors above and beyond how the type of water use, and the distinction between cost, value, and price, affect the interpretation of valuation data. In this section, we evaluate additional components of the value of water associated with physical characteristics, such as its potential for delivery at the location of interest (e.g., at the tap); at the quality required (i.e., in compliance with all federal and other applicable public health regulations and with acceptable taste, odor, and other aesthetic characteristics); and in reliable quantities. Also considered are nonmarket values such as associated with environmental and recreational uses of water.

Physical Components of Value

Water Quality

Water that requires a high level of treatment (e.g., disinfection and contaminant removals) requires that additional expense be invested to deliver the product (relative to waters requiring minimal treatment). This increases the cost of supplying the water, above and beyond the opportunity cost of alternative resource uses. Once purified, the water has the same value to the customer as any other potable source; it simply costs more to supply it (driving the price higher in the market).

From the utility perspective, the price it is willing to pay for the low quality water will reflect the higher costs of treatment. A utility will prefer to buy higher quality waters instead, if they are available for purchase at a comparable price. The value of lower quality water (e.g., high total dissolved solids [TDS] groundwaters) can also be inferred as being the same as that of high quality water, but *minus* the added expense of required treatment (all other things equal).

Timing and Permanence

The value of water, as for any other asset, is affected by the timing of when it is available for use and the duration over which it is available. Water assured indefinitely (permanent water rights) will command a higher purchase price in the market than water available only during a limited lease or transfer period. Likewise, water available for use a decade or more into the future is less valuable (other things equal) than water that is available today—an acre-foot assured in 2025 needs to be converted to its present value equivalent, using the same discounting (e.g., interest rate) principles that apply to the future value for any good or financial asset. Also, the observed price of a water transaction will reflect whether the water is acquired for definite use or is purchased simply to preserve the *option* for future use (i.e., acquired for use in a water shortage emergency, such that the price reflects an "option value" type insurance premium rather than the value of resource use per se). Other temporal aspects that affect water values are discussed later in this chapter.

Reliability and Aesthetics

Other factors also are important to the value of water, including its reliability to the utility or to a customer (for example, see the review of the empirical research on willingness to pay for more reliable water supplies in chapter 5, and the San Diego recycled water case study summarized in chapter 8). Many other factors also can influence the value a user might place on water. For example, a residential customer may value improved taste and odor of his or her tap water, and will express their demand for these attributes in public meetings or to customer service agents. Such customers might also reveal their value through ancillary purchases they make (e.g., buying an in-home filter device or bottled water if that provides them water with aesthetic properties they prefer over tap water), whereas other customers may not see sufficient value in these alternatives to motivate such purchases.

Nonmarket and Externality Components of Value

Two very important economic concepts relevant to the value and pricing of water are the related (but distinct) concepts of "nonmarket" values and "external" costs The discussion above focused on what are known in economics as internal costs and values. The term internal refers to the fact that the costs are those that the entity involved (e.g., a utility) faces as an out-of-pocket expense. These internal costs are faced by the utility because the goods and services being used are traded in relatively efficient markets. For example, labor services must be purchased at a prevailing wage rate and total compensation package—labor is not furnished to utilities for free. Internal costs are costs associated with payments for items such as acquisition of the resource, transportation, treatment, and distribution. These are "values" that can be inferred from prices because they pertain to goods and services that are traded in efficient markets. In these instances, values can be inferred directly from market behavior and associated prices.

Nonmarket Values

There are many goods and services valued by people, that are not traded in the marketplace. One example is recreational fishing or rafting. Some people place a high value on a day of game fishing along a stream bank, but there is no market in which such angling days are bought and sold. The water in the stream—by providing habitat for the targeted fish—provides an important use value to the anglers, but it is a "nonmarket" value because it is a value that is generated outside of the marketplace. The fact that the value is for a good or service that is not exchanged in a market does not mean that the commodity has little or no value—instead, it simply means it is more difficult to ascertain the value of the water as an input to providing nonmarket goods and services that are valued by individuals and society as a whole.

Many types of nonmarket values can apply to water resources. Some pertain to direct uses of the resource (e.g., recreational fishing or boating that involves a direct physical connection between the instream or near-stream activity and the values generated for the user). Use values are values that individuals hold for water because of some specific current or near future direct use of the resource. For example, some direct use values for water might be associated with individuals' desires to bath, cook, or water their lawns. At the other end of the spectrum, use values are associated with cooling water used in the production of electricity. These values may include a value to keep open the option of using water some time in the future, often called "option value."

Nonuse (Passive Use) Values

There are also values that individuals hold for some goods and services independent of any direct use. These values fall in the category of nonuse or passive use values and are often thought of as "intrinsic" values. For example, some individuals care about the preservation of threatened and endangered species that rely on waters for critical habitat. People may value the preservation of such species, even if they never expect to visit the site or see the species themselves. In this case, the person holds a value for instream flows, but he or she does not actually use the water. Nonuse values are values derived independent of any direct current use or option to use the resource in the future. These nonuse values are sometimes called "existence" or "bequest" values because some people may hold value for water, not because they individually want to use the water, but because they simply want it to exist or they want to pass it to the next generation.

The presence of nonmarket values for water makes it difficult to use markets to guide the efficient allocation of this resource, because these nonmarket values are not readily reflected in the prices paid for source water (if any price is paid) as an opportunity cost of alternative allocations.

Externalities

The concept of externalities is closely tied to the presence of nonmarket values for various types of alternative water uses and allocations. There may often be impacts or costs associated with diverting waters for public water supply that are not reflected in total (or at all) in the marketplace faced by a utility. These "externalities" are costs imposed on third parties (e.g., downstream residents) associated with the resource or use, but not explicitly paid for by the utility.

Some typical external costs that may arise would be the impact of removing water from a stream when the resulting lesser instream flow adversely affects individuals or entities that hold values for the water downstream. Removing water and leaving lower quantities downstream may, for example, adversely impact agricultural irrigators, resulting in a market-valued loss in agricultural output and farm income. This is an externality involving market-based goods and services. If the water diversion adversely impacted downstream anglers, then this would also be an externality, but one involving nonmarket services. In either case, there are impacts or costs imposed that are external to the financial accounting faced by the utility; the externalities are costs that are borne by society as a whole, not by the utility per se. From society's viewpoint, these externalities are overlooked opportunity costs and should be reflected in the costs borne by entities as they make their decisions.

Society has derived several mechanisms to limit and account for externalities. Some mechanisms are institutional controls such as water rights or instream flow requirements that preclude (or at least limit) over-extraction upstream at the expense of downstream users and resources. Another mechanism to account for or correct externalities is to set up institutional arrangements under which those who suffer the adverse impacts must be fairly compensated by the entity that creates the externality. This approach has the effect of "internalizing" the externality by creating a cost to the responsible party, and the cost of compensation thus serves as the price signal that becomes part of the basis for how the upstream party will henceforth consider its actions. For example, if an upstream polluter is forced to pay a fine or compensate downstream parties adversely affected by their discharge, then this "polluter pays" approach will motivate a reduction in future discharge levels.

Clearly, consumers of water may hold any or all of the many types of values when thinking of their total value for water. Specific values may be more strongly associated with specific types, qualities, or locations of water. For example, tap water is typically associated with the direct use of water for residential use, whereas instream water supporting aquatic ecosystems could be associated with nonuse and other nonmarket values. Thus, when thinking about the value of water, one must consider the type, quality, and location of that water. For efficient allocation of water, the total value of water to consumers, including use and nonuse values and any externalities, must be included in understanding the total demand for (and opportunity costs of) water.

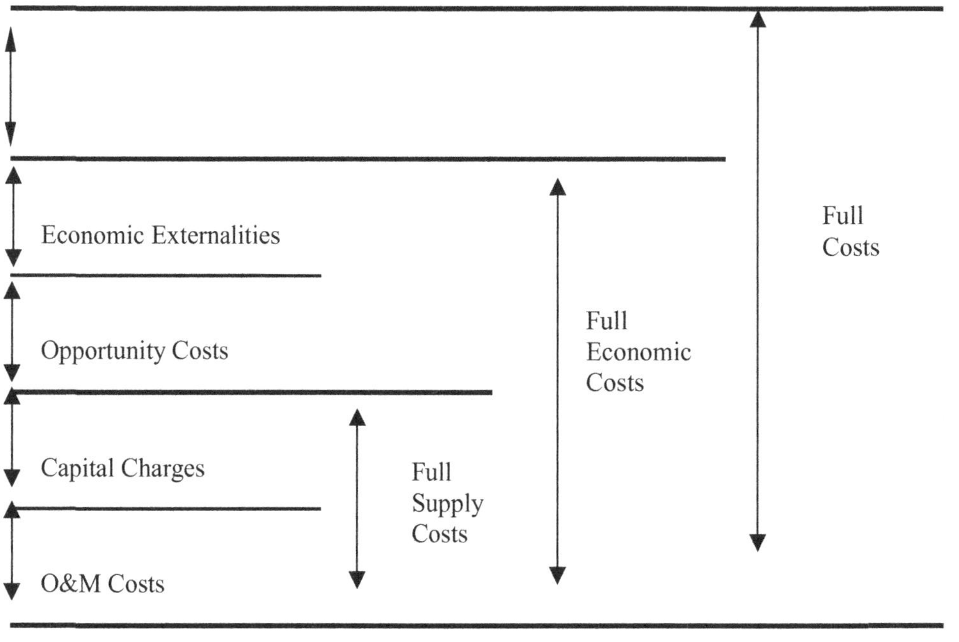

Source: Adapted from Roger, Bhatia, and Huber (1988)

Figure 4.4 Components of the full cost of water (not drawn to scale)

Components of the Full Cost of Water

Figure 4.4, adapted from Roger, Bhatia, and Huber (1998), describes the various subcategories of costs and values associated with water use. Figure 4.4 graphically represents the categories of cost that, when combined, equal the full cost associated with obtaining and supplying water. The figure is divided into three main sections: full supply costs, full economic costs, and full costs.

Full supply costs are direct financial costs associated with obtaining, treating, and transporting water. Included in this cost category is the actual cost of source water. Source water cost can fall into either operations and maintenance (O&M) costs, if obtained as an annual payment of a water lease, or capital costs, if the water right is obtained as an outright purchase of a permanent water right. Other costs included in the full supply cost include plant and distribution system O&M costs and any expenditures on current or future capital improvements and debt servicing. These direct financial costs that make up the full supply costs directly relate to the utility's ability to supply water and thus approximate the long-run marginal supply curve described in Figure 4.3.

Full economic costs include those costs associated with the full supply costs, and in addition include opportunity costs and economic externalities. Opportunity costs are cost associated with alternative uses of water. This cost accounts for the fact that in limited supplies, 1 AF of water applied to a specific use, say residential uses, cannot be applied to other uses, for example, in the industrial sector. If the productivity of water in the alternative sector would have been higher, then there is a real lost opportunity. The cost of that lost opportunity is termed opportunity cost. An economic externality is a third party effect, either positive or negative, associated with the use and allocation of water in a specific manner. For example, in the case of return flows of surface water, there can be negative effects such as reduced downstream water quality associated with contaminants added by activities using the water upstream. The decline in water quality downstream can have negative effects, and thus additional costs, such as lost crop production or higher treatment costs, on downstream users of the river. While these are real costs to other water users associated with the upstream water uses, they are external to the decision-maker extracting the water upstream, and are thus called externalities. Both the opportunity cost and economic externalities should be accounted for when determining the full economic costs of specific water supply allocations

Finally, to complete an accounting of the full cost of water, costs associated with environmental externalities need to be included. Environmental externalities, similar to economic externalities, are effects on the environment, outside of the common decision-making process. These environmental externalities include items such as the effect of surface water extraction on downstream ecological health. These downstream effects can impose real costs on society, and thus, when taking full account of water use, should be included in the overall calculations.

Components of the Full Value of Water

Figure 4.5 describes the components that make up the full value of water. Similar to the full costs of water, the full value of water can be divided into two major categories of value: economic value and intrinsic value. Economic value consists of: value to water users, positive economic externalities, positive environmental externalities, and societal objectives. The full value of water includes the economic value of water plus the intrinsic value of water.

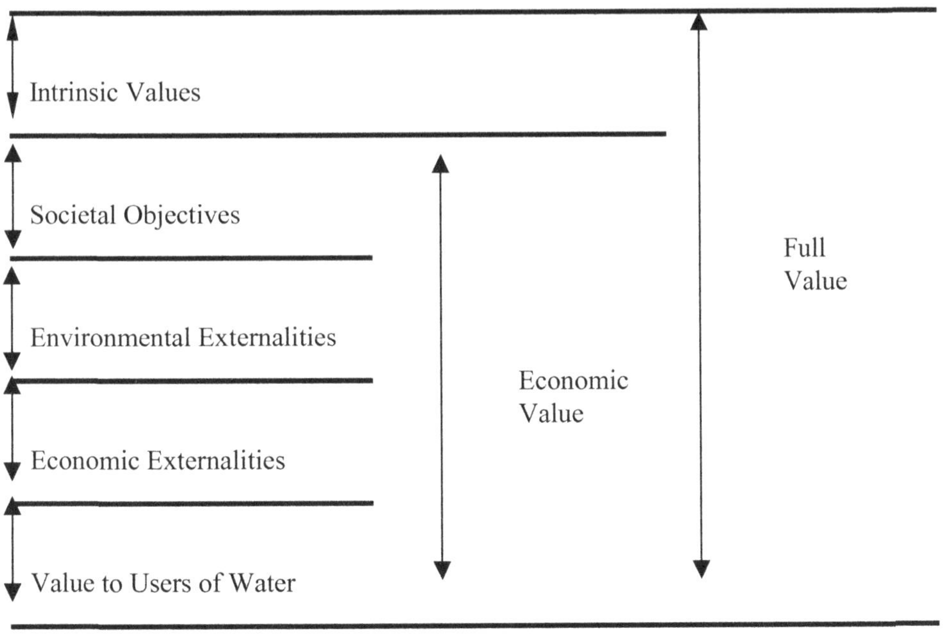

Source: Adapted from Roger, Bhatia, and Huber (1988)

Figure 4.5 Components of the full value of water (not drawn to scale)

The value of water to users is the amount that individuals, or utilities, are willing to pay to acquire the additional quantities of water. For example, to a farmer, this might be the additional value of crop production that might be generated from additional water applications. As expected, this value is specific not only to the sector but also to a specific use of the water. The value of water to a residential user for outdoor garden irrigation might be very different from the value of an additional quantity of water to a commercial user.

Economic externalities, as described above, are those economic benefits or costs that are outside of the individual decision-making process. Often these benefits are associated with return flows of water applications. For example, application of crop irrigation water can have a positive effect on users of a groundwater aquifer that is recharged by the application of irrigation water. When farmers are deciding how much water to use to irrigate their crops, they are not typically including the potential external benefits to groundwater users of downgradient aquifer recharge.

Like economic externalities, there can be positive environmental externalities associated with individual decisions on water use. Examples include the augmentation of ephemeral stream flows and the ecological function of the stream provided by irrigation return flows, and the in-stream use of water for recreational activities such as kayaking when those flows are provided for improving the overall health of the fish population.

The final component of the economic value of water includes societal objectives. There are often societal objectives, and benefits, associated with water use decisions. The value of community development and growth is a common objective when deciding on developing additional water supplies. A major component of the value of creating large water transport infrastructure such as the California Central Valley Project was the development and support of agricultural communities in the central portions of California.

Both externalities and societal objectives can be significant portions of the overall value of water. While these values are often not included in an individual's decision on water use and application, they can have effects on society's measure of the overall value of water and significantly influence water policy decisions.

The final component to make up the full value of water can be labeled intrinsic value. Intrinsic values for water are those values an individual, or society, places on water independent of any direct use. For example, many people hold value for water as an asset that could be passed on to future generations. Like many components of the value of water, the magnitude of this bequest value may be very situation specific. Another type of intrinsic value is existence value. Existence values are values that people place on water simply because it exists, not as a good to bequest to future generations or use in any manner. For some, the existence of ample instream water flows for ecological purposes provides a value to them. While perhaps difficult to measure, intrinsic values should be incorporated into any calculus on the overall value of water.

Temporal Aspects and Present Values

Benefits and costs from water projects often occur as a stream of values over time. That stream may change in magnitude over time. Water projects usually have large upfront capital costs that are paid for either up front or more likely over an amortization period at the beginning of a project, whereas benefits may accrue over the economic life of the project, which for some projects can be substantially longer than the amortization period. Values that occur in different time periods need to be adjusted to their comparable "present values" in order to compare them or make calculations with them.

There are two interrelated factors to consider when comparing values from different times—inflation and the "time value of money." When inflation is accounted for in recording or projecting values over time, the values are said to be in "nominal" terms. Many financial analyses are conducted in nominal dollars. However, for economic analyses, benefits and costs are normally *not* entered in inflation adjusted dollars. The use of "real" (i.e., not inflation-adjusted) dollars makes analyses easier and keeps inflation-related projections from clouding the analysis. In real dollars, a dollar today has the same purchasing power as a dollar 10 years from now.

The second factor to account for in comparing values over time is the fact that most people would not treat an inflation adjusted dollar today equal to an inflation adjusted dollar available in the future. Most prefer to have a dollar today instead of in the future because they prefer to use that dollar to consume today or they prefer to invest that dollar today to yield a return. This near-term consumption over deferred consumption preference is referred to as the "social rate of time preference" or the "time value of money." The annual rate at which present values are preferred is known as the discount rate (and is similar to an interest rate). The greater the preference for immediate benefits, or the greater expected rate of return on other investments today, the greater the discount rate. The discount rate can be expressed in nominal or real terms. A real discount rate is the nominal discount rate with the inflation rate subtracted out. The remaining real rate is equivalent to the risk-free and tax-free premium for investing money or resources for future use instead of consuming it today. The key is to use a real discount rate when analyzing dollars in real terms, and a nominal discount rate when analyzing values in nominal terms.

Economic theory suggests that in a world with no taxes, no financial transaction costs, and zero risk, there would be a clear signal about what discount rate to use. If consumption today would come at the expense of investments in the future, then the opportunity cost of capital should be used to discount the stream of future benefits and costs. In that case, the discount rate should be equal to the rate of return that could be earned by investing the money. If inflation is expected to be 4% in the future, and there is a 3% risk real return on capital, then the nominal discount rate should be 7%, and 3% is the real discount rate. If instead the use of funds or resources today predominantly displaces future consumption (instead of investments), then a social rate of time preference is more suitable as the discount rate.

There are philosophical and practical aspects to the choice of discount rate, and there is not general agreement among economists or policymakers about the correct discount rate to apply to evaluating projects. Justifications can be made for a range of rates, from a zero discount rate to a discount rate reflecting the private cost of capital. The argument for a zero discount rate is that discounting distorts project benefits that may occur far into the future and thus affect future generations, or that include irreversible outcomes (e.g., species extinctions). Others suggest that the discount rate should reflect prevailing interest rates on low-risk bonds because such risk-free, net of tax rates best reflect the rate of social time preference. This might be reflected by the real cost of capital to municipal agencies in raising capital through bonds, or the cost of long-term federal government bonds. Another argument is that for projects that will be paid for through water rates, that the cost of capital to ratepayers is the appropriate measure. This cost of capital might reflect an average of credit card debt rates, home and automobile finance rates, and other consumer rates—and may average around 8 to 10% in nominal terms. The argument for using the private cost of capital is that the project's funds might be otherwise invested in private ventures, and therefore reflects the true opportunity cost.

Various governmental entities have specified discount rates to be used in analyses. The federal Office of Management and Budget (OMB) regularly updates discount rates in Appendix C to its Circular Number A-94 on Guidelines and Discount Rates for Cost Benefit Analyses of Federal Programs (OMB 1992). OMB recommends using real interest rates on Treasury notes and bonds matched to project time period for the real discount rate. The real interest rate on a 30-year note as of February 1994 was 3.5%. OMB also mandates that federal agencies apply a 7% real rate of discount when evaluating the costs and benefits of federal regulatory actions (U.S. EPA 2000a).

To compare streams of value over time from different projects, the stream of values for each project is discounted to "present value" using the discount rate. If both benefits and costs are involved, the present value of the costs is subtracted from the present value of the benefits to get the net present value (NPV) of the project. If the NPV of a project is greater than zero, then the present value of the benefits is greater than the present value of the costs. The NPV of different projects can be compared if they are adjusted to be in the same year's dollars. Comparison of NPV of projects allows apples-to-apples comparisons of project values regardless of possible differences in the timing of benefits and costs for each project.

THE ROLE OF THE MARKET

Ideally, Markets Reveal Values and Allocate Resources to Highest Valued Uses

Economics is the efficient use of scarce resources to maximize the welfare of society as a whole. Economic efficiency dictates that water be allocated according to where the greatest values will be derived from its use. Where fully formed (competitive and well informed) markets exist, the working of these markets provides the mechanism through which those resource allocation decisions are optimized—a market clearing price (and associated quantity) of the good is established at the point at which the marginal costs of supplying the good are equal to the marginal value of the product to its buyers (consumers).

Under the market system, prices determine resource allocations, and water is provided to those buyers for whom value exceeds or equals the price (the amount up to Q* in Figure 4.3). Water beyond that quantity remains with those potential suppliers for whom the opportunity cost of giving up the water exceeds the price they could obtain in the market. Accordingly, utility customers need to place high enough value on those waters such that their willingness to pay is sufficiently great that they bid those waters away from their next best alternative uses. If utility customers place the waters' value (WTP) at less than the opportunity costs, then the waters should instead be used in their alternative modes (i.e., not be provided to utility customers).

These principles are evident where fully formed competitive markets determine the pricing and allocation of goods and services (e.g., where water rights are actively bought and sold without undue restrictions or price distortions). However, these pricing and allocation efficiencies may be more evasive where markets are not fully competitive (as is more typical for most water rights transactions because significant constraints on trading often exist). Therefore, data on water costs and prices need to be interpreted carefully when trying to infer values.

Interpreting Value From Water Markets

The market for water in any region where rights are exchangeable tends to be limited in the scope of its potential buyers and sellers, and at any one time may be in a period of disequilibrium. Therefore, the pricing information observed from recent transactions needs to be interpreted with appropriate caveats. Several approaches and data sources can be used to infer a range of costs and values for water resources in a given region—most approaches and data sources provide useful information on the *value* of water, but each also has its limitations (e.g., cost data provide a lower bound estimate of the value of the resource). The information therefore needs to be interpreted from a broad perspective, with the accumulated data providing a weight of evidence from which to infer a reasonable indication of the true value of water in the region.

Among the approaches used in this report to consider water acquisition pricing are those that rely on information obtained from recent transactions in the regional market for water (both source waters and treated effluent). While these transactions offer a glimpse at pricing, they also need to be interpreted within the context of several other important factors that have an impact on the price paid for a particular water right. For example, the permanence of the arrangement needs to be considered: temporary leases and transfers are less valuable than permanent transfers of the water rights (i.e., where a land purchase or an outright acquisition of the water rights themselves has been made). These and other factors make interpretation of past transactions of limited use for inferring a "fair market price" for a water utility, or for inferring the true valuation of water in the near and distant future.

Approaches used to value water resources are based on examining the "value added" or "marginal value of product" of water (or the guaranteed access to it) in various relevant uses in a region. These include examining the opportunity costs of diverting water from current uses (e.g., the value of agricultural yields produced when the waters are kept in irrigation). Also examined is the value added to potential residential development properties when service with reliable potable water is assured. The value of water to commercial activity in the region also is examined briefly (based on the cost of obtaining water from the cheapest or most feasible substitute source). As with other approaches, the information derived from these explorations of the data needs to be interpreted with caution. While marginal value added is the appropriate economic concept, the data can be difficult to interpret in isolation of other factors that affect overall property or agricultural values. Likewise, the use of an opportunity cost basis is likely to understate the value of use when water is deflected to other sectors.

Market Conditions

Markets in which restrictions in trade apply or price distortions are introduced (e.g., taxes or subsidies) do not provide price data that reveal reliable valuation information. Further, a setting in which the quantity demanded by one buyer overwhelms the demands of other market participants, or is large relative to the total available supply, provides data that deserve scrutiny in terms of their usefulness in inferring values. The anticipated growth of utility demand for surface water in a regional water market (e.g., relative to the amount of rights held by irrigators who may be willing to sell) may cause future prices to increase substantially.

Externalities

Costs (and hence prices) observed in the market may not include or reflect third-party impacts related to nonmarket goods and services (e.g., environmental and equity concerns). Such impacts may arise from water transactions or activities related to utility water use. These uncompensated nonmarket impacts are referred to as externalities (i.e., they are external to market and hence are not reflected in prices). Any significant external costs associated with water acquisition and delivery by a utility need to be identified and included in the assessment of value. These may include, for example, brine waste disposal issues (associated with desalination of high TDS groundwaters), concerns for instream flow protection, "environmental justice" related to impacts on low income families, or social values such as a desire to preserve an agricultural lifestyle and land use character for parts of the broader community.

MARKET DISTORTIONS AND WATER: SUBSIDIES RELATED TO WATER SUPPLY AND DEMAND IN THE UNITED STATES, AND HOW THEY AFFECT PRICES AND PERCEIVED VALUES

There are many aspects of water supply and demand in the United States that receive subsidies which in one way or another distort the value of water communicated to users through prices. These subsidies range from the federally subsidized supply of irrigation water to western agriculture to federal crop price support programs, subsidized crop insurance, and a variety of water rate distortions in municipal water supply.

Effects of Federal Water Resource Development Policies

Irrigation of agricultural land is a large portion of the water used in the United States. Of the 383 million acre feet (MAF) diverted annually from the nation's surface- and groundwater systems, irrigated agriculture accounts for 150 MAF. Of this, 61%, or 91 MAF, per year are consumed. Nearly half of this use occurs in the four states, California, Idaho, Colorado, and Texas, with 97% used to irrigate land in 20 mostly western states (USDA 2002). The Bureau of Reclamation supplies water to over 20% of irrigated acreage in the 17 western U.S. states. The Bureau program initially supplied water to only agricultural users, but today supplies drinking water to approximately 20 million domestic residents in the western United States.

Prices for irrigated water are not determined by markets, but usually are set administratively to recover historical water delivery costs and contract charges from federally subsidized irrigation projects. These costs in 1998 ranged from $10 to $85/acre, with an average of $41/acre, or an average of $16/AF (USDA 2002). This particular focus on historical cost recovery, coupled with institutional insulation of irrigators from market forces that would otherwise signal resource scarcity through the value of water from competing sectors, produces a significant distortion in how the value of water is perceived. This significantly distorts how the value of water can be deduced from observed pricing and use patterns, especially with respect to cropland irrigation.

The Reclamation Act of 1902 initiated the federal role in water resource development in the West. The Reclamation Act directed the newly created Bureau of Reclamation to develop water storage facilities and deliver the water to irrigators under contract. Seen as a policy to encourage settlement and economic development of western lands, the Act never intended to fully recover project development costs. Subsidized through no-interest, long-term contracts to irrigators—and in some cases capping these contract costs based on producer's ability to pay—federal policy aimed to lower irrigation water costs and increase the quantity of water demanded by the agricultural sector. Those contracts were structured to allow the irrigators to repay their share of the project cost over time from the proceeds derived from their newly irrigated land. Congress always intended that projects would be partially subsidized—irrigators were given 10 years to repay project costs interest free. However, many irrigators wanted larger subsidies, and the Bureau supported enlarging the subsidy. Congress extended the interest-free repayment period to up to 50 years in the 1939 amendments to the Act. This repayment period applies to the 50 years after final completion of the project. Congress also stated that project costs that could not be paid by irrigators could be shifted to other project beneficiaries such as consumers of hydropower generated by the project. The Bureau is authorized to enter into contracts to reimburse O&M costs and a share of the capital costs that it believes appropriate. Until recently, contracts for covering O&M costs were not indexed for inflation, and so set fixed or maximum rates ranging from $2 to $8/AF for the length of the contract. Also, irrigated acreage owned by an irrigator that could be served by a project was set at 160 acres. However, farmers found ways around this limitation by leasing lands in excess of 160 acres to technically comply with the acreage limit while serving water to more than the limit. The 1982 amendment to the Act allowed farmers to own up to 960 acres of federally irrigated land, and to lease an unlimited number of acres.

Between 1902 and 1994, the Bureau of Reclamation constructed 133 projects for irrigation water at an estimated accumulated cost of $21.8 billion. Of this, $7.1 billion of these costs were allocated to irrigation, however, 53% ($3.7 billion) were determined to exceed the ability to pay of irrigators and were shifted to other sectors (USDA 2002). A 1981 GAO study found that the subsidy supplied for six reclamation projects ranged from 92 to 97.8% of total government costs. According to the Department of Interior, reclamation subsidies to all western farmers in 1986 totaled $534 million, and from 1902 to 1986, the subsidies totaled almost $10 billion. Other reports put the amount higher (e.g., Sax, Abrams, and Thompson 1991).

It is difficult to fully account for and measure the long-term impacts of the subsidies on water allocations and economic efficiency, which certainly vary widely across projects. These effects depend on the magnitude of the subsidy, the availability of alternative water, the profitability of crops, and competing demands from other sectors. In general, these subsidies encourage western farmers to grow crops that may be more efficiently produced elsewhere, and discourages water use efficiency. Subsidies may have allowed projects not likely to produce benefits greater than costs to be built. These subsidies also encourage projects that may be environmentally damaging compared to alternatives.

The Role of Power Generation in Water Supply Projects

A considerable amount of the payment obligation for irrigators has been shifted to power producers; however, irrigation has repaid a much a larger share of debt than power users to date. About $3.4 billion of the $7.1 billion in reimbursable Bureau project costs assigned to irrigation was shifted to power users after being defined as beyond the ability of irrigators to pay. Irrigators have been relieved of requirements to repay about $373 million through legislative charge-offs, leaving a repayment responsibility of about $3 billion. Irrigators had repaid $945 million of the $3 billion as of September 30, 1994. Western power agencies—Western Area Power

Administration and the Bonneville Power Authority—have repaid a total of $32 million of irrigation debt assigned to them, leaving $1.5 billion on the books as of 1994. So irrigators have repaid roughly 30 times more irrigation debt than power users have (Driver 1997). Taxpayers have had to step in to cover costs for reclamation projects that have not been assigned to or repaid by irrigation and power users.

Power generated by a reclamation project not only can help relieve project repayment burdens but also can provide a separate subsidy to project irrigators by reducing the cost of power needed to pump groundwater. Irrigation project participants are often allowed to purchase a fixed quantity of power generated by the project at lower than market rates. The rate for this preferential power is often set to recover only O&M costs of power production, not capital costs. This low-cost power can be a long-term disincentive to select water efficient or high value crops. Low-cost power has been one factor delaying irrigators switching from nonsustainable mining of groundwater resources to more sustainable use of surface water supplies, by keeping the cost of groundwater artificially low compared to the price of surface water supplies.

Effects of Federal Commodity and Trade Policies

Agricultural commodity prices are often strongly influenced by government programs and policies. Through price support programs, import restrictions, and export programs, the U.S. government is active in agricultural markets. The federal 2002 Farm Bill continues direct support of producers of a variety of crops, including wheat, corn, grain sorghum, barley, oats, cotton, rice, soybeans, oilseeds, peanuts, dairy, and sugar. Many of these crops are grown on the more than 55 million acres of irrigated cropland, as shown in Figure 4.6. The 1997 Census of Agriculture reports a market value of $192 billion for all agricultural products produced in the United States. Of this, nearly $50 billion (over 25%) is grown in the West, where irrigation is concentrated. Figure 4.7 shows average water use in inches of depth of water applied for selected crops in the United States in 1998. The figure shows that crops such as rice and alfalfa are more water intensive users of irrigation water than others such as corn, sorghum or soybeans.

Policies aimed at supporting agricultural incomes, providing stability to agricultural markets, and increasing food security (e.g., by lowering domestic food prices) generally increase agricultural water demand. For example, raising crop prices generally leads to increased agricultural water demands as farming intensity rises along with pressure to expand cropped acreage (Boggess, Lacewell, and Zilberman 1993). However, the 2002 Farm Bill also decreases water use and provides money for improved water management by allocating money to agricultural water conservation programs and earmarking funds for wetland and stream flow protection programs.

Possible Effects of Federal Commodity and Trade Programs

Distorted product and input prices alter farm-level decisions in significant ways and, thus, alter resource allocations in largely unintended ways. Lichtenberg (2001) discusses the adverse effects of improper resource prices on, among other things, agricultural water use. He cites several examples of unintended consequences arising from policies and technological changes. "Price support policies also tend to induce farmers to irrigate more intensively. This can mean a scarcity of surface water and/or groundwater depletion, in addition to water quality problems"

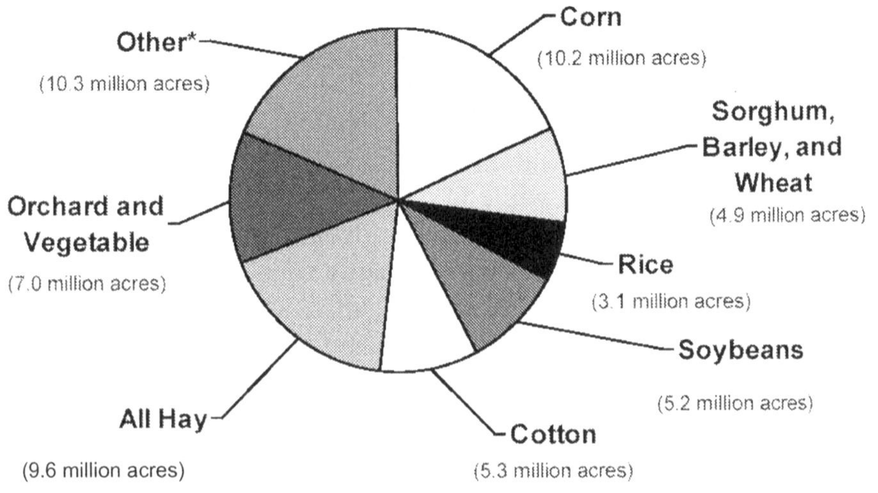

Source: USDA 2002.

Figure 4.6 Area irrigated by crop (2000)

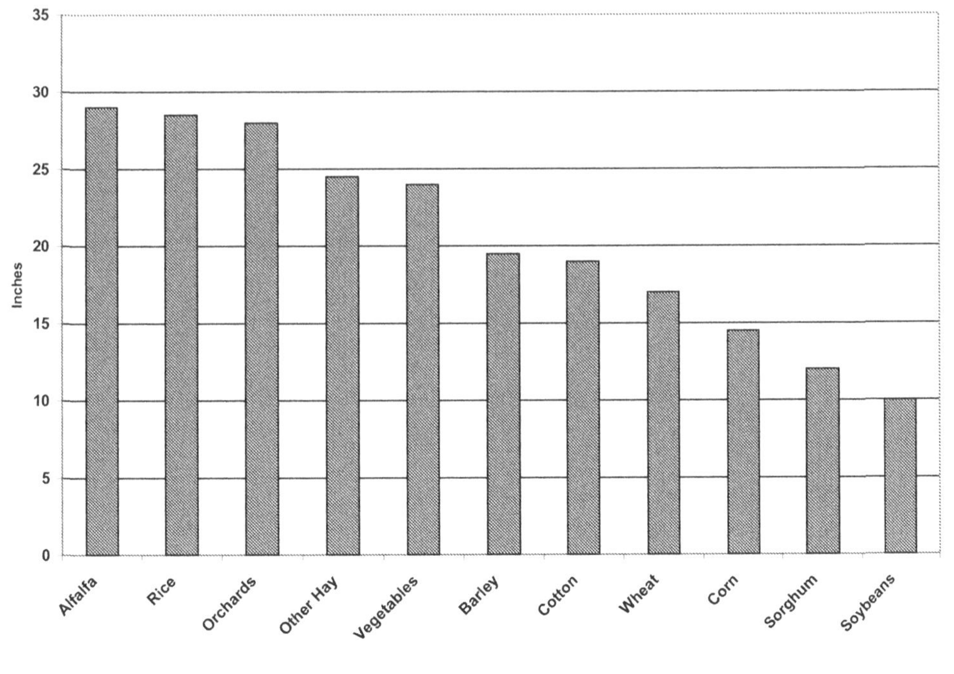

Source: USDA/NASS 1998.

Figure 4.7 Average depth of irrigation water applied by crop, 1998

(p. 6). He also notes that this problem can affect programs that support farm incomes directly as well as price stabilization programs because these programs tend to maintain farm output above desirable levels, thus distorting cropping patterns "in favor of crops which use more pesticides, fertilizers, and, in some cases, water" (p. 6).

By stimulating agricultural production, commodity support and trade programs attract resources, including water, into agricultural use. In some cases, water and other inputs are used more intensively to heighten yields, while in other cases, farmers may expand acreage and the extent of land under irrigation. For example, the advent of low-volume drip irrigation systems permitted irrigation to expand onto hillsides and sandy soils where gravity systems do not work well. Similarly, in the High Plains overlying the Ogallala aquifer, center-pivot irrigation systems have resulted in expanded production and water use—at least for the time being (Lichtenberg 2001).

Federal agricultural support programs, intended to provide farmers with sufficient and stable incomes, often distort product prices, while federal reclamation policy has significantly lowered the price of water, and prevailing institutions insulate that price from any reflection of water's opportunity cost or "real value" based on its relative scarcity. Moore, Negri, and Miranowski (1991) examine the effects of federal policies, including reclamation, commodity support, crop insurance, trade, and environmental policies on farmed acreage, production, and product prices in California. Specifically, they simulate changes in federal water supplies and commodity programs on cropping patterns. Their results show the often-conflicting nature of federal policies and suggest that "policy coordination is required to integrate diverse policies . . . to achieve a variety of goals" (p. 593). For example, in simulating a reduction in acreage set-asides for California farmers, the authors show that participation and acreage in the commodity program increase and change the land allocated to various crops. Though cropping patterns change, the net effect on water use is, however, indeterminate. This results from their analysis because of variations in crop-irrigation requirements and the authors' assumption that water use is fixed in total supply because of institutional rigidities that largely prevent transfers to, for example, urban uses.

Moore, Negri, and Miranowski (1991) also illustrate the potential for program and resource changes to affect national food prices by simulating the effect of a decrease in water supply from reclamation projects and the effects of support program changes. For example, they show that reducing federal project water by 10% in California induces an upward shift in the national supply curve for rice and vegetables (i.e., it costs more to supply each unit of farm output), resulting in higher consumer prices of more than 1.8% for rice and 1.2% for vegetables.

Possible Effects of Federal Crop Insurance Programs

Federal support programs also include subsidized crop insurance, intended to provide a safety net for farmers whose crops fail due to adverse natural events. According the USDA's Risk Management Agency: "Policies insure producers against yield losses due to natural causes such as drought, excessive moisture, hail, wind, frost, insects, and disease" (USDA-RMA 2004). The government pays the full cost of the premium for catastrophic coverage and pays a portion of the premium for higher levels of coverage.

Crop insurance, like other types of insurance, can distort decisions and behaviors in inefficient ways. Termed "moral hazard" effects, these distortions may arise from the insured limiting the use of mitigating or precautionary strategies—because the insurance coverage makes such strategies uneconomical. As federal crop insurance programs alter the planning environment of the farmer, it is reasonable to expect distortions in the behavior and allocation of resources by farmers by changing both input levels and cropping patterns. For example, even though drought conditions may be likely to persist, it may be quite rational for a farmer to plant and purchase crop insurance—using what water may be available at the time—even though the farmer may not expect to ever harvest the crop. Thus, the crop insurance program creates a "moral hazard" in which the farmer alters his/her behavior entirely as a result of the security afforded by the insurance.

Direct evidence of the effects of crop insurance on water use is unavailable except through analogy to studies of the effects of crop insurance on other inputs. Lichtenberg (2001) and Horowitz and Lichtenberg (1993) suggest that by reducing risk, subsidized crop insurance and disaster assistance programs can encourage farmers to intensify their production and increase resource use. Horowitz and Lichtenberg (1993) examine the moral hazard problem in which insured farmers are more likely to take greater risks. They observed that Midwestern corn growers' increased use of fertilizers and pesticides was highly correlated with crop insurance participation. They found that those who purchased federal crop insurance applied 19% more nitrogen and 21% more pesticide and treated more acreage with agricultural chemicals. Their results may not be directly applicable to water use, but are suggestive that crop insurance can clearly affect resource allocation at the farm level in terms of both input use and cropping patterns. It appears quite reasonable that if crop insurance can affect chemical inputs and cropping patterns in the Midwest, it may have effects on irrigation use in the West, where irrigation is a key input.

Municipal Water Pricing Practices and Subsidies

The price of water faced by municipal water customers often does not reflect the true value of water. There are several aspects influencing the price of municipal water. First, municipal water utilities have traditionally priced water to recover the average cost of supplies. In an environment in which the marginal cost of the next source of water is increasing, this practice can undervalue current supplies and does not reflect water scarcity through the anticipated cost of the next water source.

Second, utilities sometimes spread costs across multiple revenue sources in addition to water rates. These revenue sources can include grants, tap connection charges, taxes, and other charges. Partially in response to strong public opposition to water rate increases, water utility managers and city councils commonly avoid water rate increases by raising revenues from sources other than water rates.

Third, allocation of costs between water utilities and other parts of city government can become blurred. Diversion of funds from the water utility to support other city government functions are generally not allowed, especially if municipal water agencies are set up as an enterprise zone to preclude cross-department transfers. However, creative methods for extracting water agency revenues to support other local government functions are commonly used and can represent a significant portion of the water utility budget. The transfer of resources does not always flow away from water utilities, as costs can be incurred by other city departments to support water utility functions, such as a loan of a backhoe from the highway department for use on a water utility project.

For all these reasons, the rates used to price public water supply may not reflect the actual marginal costs of the water resource itself, or of pumping, treating, and delivering that water to utility customers.

ADDITIONAL KEY ECONOMIC TERMS AND CONCEPTS

To better understand the roles of value and cost in determining a socially efficient allocation of water, understanding of a few basic economic concepts is helpful. These concepts are addressed in this section.

Consumer and Producer Surplus

Two of the most fundamental concepts in determining the benefits and costs of alternative allocation of resources are consumer surplus and producer surplus. Consumer surplus is the basic measure of the benefits that an individual receives from specific goods or services. It is defined as the difference between what an individual would be willing to pay and what they actually have to pay to acquire the good. For example, if a consumer were willing to pay $20 per thousand gallons ($20/kgal) for residential water, but the market price is only $5/kgal, then that individual would be receiving a surplus of $15/kgal. The effect of different policies that either change the market price of water or shift demand, say through conservation education programs, is evaluated through the policies' effects on the resulting change in consumer surplus.

On the water provider's (producer) side, there is an analogous concept to consumer surplus called producer surplus. Producer surplus is the measure of the difference between what it costs a supplier to provide water and what they receive in terms of the market price. It is similar to (but not quite the same as) profit. Thus if the current market price is $5/kgal and it costs the utility $3/kgal to provide a certain quantity of water, the producer surplus is $2/kgal for that given quantity. The producer surplus is the sum of any quantity-specific surplus, aggregated across the quantities sold. There are a number of factors that contribute to producer surplus, such as profit, and what is termed economic rent.

Figure 4.8 shows graphically the concepts of consumer and producer surplus. For each unit of water consumed between 0 and 10, there is positive consumer surplus: WTP is greater than price. Thus for a consumer to maximize total consumer surplus, or welfare, they would consume 10 kgal of water. Beyond 10 kgal of water, the benefit of the additional water is less than the cost, so it does not make sense to buy more.

From the water provider's viewpoint, between 0 and 10 kgal, the cost of providing water is less than the market price. Thus for each unit of water between 0 and 10, the producer surplus is positive, and a utility would maximize producer surplus by providing up to 10 kgal of water. Beyond 10 kgal, the cost of providing water is greater than the price, and thus a utility would be losing money on each unit sold beyond 10 kgal.

Thus maximizing producer and consumer surplus is analogous to providing and consuming water in an economically efficient manner, which maximizes the new welfare of both providers and consumers of water.

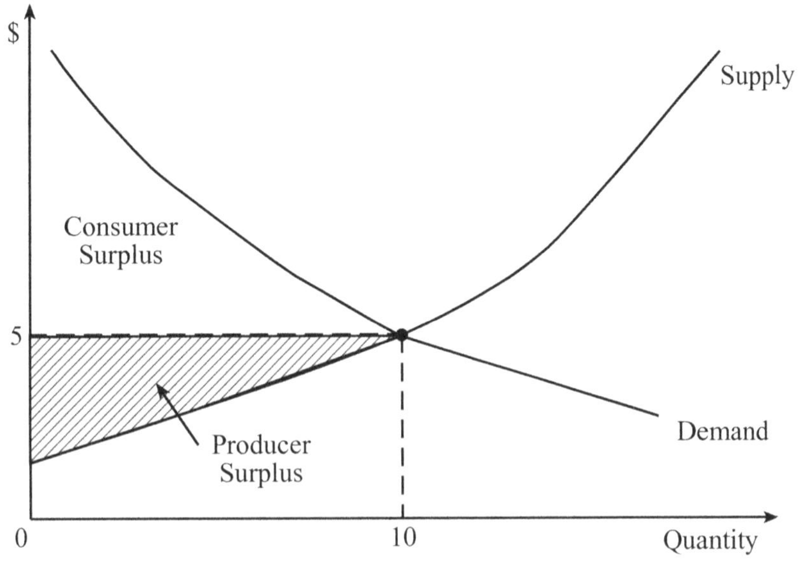

Figure 4.8 Consumer and producer surplus

Market Versus Nonmarket Goods and Services

Concepts applied in the water sectors often draw distinctions between market and nonmarket goods and services. Market goods and services are those items that are exchanged in the marketplace, and where a market price is applicable. For example, bottled water is provided in the marketplace. At the going market prices, individuals can determine if their consumer surplus would be positive, in which case they would purchase the bottled water. Water at the tap is also provided in a market setting, where individuals can determine how much to use based on the going price.

Many goods and services associated with water are not specifically traded in the marketplace. Ecological benefits, recreational fishery, and swimming are good examples of services provided by instream flows. While one may incur some costs to go fishing or swimming, you cannot typically buy those experiences in the marketplace like you can bottled water. In some cases, the nonmarket value can be quite substantial, and as discussed below, techniques have been developed to estimate these nonmarket values so they can be adequately incorporated into the estimate of the total cost and benefit of alternative uses of resources such as water.

Consumptive and Nonconsumptive Uses

Another distinction often made about resources is the manner in which they are used. Water can provide benefits both through direct consumption (e.g., drinking, irrigation) and through nonconsumptive activities such as swimming or fishing. Nonconsumptive uses often are associated with nonmarket values, but there is no direct requirement that nonmarket values be associated with nonconsumptive uses. For example, recreational fishing is typically valued through nonmarket methods; however, it can be a consumptive use of the resource.

One of the issues with consumptive versus nonconsumptive uses is that when one consumes a good, it is not available for others to use, whereas nonconsumptive uses do not necessarily preclude others from also using the resource. Water is a unique resource in that sometimes it can be used in a consumptive manner such as for irrigation, but in fact not all of the water is consumed by plant uptake or evaporation. Some applied irrigation water can re-enter the watershed either through direct surface runoff or as recharge to the local groundwater aquifer. While there may be less net consumption in the watershed, when closely examined, the act of extracting water from a stream for irrigation application keeps others in the area, or biological resources that rely on the water, from not being able to rely on the water at the current location. Thus when evaluating the actual consumptive use of water resources, it is important to evaluate the net consumption and the location (and quality) of any return flows.

Elasticity of Demand

One of the most common approaches used to evaluate effects of price changes on water demand is the use of percentage change in the demand for water in response to a specified percentage change in the price of water. The economic term used for this measure is price elasticity. Price elasticity is a measure of the percentage change in the demand for water for a 1% change in the price of water. If water demand is very responsive to price changes, then demand for that water use is considered "elastic," whereas if the demand is relatively unresponsive to price changes, it is considered "inelastic." The formal relationship is that if a 1% change in price causes a greater than 1% change in demand, then the demand is considered elastic, and if the change in demand is less than 1%, then demand is considered inelastic. Chapter 5 presents available empirical information on the estimated levels of price elasticity for water demand throughout the United States.

In addition to responsiveness to price, utility managers often care about the responsiveness to the demand for water for relative changes in a customer's overall income. Income elasticity is defined similarly to price elasticity, as the percentage change in the demand for water for a 1% change in income. Income growth can be a very important factor in predicting changes in water demand, because as households become wealthier, they tend to purchase and use more water-using appliances, purchase homes with larger yards to be irrigated, acquire pools, and so forth.

Clearly price and income elasticity will vary greatly by regions, and by the types of water used. Residential in-house water demand may be relatively price inelastic, because many of the uses of indoor waters are essential or highly valued (e.g., drinking, cooking, bathing). In contrast, residential water demand for outdoor uses typically is more responsive to changes in price because the outdoor uses are of relatively lower value to most residential customers. Hence, outdoor water use demand is more price elastic than most indoor uses.

It is important to note that caution should be used when interpreting reported price and income elasticities. Typically, elasticities are estimated for relatively small changes in prices or income. Applying these estimated elasticities to evaluate large (i.e., nonmarginal) changes will often misrepresent the expected effects of a price or income change. Additionally, estimated elasticities are based on an assumed or estimated base level of consumption. For example, a 10% change in the quantity of water used equates to a very different volume of water depending on whether the base estimate was 100 AF or 1,000 AF. Therefore, while elasticity estimates are useful to help explain expected consumer responsiveness to price or income changes, it is important to understand the range of price or income changes that the elasticity applies to, and the overall base level of demand implicit in estimated values.

METHODS FOR ESTIMATING VALUES

A number of methods are available to utility managers and researchers to measure the value of water to both consumers and providers of water. The basic distinction in the available methods is the use of data from some form of market transactions or the use of nonmarket valuation. In evaluating valuation methods, it is important to remember the distinctions between value, cost, and price.

Market Based Methods

Market Price

One of the most direct methods to estimating the value of water is to look at the market price for water. While a market price may be observed for water, the simple single observation of what people pay for water, or the price it is sold at, does not allow one to develop an estimate for the overall value of water. To use market price to estimate the value of water to consumers, market transactions for water must be observed across a number of different price levels and demand situations. By tracing the relative amount of water demanded at different prices, one can map out the demand curve for water. This demand curve for water is the consumers' willingness to pay (i.e., value), which can then be used to develop the change in consumer surplus for a policy change that changes either the demand or the supply of water.

Production Function/Factor Income

A common method used to measure the value of water in agriculture is to evaluate the benefit water provides to the overall production of crops. This approach evaluates water as one of the necessary inputs, along with seeds, land, labor, and other factors of production, to produce crops. The approach estimates the marginal value of water in the overall crop production process. This approach has two steps. The first step evaluates how additional water increases crop yields, and develops the "marginal productivity" of water in crop production. The next step evaluates the "value of the marginal product" by looking at the increase in crop value, which incorporates the price of the crops. If the addition of another acre-foot of water increases crop production by 5%, and that 5% increase increases crop revenue by $100, then if the cost of the acre-foot of water is less than $100, the value of the marginal product of water is greater than the cost.

This approach of estimating the value of the marginal product of water is also commonly applied to the commercial and industrial water sectors. There are several technical issues and nuances to using the "value of marginal product" approach to valuing water in agricultural and other such uses, and these are discussed in greater detail in appendix A.

Nonmarket Valuation Methods

Many goods and services associated with water are not traded in markets. As discussed above, often there are not well-defined markets for many water-based recreational activities.

While there may not be active markets, there are often costs that individuals incur to undertake direct use activities. For these types of uses, often we can use these costs to develop proxy "prices" for the activity, and use that information in developing the demand, and thus value, of water. This approach uses observations on people's behavior, or their associated expenditures, as indications of "revealed preferences" for the good. Methods have been developed, and are discussed below, that use these revealed preferences to develop estimates of the value of nonmarketed goods and services such as many water uses.

For other activities, where there is no direct use of the resource, and thus no behaviors or expenditures available as measure of people's preferences, methods have been developed to directly elicit preferences and estimate value. These direct methods are often described as stated preference methods because they most commonly elicit value through direct statements on value rather than using observations on behavior or expenditures to infer value.

Revealed Preference Methods

Hedonic Price Method. Almost all goods or services can be thought to consist of bundles of goods or services, some of which are priced on the market and some of which, such as environmental services, are usually not marketed. The most common example is residential housing. A house can be considered a bundle of goods and attributes such as the physical structure (bedrooms, bathrooms, square footage), relative location to schools and shopping, and the environmental factors such as views and the relative quality of the local environment (e.g., clarity of the nearby lake). The price paid for a house can be thought to incorporate the purchaser's value for all these characteristics. The hedonic price method estimates the value of each of the house's characteristics as a function of the total price of the house. This method relies on large amounts of data on the purchases of homes with a wide range of characteristics, including the nonmarket services such as environmental quality being valued. The data are evaluated to develop a relative value for each of the bundled goods and services that make up the house.

While this approach is commonly used to develop the value of nonmarket goods and services such as ocean views and the impact of hazardous waste sites on housing prices, it measures only a subset of the total value of the good or service of concern. The reason that it measures only a subset of the total value is that there may be value associated with preserved views, or removal of hazardous waste sites, above and beyond those incorporated into the local housing market. Thus, while this is a common approach to estimating the value of some nonmarket goods and services, it is not a complete estimate of value.

Travel Cost Method. The travel cost method is a revealed preference approach commonly used to estimate nonmarket recreational or environmental values. The basic concept in the travel cost method is that people expend time and money to travel to locations for recreational experiences. These expenditures of time and money can be used to develop proxy prices for the recreational experience. The "prices" can then be used to develop the overall demand for recreation and thus be used to estimate the value of the recreational experience. For those activities that are highly dependent on water resources (e.g., fishing), the value of water can be estimated through the value of the recreational experience.

Stated Preference Methods

Contingent Valuation Method. The most well-known stated preference method is contingent valuation (CV). In CV people are asked to reveal their preference, or willingness to pay, for specific nonmarket goods. CV has been used to estimate both direct use and nonuse values. Implementing CV requires developing a survey designed to elicit respondents' values for the good. A typical CV will develop a clear and focused discussion on the specific good to be valued. An example might be the protection of groundwater from reduced quality due to nitrate contamination. Respondents are then asked to provide (reveal) their willingness to pay to prevent the contamination. The responses are evaluated to develop an estimate of the demand curve for the good in question and thus their value—in this case, their value for preventing groundwater contamination.

Utility managers should use caution when using estimates from CV studies to evaluate the value of water in other situations. While CV studies may accurately estimate the values elicited in the specific scenario described in the survey, the values are specific to the scenarios used. As one moves away from the specific scenario described in the study, the transfer of those values to other situations becomes less accurate.

Conjoint/Stated Choice Method. Conjoint or stated choice methods are similar to contingent valuation, except that instead of responding to a single WTP question, survey respondents are given a set of choices. The surveys describe a set of attributes of the goods in question, and then vary those attributes across the choice questions. Respondents are asked to either rank the choices or choose a specific option. The set of responses is then analyzed to develop an overall estimate of the value of the various attributes of the choices.

Conjoint studies are commonly applied to estimate nonmarketed goods and services such as the value of cleaning up water supplies. As with CV studies, caution should be used when applying the estimated values from a conjoint study to other situations.

Benefits Transfer

Benefits transfer is not a specific method, as those described above, to estimate the value for goods and services. Rather, it is a "secondary" approach, using existing value estimates derived from prior primary research, such as using values from publications of hedonic or conjoint studies. Benefits transfer entails finding a "suitable" set of value estimates from the literature and transferring them to a new policy being evaluated. Benefits transfer is probably the most common method used in benefit-cost analysis because of the relative ease of application.

An example of a benefits transfer approach would be to use the value of residential water estimated through a market price analysis in Albuquerque, N.M., to determine the value of providing additional water to Tucson, Ariz. At first blush, this may be an appropriate transfer of values. However, there may be many differences between the two scenarios that would need to be accounted for so that the value estimated in New Mexico provides a credible estimate of the potential values in Arizona.

While benefits transfer may be commonly applied, as with the other methods, caution should be used when developing estimates of value. The accuracy of the values transferred and the ability to transfer values from one scenario to another depend on a number of factors specific to the original study, and to the new policy question. These include issues such as how comparable the new situation being evaluated is to the scenarios used in developing the original study, and the relative size of the changes in value of the specific goods being compared.

CONCLUSIONS

This chapter has provided an overview of the key economic concepts, definitions, and methods associated with valuing water. In the next chapter, empirical evidence is presented on the value of water in various key water use sectors (e.g., agriculture and residential uses) and settings, as derived from past empirical research efforts.

CHAPTER 5
EMPIRICAL EVIDENCE OF WATER VALUES

"Not everything that can be counted counts, and not everything that counts can be counted." (Albert Einstein, attributed)

The valuation of water in its many uses has been a major focus of discussion and research for decades. This chapter provides an overview and discussion of recent literature on the use and value of water in a broad range of water sectors in the United States. The focus of this chapter is to provide the reader with a basic understanding of the major issues in estimating water use and values for each of the sectors. However, a complete review of all the water use and valuation literature is beyond the scope of this report.

Growing populations, reductions in water quality, and increasing recognition of the many and varied uses of water, including instream flow requirements for ecological functioning, are increasingly challenging the primacy of historical water uses, values, and institutions. The reviews in this chapter are good examples of the types of water value estimates currently extant in North America. Future trends and conditions will certainly continue to alter the absolute and relative values, and thus require the reassessment of water allocation among competing demands.

While this literature review provides a good representative overview of the available information on the value of water in and across the various sectors, interpretation and applications of these values to other sectors or locations should be done with caution. While there are a significant number of estimates of the cost and value of water provided in this section, it should be noted that there is relatively little information on the actual value of water, as measured by willingness to pay by consumers, for many of the sectors of water use. Even within a sector, the value of water will be highly specific to regional and temporal factors. Given that water use and values are highly specific to location, timing, and quality characteristics, the presented values should be seen as representative of the potential relative magnitudes of the values of water in each sector rather than as specific estimates of value applicable for transfer to other situations.

The reviews in this chapter cover the major water use sectors: agriculture, residential, commercial/industrial, instream, cultural, and community economic development. Each section includes a discussion of specific issues related to estimating the value of water in the sector, and a review of current literature that estimates specific values. Each sector's results are summarized in a table, and a discussion of specific caveats when interpreting the values completes each sectoral section. (Appendices A and B contain detailed technical information on the agriculture and instream sectors.) The chapter concludes with a summary of the individual sectors and a comprehensive table of values. All monetary values reported in this chapter and the rest of this report are provided in 2003 U.S. dollars (USD), unless otherwise specified.

VALUE OF WATER IN U.S. IRRIGATED AGRICULTURE

"While irrigated land is but one-seventh of all cropland, it provides more than one-fourth of total crop value." Gibbons (1986, p. 23)

The implication of this observation is that water's value to U.S. agriculture is indeed significant, contributing much to the nation's food supply and security. Irrigation water also contributes to the variety and diversity of foods because of its importance for nongrain crops in many regions such as vegetable, fruit, and specialty crops in California, Arizona, New Mexico, and Oregon, among others. The contribution of water to the U.S. agricultural economy—and to many of its regional economies—is quite substantial. Perhaps of even greater consequence to society in general, however, is the overall extent and magnitude of agriculture's water use. Agriculture dominates water use patterns in many regions, most profoundly in the West and in parts of the Southeast.[1] In examining the key question posed in this section, that of identifying the sources and magnitudes underlying the "value of water" in the United States, agriculture's importance derives not from a special or relatively high value (or WTP) but rather from the sheer scale of its historical water use and its primacy within water-allocating institutions. This primacy persists because of the property rights structure underlying water use, the elemental role of agriculture in the nation's food supply and public perceptions of food security, and—until recently—a relative absence of widespread and significant competition and conflict over water.

This section reviews the literature on the contribution of water to the U.S. agricultural economy. Appendix A provides details of the specific methods used to measure the value of water in agriculture, the strengths and weaknesses of alternative approaches, the typical range of estimated values across crops and regions, and key issues that give rise to bias and uncertainty in estimated values. Key issues discussed that affect the estimation of water value in agriculture include effects of risk and uncertainty in water use decisions, the role of market distortions on agricultural water demand, impacts of technology and changes in water use efficiency, and indirect benefits and costs associated with agricultural water use.

Valuing agricultural water use is vastly more complex than it may appear at first. Figure 5.1 identifies many of the sources of complexity that confront farmers and the agricultural sector. For example, farmers are often economically squeezed by falling output prices and rising input prices, pressure that is exacerbated by government policies and regulations and weather conditions. Furthermore, farmers who are dependent on irrigation are increasingly mindful of the potential threats (and in some cases opportunities) from competing water users and political interests, especially in periods of severe drought. Each of these factors can alter water demands and values. The challenge in estimating the value of water in agriculture is to develop an assessment approach for valuing irrigation water that is attentive to the role of these and other factors.

Many of these sources of complexity, as well as a few that are not illustrated, are described in appendix A.

---

1. Eighty percent of all irrigated cropland is in the western United States, where it is the principal user of water, accounting for 80%-90% of water use. In some states to the east, irrigation is also a heavy user of water. States such as Louisiana, Florida, Georgia, Mississippi, and Michigan supplement rainfall with regular irrigation, accounting for 40-60% of consumptive water use (Gibbons 1986).

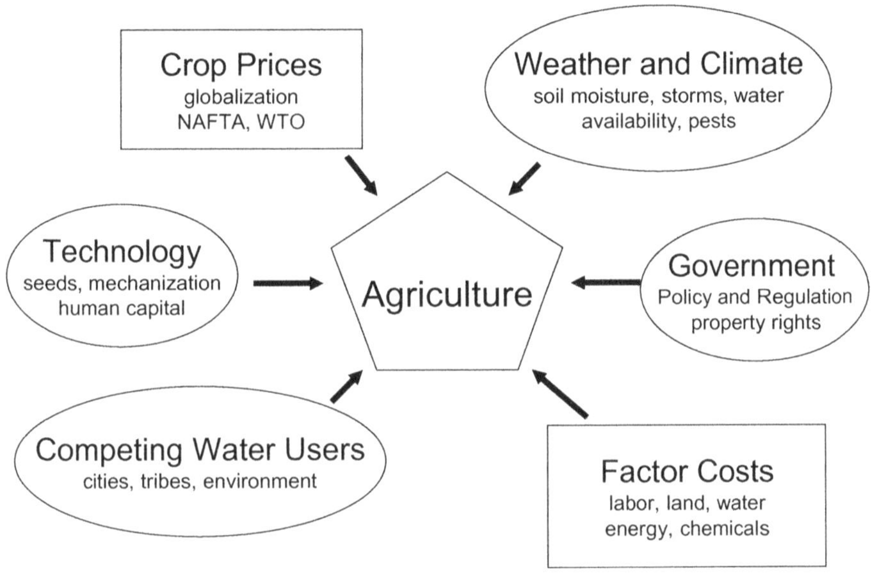

Figure 5.1 Water and agriculture: A complex environment

Table 5.1
Marginal water values from crop-water production functions (2003 USD/AF)

Crop	Idaho	Washington	California	Arizona	New Mexico	Texas
Grain sorghum				14		109
Wheat		57		21		34
Alfalfa				24	24	
Cotton			96	54	59	
Corn					50	55
Sugar beets		139				
Potatoes	672	271				
Tomatoes			375			

Source: Based on information provided in Table 2-2 in Gibbons (1986), p. 34. 1980 values in Gibbons adjusted to 2003 values using the producer price index for farm commodity prices. Farm prices in 2002 are about 96% of what they were in 1980.
Note: Values have been calculated at 10% reduction from yield-maximizing water levels assuming average efficiency and medium-textured-soil functions.

Value-of-water estimates for irrigation are widespread and varied. Frederick, VandenBerg, and Hanson (1997) have collected and presented estimated values (as well as values for other uses). Their estimates draw widely from the crop-water literature and from the estimates that Gibbons summarized in her 1986 review. Updating these estimates to 2003 values, Tables 5.1 and 5.2 highlight the variation within and across regions by crop, and the variation in regional estimates of agricultural water value, respectively. These estimates provide a good summary of the type and range of available estimates, and they provide a good "benchmark" for the relative value of agricultural water use. Caution, however, should be exercised in applying these values to site-specific applications, as highlighted in the following section. Table 5.3 provides water values by crop type.

Table 5.2
Water values for irrigation by region or watershed (2003 USD/AF)

Resource region	Average	Maximum	Minimum	Number of values
Mid-Atlantic	233	233	232	1
S. Atlantic-Gulf	23	67	0	5
Tennessee	22	22	22	1
Upper Mississippi	12	48	0	4
Lower Mississippi	24	59	0	3
Souris-Red-Rainy	0	0	0	1
Missouri	21	90	0	17
Alaska-White-Red	57	133	0	10
Texas-Gulf	95	234	0	20
Rio Grande	39	126	0	8
Upper Colorado	6	21	0	4
Lower Colorado	103	1,262	0	60
Great Basin	0	0	0	4
Pacific Northwest	169	1,447	0	18
California	131	890	0	21

Source: Based on information provided in Table 4.6 in Frederick, VandenBerg, and Hanson (1997). Values converted from 1994 to 2003 dollars using the GDP Implicit Price Deflator.

Table 5.3
Water values by crop (2003 USD/AF)

Crop	Average	Median	Maximum	Minimum	Number of values
Alfalfa	60	52	199	21	13
Apples	178	178	178	178	1
Barley	39	46	72	10	7
Beans	68	68	84	52	2
Carrots	648	648	648	648	1
Corn	187	115	157	52	7
Cotton	134	121	345	33	18
Grain sorghum	67	52	234	6	11
Hay	42	42	57	27	2
Hops	21	21	21	21	1
Lettuce	245	245	245	245	1
Melons	63	63	82	44	2
Onions	47	47	47	47	1
Potatoes	837	925	1,447	54	4
Rice	101	101	101	101	1
Safflower	62	68	81	30	4
Soybeans	142	149	209	70	3
Sugar beets	142	140	299	46	8
Tomatoes	809	809	809	809	1
Vegetables	242	242	242	242	1
Wheat	60	55	123	16	13

Source: Based on information provided in Table 4.7 in Frederick, VandenBerg, and Hanson (1997). Values converted from 1994 to 2003 dollars using the GDP Implicit Price Deflator.

Caveats on Accuracy and Bias in Valuation Approaches

Noted water economist Robert Young (1996) writes:

> Estimates of the value of irrigation water are often seriously flawed. In terms of direction of bias, the estimates are much more often overstated than understated. The pervasive bias toward overestimates arise[s] from a number of sources, not the least of which is political pressure on analysts to find the "correct" benefit-cost ratio for a project which already has strong political support.

Young finds support in his claim of bias in the observations of Franklin and Hageman (1984), who note the difficulty farmers find in repaying irrigation project costs. Young states, "While it is clear that self-interest colors many farmers' statements concerning their ability to pay for water, if the benefits were all that they are claimed to be, farmers would find it easier to repay at least a small part of the costs of irrigation projects."

Young and Haveman (1985) and Young (1996) identify the following potential sources of bias specific to agricultural water demand and valuation studies:

- Inadequate accounting for the opportunity cost of family labor and management
- Unreasonable crop price forecasts that tend to
  - Overstate scarcity of future food supplies
  - Not account for falling general agricultural price trends
  - Not account for the price and quantity effects of improved technology and increased efficiency
- Treatment of high-margin specialty crops
  - Overestimating the acreage and share of water used
  - No accounting for added management and risk costs with these crops
- Inadequate recognition of negative external costs (e.g., saline return flows)
- External (social) benefits should be net of crop price subsidies
- Failure to consider potentially less expensive, perhaps nonstructural alternatives

Each of these sources can hinder an accurate portrayal of the value of irrigation water, and it is prudent to consider these limitations and assumptions when identifying and using a specific value estimate. Space does not permit an explicit examination of each source of bias, but Young (1996) and to a lesser extent Young and Haveman (1985) offer a very clear and detailed treatment.

VALUE OF WATER IN U.S. RESIDENTIAL USE

A relatively extensive body of literature deals with the economic aspects of residential water demand. Unfortunately, there is very little literature on WTP for residential water uses, and most of these studies focus almost exclusively on developing countries. A small, yet growing body of economics literature examines WTP for specific attributes of water for residential use such as reliability. Most of the economics literature on residential water focuses on estimating demand and the factors influencing demand, particularly the price responsiveness (or price elasticity) of demand. A subset of this literature concentrates on measuring the relative effectiveness of price and other nonprice instruments (such as retrofit subsidies, rationing, and landscape irrigation controls) in reducing demand.

Given the preponderance of literature focusing on elasticities of demand, we examine this body of literature first. Chapter 4 described the concept and details of elasticities. The discussion here centers on what elasticities actually measure and some common misconceptions about elasticities. Looking at the findings from the literature provides guidance on the appropriate role of these studies in water policy and planning decisions. The discussion then turns to existing WTP studies, and the implications of these findings for decision making and future research needs.

Price Elasticity of Residential Demand

The own-price elasticity of demand—frequently referred to as the price elasticity of demand—measures the relative responsiveness of the quantity of a good demanded to a change in its price. Technically, the own-price elasticity of demand equals the percentage change in quantity divided by the percentage change in price. As a result of the negative relationship between prices and quantity demanded, own-price elasticity of demand is always less than or equal to zero. It is nonpositive because as prices increase, the quantity demanded will never increase (all else equal). Depending on the magnitude of the own-price elasticity, a change in demand is referred to as either "elastic" or "inelastic." "Elastic" means that for a given percentage change in price (e.g., 1%) the corresponding percentage change in demand is greater (e.g., > 1%). "Inelastic" means that for a given percentage change in price (e.g., 1%), the corresponding percentage change in demand is less (e.g., < 1%). Thus, elasticities are a relative rather than absolute measure. An own-price elasticity estimate provides information only about small (or marginal) changes in prices (i.e., up to a few percentage points). It cannot predict the effect of a large, say 30%, increase in price. Additionally, elasticities are relative to the total quantity of water currently being consumed and the prices paid, so comparisons across households, let alone regions or sectors, should be done with caution.

It is often stated that residential water demand is price inelastic, and most of the empirical estimates support this argument (own-price elasticity of demand is usually between 0 and -1.0). This information is commonly used to argue that consumers do not respond to water prices. This is a false statement; consumers do respond to price but at correspondingly smaller rates. If elasticity is -0.30, for example, a 10% increase in prices results in a 3.1% decrease in demand. In addition, relatively small price responses may be because the prices of water are typically extremely low to begin with, and this is frequently ignored.

Table 5.4 shows estimated own-price elasticities demand from a number of studies sorted by geographical area, from west to east. Perhaps the most striking feature is the sheer range in estimates. There are many reasons why these estimates diverge, including extent and quality of information on the determinants of demand, specification of variables and functional form, and appropriate application of statistical techniques. For a more thorough discussion of why price elasticities of demand differ, see Espey, Espey, and Shaw (1997), Arbues, Garcia-Valinas, and Martinez-Espineira (2003), and Dalhuisen et al. (2003).

Table 5.4
Estimated own-price elasticities of demand

Study	Location	Time period of analysis	Own-price elasticity
Nieswiadomy and Cobb (1993)	US	1984	-0.64 to -0.17
Nieswiadomy (1992)	US	1984	-0.60 to 0.02
Williams and Suh (1986)	US	1976	-0.48 to -0.18
Williams (1985)	US	1970	-0.62 to -0.22
Howe (1982)	US	1963-65	-0.57 to -0.03
Howe and Linaweaver (1967)	US	1961-66	-0.23 to -0.21
Foster and Beattie (1979, 1981)	US	1960	-0.76 to -0.12
Conley (1967)	US	1955	-0.35
Moncur (1987)	Honolulu, Hawaii	1977-85	-0.68 to -0.03
Corral, Fisher, and Hatch (1998)	California	1982-92	-0.30 to 0.00
Renwick and Green (2000)	California	1986-96	-0.16 to -0.20
Cassuto and Ryan (1979)	Oakland, Calif.	1970-75	-0.30 to -0.14
Renwick (1996)	Santa Barbara, Calif.	1985-90	-0.33
Renwick and Archibald (1998)	Santa Barbara, Calif.	1985-90	-0.53 to -0.11
Billings (1990)	Tucson, Ariz.	1974-80	-0.72 to -0.57
Agthe et al. (1986)	Tucson, Ariz.	1974-80	-0.62 to -0.27
Agthe and Billings (1980)	Tucson, Ariz.	1974-77	-2.23 to -0.18
Billings and Agthe (1980)	Tucson, Ariz.	1974-77	-0.61 to -0.27
Young (1973)	Tucson, Ariz.	1946-64, 65-71	-0.65 to -0.41
Griffen and Chang (1990)	Texas	1981-86	-0.38 to -0.16
Hewitt and Hanemann (1995)	Denton, Texas	1981-85	-1.59
Hewitt (1993)	Denton, Texas	1981-85	-1.23 to -1.12
Nieswiadomy and Molina (1989)	Denton, Texas	1976-80, 81-85	-0.86
Nieswiadomy and Molina (1991)	Denton, Texas	1976-80, 81-85	-0.94
Hansen and Narayanan (1981)	Salt Lake City, Utah	1961-77	-0.51 to -0.47
Jones and Morris (1984)	Denver, Colo.	1976	-0.34 to -0.07
Schefter and David (1985)	Wisconsin	1979	-0.13 to -0.11
Gottlieb (1963)	Kansas	1952-57	-1.24 to -0.66
Chicoine and Ramamurthy (1986)	Illinois	1983	-0.47
Chicoine, Deller, and Ramamurthy (1986)	Chicago, Ill.	1982	-0.42 to -0.22
Wong (1972)	Chicago, Ill.	1951-61	-0.82 to -0.02
Lyman (1992)	Moscow, Idaho	1983-87	-3.33 to -0.4
Schneider and Whitlatch (1991)	Columbus, Ohio	1955-77	-0.44 to -0.11
Stevens, Miller, and Willis (1992)	Massachusetts	1988	-0.69 to -0.10
Carver and Boland (1980)	Washington, D.C.	1969-73	-0.70 to -0.02
Hogarty and Mackay (1975)	Blacksburg, Va.	1971-72	-1.41
Danielson (1979)	Raleigh, N.C.	1969-70	-0.31 to -0.27
Gibbs (1978)	Miami, Fla.	1973	-0.51 to -0.62

Source: Developed in part based on information from Dalhuisen et al., 2001 and 2003.

Caveats on Price Elasticity Estimates

As discussed in chapter 4, decision makers who are considering using existing own-price elasticity estimates are advised to use extreme caution. Own-price elasticity estimates are site-, time-, and price-specific; they measure responsiveness of demand to *small* changes in price for a given location, time period, and apply only to a narrow band of prices and quantities demanded (i.e., they apply only to small changes from current prices and quantities consumed). Thus, an own-price elasticity estimate for Northern California is likely to be unsuitable for decision making in Southern California, and an own-price elasticity estimate for Tucson from 1980 is most likely no longer accurate in 2003. In addition, elasticities are valid only for the range of prices used to estimate demand and only over relatively small changes in price. For example, if a demand function were estimated using observed prices of between $1.25 and $1.50 per thousand gallons (kgal), then it would be inappropriate to consider price changes using a price significantly higher than $1.50/kgal, say $2.50/kgal.

THE VALUE OF WATER RELIABILITY

The reliability of a water supply refers to the ability to meet water demands on a consistent basis, even in times of drought or other constraints on source water availability. Residential and CII customers seem to value supply reliability quite highly, as indicated by the literature reviewed in this section.

The goal of any water supplier, irrigation district, or city utility is to deliver a reliable water supply. Of course there is a cost associated with the development of greater reliability for a given water supply. The question then becomes, What is the value of water supply reliability and are the associated costs justified? The managers of water supplies are constantly confronting decisions that require tradeoffs between cost and risk, especially where drought and other hard-to-predict factors affect both the demand and supply of water. The monetary benefits of risk reduction are often not known because they are difficult to quantify; therefore, it becomes difficult to determine the optimal level of risk reduction. In general, valuing nonmarket goods requires special attention and can be complex, but attempting to ascertain transferable value for water supply reliability is further exacerbated because the value of reliability is dependent on the user of the water supply, its intended use, and abundance of water in the region.

Although interest in water supply reliability is increasing, few studies have directly attempted to quantify its value. The studies that have attempted to quantify the value of reliability used stated preference and revealed preference methods. Revealed preference infers the value of reliability from data obtained from choices and decisions made in the market place. For example, expenditures made to obtain higher levels of reliability (i.e., to avert potential shortages) sometimes can be used to infer the value of reliability.

Stated preference methods determine estimates for reliability on the analysis of responses to hypothetical choices in surveys. While stated preference approaches have been applied to the valuation of nonmarket goods for many years, the method does have limitations that need to be acknowledged and considered. For example, Griffen and Mjelde (2000) note that one difficulty with stated preference studies for water reliability is the notion of the "birthright" perspective. It is not uncommon for respondents to view water as an inalienable right. Consequently, while they highly value water reliability, the notion that water should be free can lead to a reduction in their stated willingness to pay for reliability. However, if the limitations are acknowledged and efforts are made to perform the studies in an appropriate manner, stated preference studies can yield informative results.

Another method for quantifying the value of reliability attempts to infer values from some available cost and price data. While cost does not necessarily equate value, the cost that a city incurs for increased storage to improved reliability can be used as a proxy for the value of a reliable water supply. Additionally, avoided costs due to higher levels of reliability sometimes can be used to infer the value of reliability.

Stated Preference Studies

In 1987, a contingent valuation study was conducted for the Metropolitan Water District (MWD) of Southern California in an effort to determine the economic value for changes in the reliability of water supply among residents in Southern and Northern California. A reliable water supply is defined in the paper as "one without the threat of periodic shortages and mandatory rationing" (Carson and Mitchell 1987, p. 1). In the study, four scenarios of reductions in reliability are investigated and households' WTP to alleviate the threat those reductions in reliability is determined. Reductions in reliability are defined in terms of magnitude and frequency. The scenarios with the associated annual median WTP are shown in Table 5.5.

The study uses the 1983 census to determine that there were approximately 5.5 million households in the State Water Project service area. Multiplying the WTP per household by the number of households yields the annual aggregate value of providing greater water reliability.

Significant attempts were made to ensure that the estimated values are conservative and represent lower bound estimates. First, the study defines the value of water reliability in terms of willingness to pay rather than willingness to accept. Studies have shown that WTA is typically 2 to 6 times larger than WTP for public goods for which there are no substitutes (Carson and Mitchell 1987). Second, the study's WTP estimates are based on median values rather than mean values. The authors note that, while mean WTP is usually used in economic valuation, mean WTP values are typically 1.5 to 4 times larger than median WTP (Carson and Mitchell 1987). Third, those respondents that refused to participate in the survey or responded "don't know," are treated as households who are truly not willing to pay the specified amount. Therefore, they are treated as respondents willing to pay $0 and are not discarded from the study as is typically done in contingent valuation studies (Carson and Mitchell 1987).

Table 5.5
Annual median willingness to pay for households under four scenarios (2003 USD) (baseline = household's current consumption of water)

Scenario	Description of scenario	Household annual median WTP	Annual aggregate value of supply reliability ($ millions)
A	A 30-35% reductions from the baseline once every five years	$186	$1,027
B	A 10-15% reduction from baseline once ever five years	$135	$751*
C	A 30-35% reduction from baseline in two out of five years	$421	$2,280
D	A 10-15% reduction from baseline in two out of every five years	$248	$1,370

Source: Based on data from Carson and Mitchell 1987.

*The results for Scenario B were given using a 95% confidence interval ($653 million to $848 million). The mid-point of the confidence interval is reported in the table.

Though the authors attempt to be sound in their methodology, there are some problems associated with the study. The study uses a referendum format in the survey. In the discrete choice referendum format, respondents are asked whether they would vote yes or no on a referendum that would alleviate the threat of a water shortage for a specific magnitude and frequency, given a specified cost to their household if the referendum were to pass. This format of questioning has been shown to be inconsistent and usually overestimates WTP values (Jenkins, Lund, and Howitt 2003). Furthermore, the fact that the survey allows for the prevention of a water shortage (rather than a reduction in likelihood or severity) indicates that the WTP values should be interpreted as upper bounds for consumer valuations (Griffin and Mjelde 2000). The elimination of shortfalls is not a realistic scenario. However, it should be noted Griffin and Mjelde use an improved survey design, not allowing for complete avoidance of shortages, and obtained inconsistent WTP values.

In 1993, the California Urban Water Agencies (CUWA 1994) hired Barakat and Chamberlin, Inc. to design, conduct, and analyze the results of a contingent valuation survey to estimate the value to residential users of water supply reliability in 10 California water districts. More specifically, they sought to estimate how much residents are willing to pay to avoid water shortages of varying magnitude and frequency. Shortage magnitudes ranged from 10 to 50% and frequencies ranged from once every 3 years to once every 30 years. Bid amounts ranged from $1 to $50 (1994 dollars) increments to monthly water bills. The survey design was very similar to the Carson and Mitchell (1987) study and faced many of the same problems.

The study found that the mean WTP for the detailed model over all counties varies from a low $14.49/month ($143/year) to avoid a 20% shortage once every 30 years to a high of $21.10/month ($253/year) to avoid a 50% shortage once every 20 years. The WTP results were not used to calculate annual aggregate value of providing water reliability as in the Carson and Mitchell 1987 study, nor is there any indication of the total number of users served by the CUWA. However, the study does indicate that the given levels of increase that residents are willing to pay per month per household on their monthly water bills demonstrate a significant value of water reliability in the area. Aggregating across all consumers in the state, additional customer payments would total more than $1 billion per year (CUWA 1994).

- As expected, WTP increases with increasing magnitude and frequency of shortages. Respondents were willing to pay to even avoid minor shortage scenarios.
- Users may make a greater distinction between "shortage" and "no shortage" than between magnitudes and frequency.
- Shortage magnitude is a more important determinant of WTP than shortage frequency.
- Individuals who indicated a desire for their community to grow have a higher WTP than those that wish for their communities to stay the same size or get smaller.
- Those respondents who considered water to be long-term problem in the area have higher WTP than those that did not.

The survey was designed and executed well, and the study is cited several times in water reliability literature. However, it is by no means perfect. Like Carson and Mitchell (1987), a shortfall in the design of the survey was their use of WTP to "avoid" a shortage. Barakat and Chamberlin's findings should be interpreted as upper bounds for consumer valuations pertaining to modified shortfall scenarios because the elimination of shortfalls is not a realistic scenario

(Griffin and Mjelde 2000). Furthermore, again like the Cason and Mitchell (1987) study, the survey asks questions in a referendum format, which may produce unreliable and usually overestimated values (Jenkins, Lund, and Howitt 2003).

Another stated preference study was conducted in seven Texas cities by Griffin and Mjelde (2000). Their first objective was to investigate the value of current water supply shortfalls (existing shortages of known strength and duration). Second, the study attempted to determine the value of future shortfalls, probabilistic shortages of differing strength duration and frequency.

Each questionnaire included two contingent valuation questions.

- The first CV question was a closed-ended WTP question that established a current supply shortfall of X% of the community's water demand expected to have a duration of Y summer days. The respondent was then asked if they would be willing to pay a one time fee of $Z to be exempt from the outdoor water restrictions.
- The second CV question was an opened ended WTP or WTA question concerning a hypothetical increase or decrease in future water reliability. An initial situation was posed to the respondent in which approximately once every U years a shortfall of V% would occur for a duration of W days. Depending on the particular survey, the question then posed a potential improvement of decline in one of the parameters and the other stayed constant. This question design is intended to be an improvement on the "avoided shortage" problem in the Carson and Mitchell (1987) and the CUWA (1994) studies.

The results of the WTP survey for households to be exempt from outdoor water restrictions due to the current shortfall are given in Table 5.6.

- For the average respondent, $32.04 is the avoidance value for a three week current shortfall of 20%. A one week increase/decrease in shortfall duration increases/decreases this value by $2.59. Every 10% increase or decrease in shortfall strength increases or decreases this value by $2.27.
- As duration increases, respondents are likely to pay more to avoid restrictions; therefore, the value of reliability increases with duration of the shortage.

Table 5.6
Respondents' WTP to avoid water restrictions from a single current shortfall event (2003 USD)

	Shortfall duration		
Shortfall strength	14 days	21 days	28 days
10%	$27.19	$29.77	$32.35
20%	$29.46	$32.04	$34.62
30%	$31.74	$34.31	$36.90

Source: Based on information provided in Griffin and Mjelde 2000.

WTP and WTA measures were obtained as means from the survey responses as well as calculated from the tobit models. Both are presented below:

- Mean WTP and WTA per respondent are $109/year and $163/year, respectively.
- The mean tobit predicted WTP and WTA per respondent are $125/year and $170/year, respectively.

While the authors attempted to leave the future shortfall scenarios open ended for improved methodology purposes, the future shortfall values appear to be inconsistent with the reported current shortfall values. When the current shortfall values are used to calculate the future shortfall values, the calculated values are much lower than the WTP and WTA from the survey results The authors believe that the future shortfall valuation is the source of the discrepancy because the current shortfall valuation is easily understood by the respondents and is a common line of questioning for contingent valuation surveys. On the other hand, respondents did not appear to understand the future shortfall query. Using frequency to convey probability may have been a bad idea because of scaling problems and because respondents became confused by the added dimension of frequencies and probabilities. Therefore, while the Griffen and Mjelde study may have been an improvement in design from previous studies, their WTP/WTA results are inconsistent and somewhat overstated for small changes in future probability shortages (Jenkins, Lund, and Howitt 2003).

A study conducted by Howe and Smith (1994) attempts to formulate a framework for determining the optimal level of water supply reliability. The study uses contingent valuation survey methods to measure customers' willingness to pay for improved reliability and willingness to accept lower water costs for reduced reliability.

This survey was conducted in three Colorado towns: Boulder, Aurora, and Longmont. Respondents were asked to consider several specific changes in their city's level of reliability (increase and decrease) and to assert whether or not these changes would be acceptable if accompanied by appropriate (but unspecified) changes in their water bills. The first decrease (scenario 1) was smaller than the second decrease (scenario 2). The questions were set up in a "yes" or "no" format. For "yes" answers, quantitative WTA and WTP values were elicited from the respondents.

The type of shortage investigated in the study is a "standard annual shortage event" (SASE). The SASE can be defined as "a drought of sufficient severity and duration that residential outdoor water use would be restricted to three hours every third day for the months of July, August, and September" (Howe and Smith 1993). The base probabilities of the SASE occurring for each city were 1/300 for Boulder, 1/10 for Aurora, and 1/7 for Longmont. Their results were as follows:

- In Aurora and Longmont, the two towns with lower levels of reliability, consumers were not willing to pay enough to cover the cost of investment necessary to improve reliability. However, in the town with very reliable water supplies (Boulder), consumers were actually willing to pay less for reduced reliability.
- A household's WTA compensation for the first scenario (decrease in reliability in a range of approximately 0.7% to 11%, depending on the city) ranges from $68/year in Boulder to a high $166/year in Longmont. The WTA compensation for the second

scenario (decrease in reliability in a range of approximately 1.7% to 40% depending on city) ranges from $81/year in Boulder to $240/year in Longmont.

- Two sets of WTP averages were developed for each scenario. The first average is based only on "yes" answers to the accompanying WTP. For the second average, "no" responses were counted as $0 and incorporated into the overall average.
  - The WTP for the first scenario (increase in reliability in a range of approximately 0.16% to 9.2% depending on city) ranged from $70/year in Boulder to $90/year in Longmont. The WTP including "no" respondents ranged from $16/year in Boulder to $28/year in Aurora.
  - The WTP for the second scenario (increase in reliability in a range of approximately 0.23% to 12.2% depending on city) ranged from $64/year in Boulder to $119/year in Longmont. The WTP including "no" respondents ranged from $15/year in Boulder to $29/year in Aurora.
- In general, as expected, larger WTAs are required for greater decreases in reliability and larger WTP are offered for greater increases in reliability.

The WTP results, when compared to the results of the other contingent valuation survey, are clearly lower. The reason that Howe and Smith's results differ from the others is that their methodology is slightly different. Carson and Mitchell (1987) and CUWA (1994) both asked respondents their WTP for complete avoidance of a shortfall with given percentage. Griffen and Mjelde (2000) questioned respondents on their WTP to reduce the probability of, not avoid, the probability of a potential shortfall. All three of these studies are determining what people will pay to maintain their current well-being. Howe and Smith approached the issue from the opposite direction. They determined respondents' WTP for a percentage increase in reliability. The lower values of their study may be attributable to the fact that respondents were already content with their current level of reliability. People may be more willing to pay for maintaining a level of service they currently have than they will pay for a potential improvement in that service, which is consistent with economic theory.

It should be noted that Howe and Smith study's emphasis on a single type of shortage, the SASE, limits the transferability of the results (Griffin and Mjelde 2000). More severe or moderate events are not considered in the calculation of the WTP/WTA results.

Another noteworthy observation about Howe and Smith's theory is that they set aside the potential role of price in managing excess demand during shortfall events (Griffen and Mjelde 2000). It often is difficult for water managers to use price as a demand management tool because of institutional controls, time lags, and local political considerations. However, districts in California have used price to help manage demand during periodic shortfall events (Fisher et al. 1995) and, as shown later, analysis of demand management using price can aid in estimating the value of reliability through avoided consumer loss due to price increases (Fisher et al. 1995).

Based on data from Michelsen, McGuckin, and Stumpf (1998), McGuckin (2000) estimated the WTP for water by residential customers in three Southwest cities—Albuquerque, N.M.; El Paso, Texas; and Las Cruces, N.M. The study estimated the economic loss if current water deliveries were reduced by 5% because of drought. To do this, the study used inverse demand functions to estimate WTP under current and 95% of current consumption levels, on a monthly basis, and found that the WTP for water at the 5% shortfall margin more than tripled in each city. The economic loss attributed to a potential drought was estimated as the difference in WTP before and after the reduction in use, multiplied by 12 months and the number of households. The findings are shown in Table 5.7. However, the research team's investigation suggests that there are several empirical limitations to this analysis and, therefore, McGuckin's empirical results should be interpreted with a high degree of caution.

Overall, while the stated preference studies discussed above are valuable in terms of gaining insight into the value of reliability, none of them are perfect in their methodology. Furthermore, the studies are unique to each location and situation. It probably is ill advised to attempt to use any single value for the transfer of benefits to other situations. However, in looking at the entire set of stated preference results, as summarized in Table 5.8, it is interesting to note the consistency in the range of values across all the studies. It appears the majority of households value water supply reliability in excess of $100 per year.

Revealed Preference Studies

Fisher et al. (1995) explored how price can be used as a tool to reduce demand during a drought. The authors note that the associated consumer surplus loss due to a price induced reduction in water consumption can be regarded as the benefit from mitigating drought through the construction of new storage capacity or operating a conjunctive use program. Using estimated price elasticities for residential customers, the loss of surplus was computed with a price-induced cutback of 25% in consumption in the East Bay Municipal Utility District (EBMUD, California) service area. A demand reduction of 25% was used in the study because it is one of two EBMUD operation parameters for maximum acceptable demand reductions in drought (the less conservative parameter is 39%). The authors produced a range of estimates using a selection of studies that most effectively addressed the econometric issues associated with calculation (Fisher et al. 1995). The selected studies, with varying demand elasticities, produce a range of welfare loss of $51 to $230/AF.

Table 5.7
Willingness to pay for water before and after drought: Three southwestern cities (2003 USD)

	Albuquerque		El Paso		Las Cruces	
	Current use	Drought use current less 5%	Current use	Drought use current less 5%	Current use	Drought use current less 5%
Residential use (kgal/mo)	14.7	13.4	13.3	12.0	16.6	15.3
WTP/kgal	$1.38	$4.40	$0.99	$3.95	$1.08	$3.55
WTP/AF	$450	$1,433	$322	$1,287	$352	$1,156
Number of households	107,000	107,000	120,553	120,553	18,840	18,840
Annual economic loss per household		$28.57		$29.38		$22.89
City total		$3,058,000		$3,342,000		$431,000

Source: Based on data from Michelsen, McGuckin, and Stumpf 1998, as interpreted by McGuckin (2000).

Table 5.8
Summary table of results from stated preference studies (2003 USD)

Source	Shortfall amount	Frequency	Probability	Annual WTP/household
Carson and Mitchell (1987)	10% to 15 %	1 in 5 years	20%	$135
Carson and Mitchell (1987)	10% to 15 %	2 in 5 years	10%	$248
CUWA (1994)	20%	1 in 30 years	3.3%	$143
Carson and Mitchell (1987)	30% to 35%	1 in 5 years	20%	$186
Carson and Mitchell (1987)	30% to 35%	2 in 5 years	10%	$421
CUWA (1994)	50%	1 in 10 years	5%	$253
Griffen and Mjelde (2000)	na	na	na	$109
Griffen and Mjelde (2000)	na	na	na	$125
Howe and Smith (1994)*	0.16% to 9.2%†	na	na	$80‡
Howe and Smith (1994)	0.23% to 12.2%†	na	na	$92§

na = not applicable.
*Howe and Smith (1994) also estimated WTA values for decreases in reliability. Mean annual WTA results per household for approximately a 0.7% to 11% decrease in reliability, depending on the city, ranged from $68 to $166. Mean annual WTA results for approximately a 1.7% to 40% decrease in reliability, depending on the city, ranged from $81 to $241.
† This percentage range does not represent the magnitude of the shortfall, as is the case in the other studies. Rather, this range represents increased probability over the base probabilities of the SASE. The actual percentage increase is dependent on the city. The associated dollar values are the annual WTP per respondent for an increase over their current reliability.
‡Value represents the average of the WTP range given in the study ($70 to $90 per year). If "no" respondents for this increased probability range are included into the data set (respondents' WTP = $0), the WTP range is from $16/year to $28/year per respondent.
§Value represents the average of the WTP range given in the study ($64 to $119 per year). If "no" respondents for this increased probability range are included into the data set, the WTP range is from $15/year to $29/year per respondent.

Inferring the Value of Reliability From Cost and Price Differentials

In 2002, the California Recycled Water Task Force was established to investigate specific recycled water issues. The economic group of the task force was charged with identifying economic impediments to enhancing water recycling statewide. The report uses a case study of the Ground Water Replenishment System (GWRS) in Orange County as an illustration for the importance of economic feasibility analysis. The GWRS was designed to recycle an estimated 70,000 AF/year of effluent and inject it into the Orange County Aquifer.

According to the Groundwater Replenishment System Financial study (Public Resources Advisory Group 2001), conducted for Orange County Water District and Orange County Sanitation District, the value of drought proofing (the value of reliability) based on drought penalties and rate increases for consumers is estimated at $179-$256 AF/year (and aggregate benefits of $7.8-$13.3 million per year for 40 years, with a total present value of $232 million, 5.5% discount rate) (2002 Recycled Water Task Force 2002).

The task force also looked at several other benefits realized from the Orange County Groundwater Replenishment System. They noted that one benefit of reusing wastewater is its availability year round, thus making it an uninterruptible supply. The lower bound value for the higher degree of reliability can be inferred through the difference between interruptible and noninterruptible supply if purchased from MWD in Southern California. In 2002, the price difference was $141/AF (2002 Recycled Water Task Force 2002). The study asserts that when the $141/AF is eliminated from future water supply cost calculations, the Groundwater Replenishment system will provide well over $10 million annually in terms of cost reductions for reliability (2002 Recycled Water Task Force 2002), but the study goes no further to show how they arrived at this estimate. Clearly, these values are unique to the GWRS; however, they illustrate the potentially high value of reliability, especially in relatively arid regions.

In a similar investigation in 1997, the National Research Council (NRC) estimated that if Orange County were to lose its reliable groundwater supply to salt water intrusion, the cost of securing water by the retail producers to serve their customers would jump from the 1997 cost of $106 million to $210 million. The $104 million increase arises because the water once pumped from the aquifer would now have to be purchased from MWD at the noninterruptible rate (NRC 1997).[2] The sharp increase in cost charged by MWD for noninterruptible water supplies highlights the fact that reliability has a key role in water pricing (Paul 2004). Water users are willing to pay a premium for water that is available to them in drought years. As actual or potential shortages worsen and demand outpaces supply, users are willing to pay more for water. Therefore, it can be assumed that water with a high reliability factor will be much more valuable in the future than in the past and should be priced accordingly.

As mentioned earlier, while cost of a water project does not necessarily equal the value of the project or program, cost sometimes can be used as a lower bound proxy estimate of the value attached to increased reliability. For example, Varga (1991) investigated the role of local projects and programs in the City of San Diego to enhance imported water supply and improve reliability. The MWD provides water to San Diego from the Colorado River and Northern California, based on availability. To encourage the use of existing local reservoir capacities and improve the reliability and yield of the imported water system, MWD and California introduced water rate credits for serviced cities. The first of two programs instituted was the Interruptible Credit Program. An interruptible credit applies to either treated or untreated water that either could be reduced or have its delivery interrupted by the MWD or another external agency. This program encourages water retailers like the City of San Diego to store water in reservoirs to overcome possible interruptions of imported water. In 1991, the interruptible credit rate was approximately $60/AF. The second program is the Seasonal Storage Credit program. A seasonal storage credit applies to water stored during seasons when imported water is available in excess of demand. This program encourages water agencies to use available local storage to increase the capacity and yield of the imported water system. The 1991 seasonal storage rate was approximately $111/AF. MWD is paying for direct increases in reliability, and therefore, the credit rates can be used as the value for an AF increase in water supply reliability.

A 1996 study by Thomas and Rodrigo attached higher levels of value to storage than the MWD credit rates that Varga (1991) cites in his paper. Thomas and Rodrigo (1996) measured the benefits of nontraditional water resource investments. The focus of the study was on MWD and its member agencies. They investigated the benefits (expected yields and cost savings) of

---

2. In 1995, 300,000 AF were pumped from the basin at a cost of $158.33/AF (pumping assessment of $97.52/AF and an energy cost of $60.81/AF). The retailers in the district purchased 130,000 AF from MWD. Of the water purchased, approximately 100,000 AF was noninterruptible treated water at $488.76/AF. The additional 30,000 AF was purchased as seasonal shift water at $328.13/AF. In 1995, the cost of purchasing all water (groundwater and imported water) by the retail producers to serve their customers was approximate $106.2 million. If the groundwater basin had not been available, the entire 430,000 AF of necessary supplies would had to have been purchased at the noninterruptible rate of $488.76/AF.

developing additional resources in the region through several alternatives: increased imported supplies (base case), the addition of significant conjunctive storage of local groundwater basins (groundwater case), and the implementation of recycled water and groundwater recovery projects (preferred case). To determine the value of recycled water and conjunctive use storage, the savings attributable to each of these resources were compared to the yield associated with the resource. Thomas and Rodrigo note that "dividing the total present value of benefits by the expected groundwater replenishment deliveries (e.g., the difference between the base case and the preferred case and the groundwater case for conjunctive use storage), yields a dollar/AF index" (Thomas and Rodrigo 1996). In the case of conjunctive use storage, the modeling revealed that carryover or drought storage, which helps ensure greater reliability during dry periods, provides a benefit of approximately $353/AF to the region. Implementing the recycled water program would increase that figure by an additional $294 per AF.

An overview of the value of reliability inferred from results of revealed preference and cost-based approaches is provided in Table 5.9.

Drawing Inferences About the WTP for Residential Water

Despite the considerable body of empirical research reviewed in the preceding sections regarding elasticity of demand and reliability values, there is a general lack of direct empirical evidence about how much residential customers of water utilities value the water they receive. This leaves open the key question of "how much are households willing to pay for the water provided by their community water system?" In this section, the research team applies a series of simple assumptions to interpret the available empirical evidence on reliability values, in a manner that provides some insight on the more basic issue of the WTP for residential water. In addition, the few studies that directly estimate WTP for residential water are reviewed.

Table 5.9
Water supply reliability values inferred from revealed preference or cost and price differential results (2003 USD/AF)

Source	Value ($/AF)	Basis
Revealed preference studies:		
Fisher et al. (1995)	$51 to $230	Welfare loss per AF due to a price induced reduction in water consumption of 25%
2002 Recycled Water Task Force (2002)	$179 to $256	The value (AF/year) of drought proofing based on drought penalties and rate increases for customer
NRC (1997)	$331	The difference in cost of local groundwater supplies versus the MWD noninterruptible rate
Cost based studies:		
Varga (1991)	$60	The rate per AF that MWD credits local water retailers to store imported water in local reservoir to increase reliability of imported supplies
Varga (1991)	$111	The rate per AF that MWD credits local water retailers to seasonally store imported water to increase capacity and yield of imported water system
Thomas and Rodrigo (1996)	$353	The benefit per AF of conjunctive use storage to ensure greater reliability

Several of the reliability valuation studies summarized above provide WTP estimates for specific frequencies and severities of water shortfalls. These scenarios imply an approximate volume of water foregone over a given time period. The WTP estimates derived from these studies can be compared to the associated water quantities, to infer a monetary WTP per AF.

For example, Griffin and Mjelde (2000) evaluated a "current shortfall" scenario of 20%, lasting for 3 weeks. To estimate how much water is at stake in this scenario, consider that the average U.S. household uses approximately 0.5 AF per year (172 gallons per capita per day [based on Mayer et al. 1999], times 2.6 persons per household, times 365 days per year, which equals over 163,000 gallons per household per year, or about 50% of the 325,850 gallons in an AF). The shortfall scenario used by Griffin and Mjelde thus may amount to about 0.0058 AF of water (3 weeks out of 52 weeks being 5.77% of the year, times a 20% shortfall, times 0.5 AF per year, which equals 0.0058 AF). Given the estimated WTP to avoid such a shortfall was $32.04 per household per year, the implied value per at risk AF is $5,553 ($32.04 divided by 0.00577 AF).

Several caveats are required in evaluating a value estimate derived from this process. First, the assumptions applied to estimate the volume of water at stake might be in error. For example, if the water shortfall occurred in summer (which is likely), and the water use in summer is 2.4 times higher than in winter (the ratio of typical total use to indoor use only, as per Mayer et al. 1999, as discussed in chapter 3), then the implied quantity of water shortfall is understated. If the outdoor water use season in California (the study location) is assumed to be roughly one-half the year, then the 0.5 AF used per home per year comprises roughly 0.15 AF used in the winter months and 0.35 AF per household used in the six months in which outdoor irrigation occurs. The 3-week shortfall of 20% is thus equivalent to 0.008 AF (3 of 26 weeks of the outdoor watering season, times 20%, times 0.35 AF). Then, the implied residential customer WTP is $4,005 per AF ($32.04 divided by 0.008 AF).

Second, the reliability-based WTP values obtained by the original researchers reflect not just the value of the water per se, but also some degree of the residential customers' aversion to risk and uncertainty. In other words, the WTP values from the reliability studies undoubtedly embody some risk avoidance premium as well as the value held for the quantity of water at risk. This implies that the inferred WTP estimate would overstate the value of the water alone. This may be particularly true for the studies that value eliminating the risk of shortfalls, rather than reducing their likelihood or severity.

Third, the WTP estimates reflect values at the margin for the households' lowest valued current uses of the water (e.g., a portion of their outdoor irrigation). As more and more water is withheld from the households, the water uses that would be affected would be of increasing importance and value to the residential customers. Therefore, the WTP estimates inferred above might be understated compared to the WTP for water used for more highly valued purposes in the home (e.g., drinking, cleaning).

Finally, the reliability estimates we are interpreting are based on stated preference surveys of households. Given the hypothetical nature of some of the survey questions and the difficulty some respondents may have had with probability-based scenarios of water shortfalls and reliability, it may be the case that the results from the original research are skewed in one direction or the other.

Based on the above caveats, the values derived here need to be interpreted with considerable caution. There are reasons why the estimates may be under- or overstated relative to the true WTP of households for utility-supplied water. With these caveats in mind, by applying the general assumptions and procedures described above to the applicable reliability value estimates, the following illustrative WTP estimates for residential water are inferred:

- Griffin and Mjelde's (2000) current shortfall scenario implies a WTP for residential water on the order of $4,005 per AF.
- Carson and Mitchell's (1987) scenarios for MWD imply a possible WTP for residential water of between $4,675 and $7,714 per AF.
- The Barakat and Chamberlin study for CUWA (1994) implies a possible WTP of over $14,500 per AF.

As noted, these value estimates may be overstated for water use at the margin (i.e., for modest cutbacks in current outdoor uses), for reasons described above. In particular, the results based on Carson and Mitchell (1987) and CUWA (1994) may be overstated because they are based on certainty equivalents of eliminating future shortfalls. However, these estimates may be on-target, or possibly understated, for more essential water uses.

To provide some additional context for these results, some existing empirical research that examines water values more directly is described below. Colby (1989b) presents a simple, utility-theoretic model of water demand and consumer surplus to present a synthesis of water value studies for different water uses. Of particular relevance is the work of Young and Gray (1972), who found that the national value of residential outdoor water use is $810 per AF, and the value for indoor use is $1,435 per AF. This study is more than 30 years old, and so the values, even as updated by the Consumer Price Index (CPI) here, may not be entirely applicable today.

Gaudin, Griffin, and Sickles (2001) used data from 221 Texas communities to estimate a utility-theoretic Stone-Geary demand function for municipal water. Based on estimated demand curves, loose lower-bound assumptions, and simple geometry, total WTP (i.e., at a zero price) for *discretionary* municipal water use (as opposed to basic necessity water use) is $267 per person per year. Assuming a typical household size (roughly 2.6 persons per household), this amounts to about $700 per household per year. Assuming typical water use patterns of outdoor use being roughly 60% of total household use (Mayer et al. 1999), and assuming outdoor uses are largely discretionary and indoor uses are largely nondiscretionary, then about 0.3 AF may be assumed to be discretionary use by a typical household (60% of 0.5 AF per household per year). This suggests a value of $2,333 per AF for discretionary uses ($700 divided by 0.3 AF). Higher values would be expected for essential water uses.

In conclusion, based on the body of empirical evidence provided above, it is clear that households hold a high value (WTP) for their residential uses of water. While each of the above value estimates has some limitations, it seems likely that values of at least $800 to $1,400 per AF apply for non-essential water uses by residential customers of water utilities. The evidence also suggests that WTP values as high as $4,000 per AF, or perhaps considerably higher, may apply for current residential water uses.

WATER USE AND VALUE IN THE COMMERCIAL, INDUSTRIAL, AND INSTITUTIONAL SECTORS

This section presents an overview of water use in the commercial, industrial, and institutional sectors in the United States and provides a brief literature review on the value of water to these sectors.

Nonresidential water use accounts for 53% of total water use in U.S. community water systems (CWSs), and of that, over 70% is delivered to commercial, industrial, and institutional water users (RMI 2003). Moreover, self-supplied commercial, industrial, and institutional facilities use about as much water as all public and private community water systems put together (RMI 2003). Commercial, industrial, and institutional (CII) sectors alone have a significant impact on water consumption, particularly in urban areas. Nearly one-fourth of the potable water demand in an urban area is made by a wide variety of CII customers who in turn devote the water to a wide variety of end uses (Aquacraft 2003).

The USGS estimated that total water withdrawals during 1995 were 402,000 million gallons per day (mgd) for all water use categories, which include domestic, commercial, irrigation, livestock, industrial, mining, and thermoelectric power (Solley, Pierce, and Perlman 1998). Nine percent of this total water use (or 37,000 mgd) is attributable to the commercial, industrial, and institutional sectors. To demonstrate the relative shares of the CII sectors, Table 5.10 displays the categories of water users and their water use and corresponding percentages during 1995.

Industrial Water Use

Industrial water use (27,184 mgd in 1995) includes water for such purposes as processing, washing, and cooling in facilities that manufacture products (Solley, Pierce, and Perlman 1998). Major water-using industries include, but are not limited to, steel, chemical and allied products, paper and allied products, and petroleum refining.

Table 5.10
Water use by sector in 1995 (all values in million gallons per day)

	Millions of gallons per day	Percent of total water use*	Percent of public supplied water†
Thermoelectric	189,900	47%	<1%
Irrigation	134,000	33%	
Industrial	27,184	7%	12%
Domestic	25,902	6%	56%
Commercial/institutional	9,724	2%	17%
Livestock	5,490	1%	
Mining	3,770	1%	
Total	401,800	100%	85%†

Source: Solley, Pierce, and Perlman 1998.
*Figures may not add to totals because of independent rounding.
†Fifteen percent of public water is unaccounted for water or public use and losses. This unaccounted for water represents 2% of freshwater use.

CWSs supply just 18% of the industrial water, while the rest of industry's water needs are self supplied (Solley, Pierce, and Perlman 1998). Fifteen percent of industrial water use is consumed, and 85% is return flow (Solley, Pierce, and Perlman 1998). Consumed water is that part of water withdrawn that is evaporated, transpired, incorporated into products or crops, consumed by humans or livestock, or otherwise removed from the immediate water environment.

Return flow applications consist of industrial water needed primarily for production purposes, such as in the manufacturing of computer chips or steel. This water typically is treated and returned to surface water or groundwater sources (U.S. EPA 2000b).

Commercial and Institutional Water Use

Commercial and institutional water use includes water for motels, hotels, restaurants, office buildings, other commercial facilities, and civilian and military institutions. The USGS Survey estimates of CII water use (9,724 mgd) also include public-supply deliveries to golf courses (Solley, Pierce, and Perlman 1998).

Similar to industrial water use, 14% of CII water use is consumptive use and 86% is return flow (Solley, Pierce, and Perlman 1998). The public supplies 70% of the CII water needs, while the rest of the CII sector meets its water needs by self supplied withdrawals (Solley, Pierce, and Perlman 1998).

In 1998 the AWWA Research Foundation (AwwaRF) funded a study on the Commercial and Institutional End Uses of Water (Dziegielewski et al. 2000). This study spent two years working with five municipalities to create a database of a representative sample of CII customers from existing information, which was verified by field studies. The study team evaluated several characteristics of 11 categories of CII customers, shown in Table 5.11. Table 5.11 includes both the average annual use by customer category and the scaled average use, which combines the intensity of use for each category with the prevalence of the category in the municipal system.

Five of the 11 categories were selected for further study. These five categories were chosen because they ranked high on the list of scaled average daily use, which means they are significant both in terms of average use per customer and as a percentage of the total CII use in the municipal system. The final five categories selected for the study were schools, hotels/motels, office buildings, restaurants, and food stores.

Schools/Colleges

Schools and colleges comprise approximately 9% of the total CII water use. They may range from very small facilities where water use is limited to hand washing and toilet facilities to large campuses requiring water use for swimming pools, showers, and food preparation and service. Water may be used for seasonal and nonseasonal purposes; however, separate metering data for these purposes are not always available. Some campuses may serve students on a year-round schedule while others may follow a more conventional schedule with minimal use during the summer months.

Table 5.11
Characteristics of significant CII categories in five participating agencies

Customer category description	Average annual daily use (gpcd)	Coefficient of variation in daily use (gpcd)*	Percent of total CII use (%)	Percent of CII customers (%)†	Percent seasonal use (%)‡	Scaled average daily use (gpcd)§
Urban irrigation	2,596	8.73	28.48	30.22	86.90	739.0
Schools and colleges	2,117	12.13	8.84	4.79	57.99	187.0
Hotels and motels	7,113	5.41	5.82	1.92	23.07	414.0
Laundries and laundromats	3,290	8.85	3.95	1.38	13.35	130.0
Office buildings	1,204	6.29	10.19	11.67	29.04	123.0
Hospitals and medical offices	1,236	78.50	3.90	4.19	23.16	48.0
Restaurants	906	7.69	8.83	11.18	16.13	80.0
Food stores	729	16.29	2.86	5.20	19.37	21.0
Auto shops	687	7.96	1.97	6.74	27.16	14.0
Membership organizations	629	6.42	1.95	5.60	46.18	12.0
Car washes	3,031	3.12	0.82	0.36	14.22	25.0

Source: Dziegielewski et al. 2000.
*Coefficient of variation in daily use: The ratio of standard deviation of daily use to average of daily use.
†Percent of CII customers pertains to CII customers in agencies that have respective category only.
‡Percent seasonal use = [(total annual use—12 x minimum month use] / total annual use.
§Scaled average daily use = average annual daily use in category x percent of total CII use attributed to the category.

Hotels/Motels

Water use in hotels and motels is 6% of the CII water use. Water use is affected by the number of rooms, the number of occupants, and services provided, ranging from icemakers and water cooling to swimming pools, spas, laundry, and restaurants. In-room use for toilets, faucets, and showers places the most significant demand on water use; irrigation, cooling, and swimming pools contribute a seasonal component of water use for many facilities.

Office Buildings

The CII water demand for office buildings is approximately 10%, the highest water user of the five categories evaluated in the AwwaRF study. Occupancy rates and worker numbers affect the water use in office buildings, as does water for seasonal purposes such as irrigation and cooling. Further, the number and types of business found on the first floor of large office buildings vary and may include restaurants, salons and florists, all requiring diverse uses of water. While disaggregation of water uses in office buildings often presents a challenge, much of the water is used for typical domestic purposes such as faucets, toilets, cleaning, cooling, and irrigation.

Restaurants

Overall, restaurants account for approximately 9% of CII water use. A 2001 study of water use for Westminster, Colorado, revealed considerable variation in water use per meal prepared based on the type of restaurant. The results of this study revealed that Chinese restaurants used significantly more water than average while fast food restaurants used significantly less (Aquacraft, Inc. and Stratus Consulting 2001). Typical water uses for restaurants include faucets for bathrooms, food preparation, and janitor closets; dishwashers; toilets; ice makers; irrigation; and cooling.

Food Stores

Food stores account for approximately 3% of CII water demand, and the water use per customer is dependent on factors such as hours of operation, amount of refrigeration, type of cooling system, the number of aisles, types of services, presence of restroom facilities, and the use of mist sprayers on vegetables. Typically, however, water used in the cooling system for refrigeration represents more than half of the water used for all other purposes combined.

Value of Water to the Commercial, Industrial, and Institutional Sectors

The value of water to the commercial, industrial, and institutional sectors has not been studied as extensively as the value of water used in other sectors such as the agriculture and residential sectors (Renzetti and Dupont 2002). Thus, while it is commonly believed that commercial, industrial, and institutional water use is a relatively high value application of water (Frederick, Vandenberg, and Hanson 1997), there is actually relatively little empirical evidence to support this (Renzetti and Dupont 2002).

Values of Industrial Water Use

It is difficult to develop empirical estimates of estimate value for industrial water use largely because most industries self-supply their water, and pay little or nothing for their raw water input (Dinar and Subramanian 1997). Where industries have had to make purchases on water markets, those prices may be examined.

Frederick, VandenBerg, and Hanson (1997) examined seven industrial water valuation studies (the most recent of which was 1982) and found the median value of industrial water intake to be $132/AF (1994 USD), with a range of values from $28 to $802/AF. Renzetti and Dupont (2002) created a model, applied it to the manufacturing sector in Canada at the margin, and found a much lower median value (less than $1/AF). The authors attribute the low estimate to the fact that manufacturing in Canada self-supplies 90% of their water needs at almost zero internal cost, and that marginal costs for the remaining water needs are very low.

Water used in the production process can also be estimated by examining its value as an input to production. The ratio of the value of the output and the quantity of water intake may offer some insight; however, this method cannot be used as a shortcut to obtaining water values because it is the value of the marginal products that is needed, which is extremely difficult to extract from typical production data. In addition, Renzetti and Dupont (2002) point out that this approach does not account for the contributions to production of non-water inputs and for differences in revenue across firms that are related to the structure of output markets. Renzetti and Dupont (2002) also mention that two studies have used this approach: Giuliano and Spaziani (1985) and Mody (1997). Table 5.12 lists the water needs of various products for illustrative purposes, but provides no values for these quantities of water because of the inherent difficulties described above.

Table 5.12
Water needs of selected products

Product	Water use (gallons)
1 day's supply of U.S. newsprint	300,000,000
1 ton of steel	62,600
1 new car	39,090
1 ton of beet sugar (water used to process)	33,100
1 ton of cane sugar (water used to process)	28,100
1 kilowatt-hour of hydroelectric power	4,000
1 barrel of crude oil (water used to refine)	1,851
1 barrel of beer (water used to process)	1,500
1 ton of cement	1,360
1 car or truck tire or inner tube	518
1 pound of wool or cotton	101
1 pound of synthetic rubber	55
1 pound of plastic	24
1 gallon of paint	13
1 chicken (water used to process)	11.6
1 computer chip	10
1 can of fruit or vegetables (water used to process)	9.3
1 board foot of lumber	5.4
1 pound of meat (water used to process)	3.6
1 quarter pound of hamburger (water used to process)	1

Source: U.S. EPA 1995.

Values of Water to the Commercial and Institutional Sector

Values of the water used in the CII sector are also difficult to estimate. Information on CII withdrawals is limited but may be available through state agencies that permit withdrawals or require permits to operate potable water supplies (Solley, Pierce, and Perlman 1998). In many cases, withdrawal estimates are based on the population of the commercial facilities, that is, the number of students attending a university, inmates in a penal institution, workers in an office building, or the average occupancy rate of a hotel, rather than actual reported use (Solley, Pierce, and Perlman 1998).

Aquacraft (2003) analyzed the value of CII water and found that it is intimately connected with the purpose for which the water is being used. The study examined water use in three CII sectors (office buildings, hotels, and dialysis centers,) and concluded that low priority end uses have low values to the customer and high priority end uses have high values. In other words, faced with an increase in the price of water, the customer would first reduce or eliminate the low value uses before considering changes to those of high value. Water can be a convenience in some uses, but in others it can be an irreplaceable commodity.

For example, the study examined the relative value of water by observing water use patterns in two comparable office buildings. Building 1 is a relatively efficient user and Building 2 is a very heavy user. They deduced that the marginal value of water in Building 1 is greater than the marginal value in Building 2, since reducing water use in Building 2 (by upgrading fixtures and repairing leaks) would be relatively more simple (quicker and less costly) than reducing water use in Building 1. The net effect would be that because it would be less costly for the owners of Building 2 to reduce their water consumption, it would take a smaller increase in the cost of water to motivate them to do so. In Building 1, on the other hand, the inexpensive conservation steps have already been taken, so additional reductions would be more costly, and the owners would require a larger increase in the price for them to be justified from an economic perspective in making these upgrades.

In a second example, Aquacraft (2003) compared the average daily water use per room in five hotels. Water use in four of the hotels was very similar (about 78 gallons per room per day), while one of the hotels used 204 gallons per room per day. This hotel is a luxury hotel with high water using fixtures in the rooms, including extra deep soaking tubs and multihead showers. The rooms also include wet bars. The study deduced that the four hotels had probably eliminated most of the low value water use in the rooms, while the high water use hotel provides a great deal of luxury water use to its guests, and would be presumably more subject to reduction in response to price changes.

These two examples clearly support the long-held notion that the efficiency with which water resources are produced and consumed can be improved considerably if the general principles of marginal cost pricing are used as a guide in evaluating water pricing policies (Hanke and Davis 1973).

A similar conclusion was reached when water use was investigated at one of the largest dialysis clinics in the Midwest. Each year this center treats 300 patients, who each require 150 dialysis treatments per year. Water is essential for this process, and one would imagine that it would have such a high value that a very high price would be paid. However, because the costs of maintaining a patient on dialysis are paid by Medicare, Medicaid, and private companies, cost increases cannot be passed along and every effort must be made to contain all costs rigorously.

The total water used at the dialysis center in a typical year averages 12.5 million gallons. Of this, the major uses are 2.25 million gallons for dialysis, another 2.25 million gallons rejected in the purifying process (explained in further detail below), 2.0 million gallons used by the laundry, and 1.0 million gallons used by the staff for sanitation and cleaning (see Table 5.13). It was estimated that 2.75 million gallons of water use per year is unknown (this is a very large volume for this type of facility, and warrants further investigation).

The reject water is produced in the purifying process as the center produces its ultra-pure water on site using city water as raw material. During this process, for every gallon of final product water, about another one-half gallon is rejected. This reject water is currently wasted to the sewer.

Next, water costs were examined, and unit water and wastewater costs amounted to \$2.31 and \$4.02/kgal, respectively (see Table 5.14). The average budgeted cost for water and wastewater at this facility, excluding the laundry, is approximately \$76,000 per year. When non-personnel costs for this facility were examined, water and wastewater charges were found to rank sixth in a list of 18 categories of non-personnel costs.

The study concluded that the value of water ranged from a value of practically zero for the reject water (and a low value for the leakage and unaccounted water) to almost infinite for the water used for the actual treatment (the value for this water relates to the value placed on maintaining the lives of the 300 patients using the center each year), with the small amount of water used for domestic and irrigation uses falling in between.

Table 5.13
End uses of water at dialysis center

End use	Amount (kgal)
Dialysis water	4,220
Reject water	2,260
Toilet, urinals, hand sinks	1,000
Cleaning	50
Irrigation	20
Cooling and cooling system bleed	200
Laundry	2,000
Unknown (leaks, other uses, etc.)	2,750
Total	12,500

Source: Aquacraft 2003.

Table 5.14
Water and sewer charges for dialysis center*

Parameter	Volume (kgal)	Water charge ($)	Sewer charge ($)	Total charge ($)
Water consumption	12,536	28,876	50,212	79,880
Unit price ($/kgal)		2.31	4.02	6.33

Source: Aquacraft 2003.
*Estimated from 1st quarter 2003 water and wastewater bills.

The Aquacraft study then pointed out that another way to gauge the value of water to a particular group of customers is to observe how their use patterns change as a function of price. This provides a measure of the elasticity of demand. The inverse of this value, sometimes referred to as the elasticity of value, provides a fairly direct estimate of the relative value the customer would place on the next unit of consumption in the face of a unit increase in the price.

Aquacraft reviewed a study (Lynne, Luppold, and Kiker 1978) on the responsiveness of a group of commercial water users in south Florida that used a multiple regression modeling approach. The use of water was modeled as one of several inputs required for the production of a specific good or service. Since this water had an easily determined cost, it was possible to create what the authors referred to as a "derived demand" model using regional data from south Florida of the response of water use to changes in its price for a range of commercial customers.

The study included 4,356 customers in the Miami/Florida Keys area. Five categories were examined: department stores, grocery stores, drinking and eating establishments, motels and hotels, and "other commercial." Price elasticities for each customer category were determined by regression techniques. All had the expected negative slope (reflecting the decrease in demand in response to increases in price), but their magnitude varied considerably (see Table 5.15). Department stores had the largest absolute value of price elasticity and therefore are expected to have the largest response of the sectors to price changes, while eating and drinking establishments had the smallest absolute value of price elasticity, and therefore would not be expected to change their water consumption much (compared to the other sectors) in response to water price changes.

Table 5.15
Price elasticities for commercial customers in Miami/Florida Keys

Category	Price elasticity
Department stores	-1.074
Grocery stores	-0.719
Other commercial	-0.480
Motels and hotels	-0.175
Eating and drinking	-0.174

Source: Based on Lynne, Luppold, and Kiker 1978.

Summary of Water Use and Value in Commercial, Industrial, and Institutional Sectors

Even though the allocation of water is not often optimized in an economic sense, available commercial, institutional, and industrial water values can be compared within their sectors and among other water uses (such as domestic and agricultural, or even instream flow values for recreation and ecosystem support) to reveal the wide range of values that exist for various water uses. Drought, natural disasters, and structural failures in water supply systems sometimes lead to temporary water shortages, and minimizing economic losses would require knowing where the least valued water is provided (Renshaw 1982).

Values among the sectors tend to vary significantly by region, yet some relative values may be consistent across regions. Another factor affecting value is time. Future market values depend on a particular industry's sustainability with regard to its inputs (water, in particular) and the fate of each industry (i.e., their economic growth or decline). Values will have less variation over longer periods of time in sectors where applied use is sustainable in the long run, such as recreation or ecosystem support.

THE VALUE OF WATER TO SUPPORT INSTREAM FLOWS

The concept of instream flows has a variety of meanings, including stream flows necessary for swimming, floating a canoe, supporting fisheries, sustaining waterfowl habitat, providing downstream water supplies, assimilating wastes, and generating hydroelectric power. These uses require different quantities of water at different times and at different levels of quality. Each use produces economic benefits. The extent of these benefits has been estimated in numerous studies that have focused on the actual economic benefits produced by instream flows (and related activities and ecological services), on the public's willingness to pay to protect instream flows, and on the adverse economic impacts resulting from a failure to protect such flows (Sherk 2002).

This section reviews the existing literature addressing the economic benefits derived from instream flows.

Use Values

As discussed in chapter 4, use values reflect the values people have for a given resource based on their direct use of the resource in question. For instream flows, use is primarily recreational activities such as fishing or boating. The total value of recreational activities is examined in the following two sections: (1) what people actually pay to participate in a given sport (e.g., equipment expenditures), and (2) what they would be willing to pay over and above what they currently pay. The first component of value can be represented simply by the expenditures incurred. However, cost is typically an underestimate of total value, and thus the second component of value requires more explanation. Recreational site visits cost money and time, and recreationists would not undertake visits unless the visits yielded net benefits, also known as consumer surplus (as defined near the end of chapter 4). Thus, even though there are often not direct markets to "purchase" recreational activities such as fishing days, there are costs, in both money and time, that individuals incur to participate in recreational activities. These costs can be used to estimate an individual's willingness to pay, and thus demand, for the recreational activity.

Recreation Expenditures

A significant portion of recreational spending is tied to fish and wildlife, both of which require instream flows and the habitats they support, such as wetlands, for survival (wildlife to a much lesser extent than fish). In 1995, total recreational spending generated sales of over $450 billion and provided jobs for over 6.8 million people (U.S. EPA 2000b). However, just a portion of this is attributable to freshwater sources. Freshwater fishing alone generated $61 billion in 2001 from expenditures and indirect economic impacts, from over 28 million freshwater anglers (USFWS 2002). Ducks and other birds that depend on clean water also generate economic activity for the recreation and tourism industry. In 2001, more than 3 million people spent over $1.4 billion hunting migratory waterfowl, most of which depend on freshwater sources (USFWS 2002). Also, over 19 million Americans watch and photograph waterfowl, which generated nearly $10 billion in total economic activity in 1995 (U.S. EPA 1996). When all birds, not just waterfowl, are factored in, the impact is even more significant: the U.S. Fish and Wildlife Service estimates that nearly 30 million people participate, and the total economic impact was nearly $20 billion in 1995 (U.S. EPA 1996). All of these activities require the protection of instream flows. Fishing is perhaps the most obvious but, in many areas of the country, hunting in the riparian corridor is of equal importance.

Floating on rivers in canoes or rubber rafts has emerged as a major recreational activity in those areas of the country with sufficient stream flows. These flows and the riparian corridors that they support are also used for nature study, picnicking, hiking, camping, and bird watching. In fact, hiking on trails along rivers is growing in popularity. For example, along the Chattahoochee River in metropolitan Atlanta, over 3 million people a year use the trails, picnic areas, and other amenities in the Chattahoochee River National Recreation Area (U.S. NPS 2003).

Canyon Lake in Comal County, Texas, operates dam releases to provide flows downstream in the Guadalupe River. This operating plan, which provides recreational opportunities on both Canyon Lake and the Guadalupe River, results in recreator expenditures of between $7 and $8 per person per visitor day (TCPS 2002). This results in economic benefits to Comal County of approximately $500 million per year (TCPS 2002), and thus it is important that management of this resource accounts for recreational activities.

The Natural Resources Council of Maine addressed the economic benefits derived from the protection of instream flows, concluding that inland fishing alone produced $332.8 million in annual economic benefits and supported at least 5,320 full and part-time jobs (Day 2001). In addition, the Council noted that "millions of dollars are spent annually by birdwatchers and other recreational pursuits that are supported at least in part by adequate flow conditions" (Day 2001).

Willingness to Pay

Most recreational fishing and boating days provide a benefit to participants. It is a benefit for which participants would, if they had to, pay more than equipment costs or the current nominal fishing license fee or launch fee. The fact that they do not have to pay a market price to participate results in the visitor enjoying a "consumer surplus" as a value above and beyond any out-of-pocket expense. Several studies have examined what participants are willing to pay to participate in water-related sports, which captures the total value: costs and fees, plus consumer surplus.

Two studies reviewed a number of other empirical research efforts that address willingness to pay for various recreational activities (Walsh, Johnson, and McKean 1988, Loomis 1999a). These reviews are meta-analyses, which are amalgamations of many individual studies to develop estimates of central tendency. Meta-analysis is used to exploit and combine the strengths of multiple studies that use different valuation methods, and to avoid being misled by a single potential outlier study. Values from the Loomis meta-analyses are summarized by activity in Table 5.16. The Loomis study is the only one displayed because it is the most recent study, and the analysis includes studies used in the Walsh, Johnson, and McKean summary.

The studies used to estimate the values listed in Table 5.16 represent the two types of valuation methods typically used in the literature: (1) revealed preference methods such as travel cost models, which use observed recreational behavior to infer values; and (2) stated preference methods such as contingent valuation, which ask people to state their values or their willingness to trade off different resource commodities. Carson et al. (1996) demonstrated that estimates of use values do not vary substantively whether revealed preference or stated preference methods are used.

The results summarized in Table 5.16 have been confirmed in numerous studies documenting a public willingness to pay to preserve instream flows and the activities associated with those flows. Daubert and Young (1981), for example, used contingent valuation to analyze public willingness to pay for recreational opportunities on the Cache la Poudre River in northern Colorado. With regard to use of the resource for fishing, their studies indicated that willingness to pay peaked at about $100 per angler day (2003 USD) when flows were maintained at 500

Table 5.16
Average willingness to pay by activity (2003 USD)

	Average value per adult user day	Number of studies evaluated
Camping	$42.85	36
Picnicking	$43.57	10
Swimming	$26.82	11
Hiking	$31.68	20
Float boating	$36.89	14
Fishing	$38.74	109
Wildlife watching	$36.74	109

Source: Loomis 1999a.

cubic feet per second. It is interesting to note that this willingness to pay declined significantly when (a) these flows were not maintained or (b) flows were excessive, thus illustrating the fact that water levels affect value. Too little or too much water can reduce the overall value of the resource.

González-Cabán and Loomis (1999) used contingent valuation to measure the economic benefits derived from protecting instream flows in the Río Mameyes, and avoiding a dam on the Río Fajardo, in Puerto Rico. The study was triggered by a proposed withdrawal of 10 million gallons per day from the Río Mameyes by the Puerto Rico Aqueduct and Sewer Authority. Annual willingness to pay was $13.25 for each of the rivers, or $25 for both. When expanded to the 1 million households in Puerto Rico over the 5-year period households were asked to pay, this amount increased to $130 million. Of perhaps greater importance was the recognition that expenditures would decline sharply if instream flows were not protected. Visitors to the Río Mameyes indicated they would reduce trips by 93% (or 22,480 trips) if 10 million gallons daily were withdrawn from the river, resulting in a loss of $298,000 (22,480 times $13.25) in annual recreation benefits alone. Both the household and recreation values can be compared to the costs of repairing water lines and in-home water conservation measures to determine if there are more net benefits to these alternative sources of water than withdrawals from the Rió Mameyes.

A similar study of the Hassayampa River Preserve in Arizona was conducted by Crandall, Colby, and Rait (1992). Again using contingent valuation, this study indicated a public willingness to pay $107 per year to restore streamflow from intermittent to perennial. Of particular note was the conclusion that public willingness to pay would be reduced by 80% if instream flows were not maintained. The authors also concluded that water-based recreation was a key element in diversifying and developing rural economies.

Douglas and Taylor (1998) also found that instream flows were important to recreation. Construction of the Central Valley Project in California and subsequent transbasin diversions reduced flows in the Trinity River by approximately 90%. Using the travel cost method, the authors concluded that restoration of the Trinity River as a recreational resource would produce benefits of $491 million per year. Depending on flow requirements, the cost of providing the necessary streamflows to restore the recreational potential of the Trinity River ranged between $20 million and $52 million per year.

Many studies have also examined the economic values of instream flows at the margin: Johnson and Walsh (1987), Ward (1989), Sanders, Walsh, and Loomis (1990), and Cole and Ward (1994). Black et al. (1998) determined that most recreational values lie in the $10 to $50/AF range for rivers that offer high-quality recreational opportunities or are used extensively for recreation, and values usual range from about $1 to $10/AF for rivers that offer conventional opportunities or receive only a modest number of visitors. However, Ward, Roach, and Henderson (1996) estimated that recreational values of water range from a low of $8/AF to a high of $790/AF for reservoirs in the California Central Valley, depending on reservoir contents and other characteristics of the reservoir at which the recreation occurs. In addition, Lansford and Jones (1995) estimated recreational economic values of water in Lake Travis, Texas, to be between $147 and $183/AF.

A case study of recreational whitewater values in Colorado, developed in this report in chapter 8 and appendix I, shows that instream flows in Gore Creek (Vail) generate values to boaters and local economies on the order of $1.8 million per year. Based on the streamflows associated with the whitewater park, this translates to an instream water value of about $1,664 per AF in the peak use months of May, June, and July.

The values discussed in this section may underestimate the value of increased streamflow because most studies look exclusively at changes in willingness to pay among current river users. In addition to this effect, increased streamflow may attract new users to the site. Adding new users increases the overall value of increased flow. The studies that developed separate estimates of quality and participation effects found that impacts on participation can add significantly to overall streamflow values (Black et al. 1998).

Ecological Values

Instream flows support numerous ecological services, each providing valuable benefits to the ecosystem. These services include dilution of wastewater, natural water purification, erosion control, and habitat for fish and wildlife.

One aspect that has been addressed in the literature focuses on instream flows needed to protect coastal and estuarine areas. Adequate flows of fresh water into these areas are critical to coastal and estuarine resources as well as all of the economic activities associated with these resources. At some point in its lifecycle, approximately 95% of marine life "depends on the wide range of salinities and abundant food and shelter provided by bays and estuaries. [T]he health of our bays and marine life in our oceans is intrinsically linked to adequate freshwater flowing from our rivers to the bays" (TCPS 2002). Disrupting freshwater flows into bays and estuaries could have tremendous impacts on the coastal economy, directly affecting commercial and recreational fisheries and other nature-based tourism activities. In Georgia, the Center for a Sustainable Coast has argued that reduced streamflows into coastal and estuarine areas could have significant adverse impacts on "fisheries and nature-based tourism activities, worth at least $1 billion annually, and supporting some 40,000 jobs" (Kyler 2001). Similar adverse impacts, particularly on the production of oysters, could occur in Apalachicola Bay, Florida, if freshwater inflows are reduced (Christensen et al. 1998).

Adequate streamflows are also important to the maintenance of water quality (Sherk 2002). The Clean Water Act requires point sources of pollutants to obtain a National Pollutant Discharge Elimination System (NPDES) permit before such pollutants are discharged into waters of the United States. In part, the requirements of individual NPDES permits are based on streamflow, specifically the assimilative capacity of the stream receiving the pollutants.

The pollution control requirements of the NPDES permit are usually based on the lowest flow occurring for seven consecutive days in a 10 year period (called a 7Q10 flow). If streamflows fall below this level, then the 7Q10 flow would have to be recalculated. If the 7Q10 flow has to be recalculated, then the requirements of the NPDES permit must be recalculated. In essence, if the assimilative capacity of the stream is reduced, then the stringency of treatment requirements must be increased (Sherk 2002). Any recalculation of the treatment requirements contained in the NPDES permit could increase treatment costs significantly, as "[e]ven a small increase in pollutant removal requirements can result in a large expense in control technology if existing systems cannot be modified" (Bailey undated).

In lieu of available estimates of values of all of the ecological services provided by instream flows, one can examine the prices that have been paid to protect instream flows for ecological purposes. On average, water rights purchased for instream uses in the Pacific Northwest sold for about $185/AF (Landry 1998). The highest sale price, $520/AF, was charged to the Oregon Water Trust in 1996 (OWT 1998). The purchase was a part of a conservation project that made water available for instream flows by improving irrigation technology. On the low end of the market, the U.S. Bureau of Reclamation paid \$71/AF in 1996 to buy back a water service contract for water stored in the Lucky Peak Reservoir in Idaho (Rigby 1997). However, these are sales prices, and they do not reflect the value of the numerous ecological services provided by instream flows; they are believed to be an underestimate of true value.

Nonuse Values

Nonuse values stem primarily from existence (stewardship) and bequest (intergenerational equity) motives, and sometimes are largely values to preserve threatened and endangered species. Many individuals who may never fish or boat still receive some benefits from just knowing that free-flowing rivers exist (Sanders, Walsh, and Loomis 1990), that salmon migrations will continue as part of the cycle of life in a region (Olsen, Richards, and Scott 1991), or that endangered species will be protected (Berrens, Ganderton, and Silva 1996).

Several studies have conducted surveys asking households what they would be willing to pay for protection of water resources. A common method is to use a hypothetical referendum, where households are asked if they would vote in favor of a particular resource protection action, if it cost their household \$X. The amount of \$X varies across households, so that a demand curve can be traced. From this demand curve, willingness to pay is calculated.

Loomis (1987a) reports results of a survey given to California households, asking them what they would be willing to pay to protect Mono Lake in California. At the time of the survey, the waters of Mono Lake were being used for hydropower generation and water supply. If the Mono Lake ecosystem was to be preserved, alternative sources of energy and water supply would have to be acquired, and thus the original implication was that providing water for fish, birds, and other parts of the Mono Lake ecosystem deprived people of economic benefits. The Loomis study revealed that California households (statewide, regardless of whether they were users of the resource or not) were willing to pay between \$100 and \$245 annually to maintain lake levels in Mono Lake. The estimated amount of the general public's willingness to pay to protect the Mono Lake ecosystem was counted dollar for dollar as equivalent to hydropower and water supply benefits and costs in the economic analysis of the different water allocation alternatives (Jones and Stokes Associates 1993). It was estimated that the cost of alternative sources of energy would be only \$6 per household per year. In the end, the state ordered that the flows into Mono Lake be increased and Los Angeles' water right be reduced by nearly half.

A few studies have examined values of instream flows for protecting threatened and endangered species. Berrens et al. (2000) found that the annual household willingness to pay for protecting instream flows specifically for the silvery minnow on the middle Rio Grande is \$28, and \$61 to protect minimum instream flows on all major New Mexico rivers to protect 11 total listed fish species. These estimates of annual household willingness to pay are comparable with estimated values identified in previous nonuse value studies of western instream flows: Colby (1993) cites a range of annual household values from \$50 to \$100.

Using the Berrens et al. estimates and applying them to a conservative figure of 500,000 households (the 2001 U.S. Census Bureau estimated population of New Mexico is 1.8 million individuals, or 692,000 households), estimates of annual statewide values are \$13 million for the protection of instream flows specifically for the silvery minnow on the middle Rio Grande, and \$30.5 million for the protection of minimum instream flows on all major New Mexico rivers and the protection of 11 total listed fish species. Applying the \$14 million annual value to a sample flow range of 72,500 to 145,000 AF implies a range in values of $95/AF to $190/AF to protect instream flows for the silvery minnow.

Other Values

Instream flows also provide services such as hydropower and supply water to areas downstream. Such services are discussed in this section. Other benefits of instream flows include improved water quality, which affects adjacent property values (Epp and Al-Ani 1979, Leefers and Jones 1996), and the attraction of new small businesses to communities (Koberstein 1997).

Leefers and Jones (1996) examined property values and selling prices in areas with "Natural River" designation in Michigan, and found that property values and selling prices are indeed higher in those areas. Epp and Al-Ani (1979), who used real estate prices to value improvements in the water quality of small rivers and streams in Pennsylvania, found similar results: water quality is positively correlated with economic value of adjacent properties.

Koberstein (1997) supports the claim that revitalized and protected rivers can produce quantifiable economic benefits. This study showed that the Missouri, Columbia, and Blackfoot Rivers provide examples of how revitalized and protected rivers can attract new small businesses and recreation and tourism dollars to communities.

Water Supply

Instream flows supporting downstream water supplies meet municipal and industrial supply needs. Large economic benefits are associated with these instream flows, especially the value of reduced uncertainty of supply (i.e., the value of increased supply reliability).

Surface waters often are used primarily for irrigation and recreation. Typically, downstream water uses limit upstream diversions. Groundwater is a major source of public and domestic water, and many future developments depend on this water supply. Groundwater and surface water resources cannot be considered separately in many areas. Discharge from some aquifers maintains the flow of many streams. Large-scale groundwater development would ultimately decrease surface outflow from the basin.

Brown and Duffield (1995) reviewed a series of estimated marginal values of water in municipal use (for a 10% reduction in current use) ranging from $37 to $233/AF, and for irrigated agriculture, approximately $105/AF.

Hydropower

Instream flows also provide the "fuel" for hydroelectric generation. Any reduction in streamflow has the potential to reduce hydroelectric production. Assuming that energy demands remain constant, any reduction in hydroelectric generating capacity could result in increased generation using fuel sources that are significantly more expensive than hydropower. "Peaking power" generation is intended to meet peak electricity demands during a typical 24-hour period. Hydropower is generated at numerous public and private facilities to meet peak power demands because it can be brought online quickly and has low marginal costs (Sherk 2002).

An alternative source of peaking power is thermal electric generating facilities powered by natural gas. Like hydroelectric generating facilities, natural gas-fired generating facilities can be brought online quickly. Both systems are well suited to meet peak power demands. The critical difference is the price of the power generated. The cost of hydroelectric power purchased from the Southeastern Power Administration is approximately $10.00 per megawatt hour; the cost of natural gas-fired generation is approximately $70.00 per megawatt hour (in 2002 USD) (Sherk 2002).

The Powell Consortium (1995) examined the marginal hydropower benefits at seven dams along the Colorado River. The study found the estimates to range from $5 to $70/AF. A failure to protect instream flows could reduce hydroelectric generation and, thus, increase the costs of both energy and consumer goods.

Finally, an evaluation by the research team of the value of hydropower at the Shoshone Power Plant on the Colorado River, east of Glenwood Springs, reveal electricity generated in peak flow months yielded power company revenues of about $435,000. The corresponding water rights (1,400 cubic feet per second) imply a value generated by the instream flow of $157/AF. This assigns all value to the water and ignores other inputs to the generation process, so the estimated value is an upperbound for the value of water alone.

Summary of Values of Instream Flows

Ranges of the various values of instream flows discussed in this report are summarized in Table 5.17. Ranges are given for use values (recreation), ecological values, values of downstream water supply, hydropower, and nonuse values. These are but a sampling of values found in the literature examined.

Estimates of willingness to pay for instream flow must be interpreted carefully. The estimates are very sensitive to differences in use levels and the underlying value per day for recreation at the site. Comparing the marginal values requires correcting for key differences in activity levels, value per day of activity, and the baseline flows on which the marginal values are based (Black et al. 1998).

Table 5.17
Summary of economic values of instream flows (2003 USD)

Category	Low ($/AF)	High ($/AF)	Source	Comments
Use values (recreation)	$1	$770*	Low: Black et al. 1998 High: Ward, Roach, and Henderson 1996	Several values are available depending on baseline flows, quality of recreation, and level of use
Ecological values	$79	$503	Low: Rigby 1997 High: OWT 1998	Based on purchase prices of agreements for the purpose of restoring/protecting instream flows for ecological reasons
Downstream water supply	$37	$233	Brown and Duffield 1995	For municipal/ irrigation use
Hydropower	$5	$70	Powell Consortium 1995	Based on dams along the Colorado River
Nonuse values	$95	$190	Berrens et al. 2000	Based on protecting instream flows of the middle Rio Grande for the protection of endangered species (and applying estimates to an assumed 72,500 to 145,000 AF)

*$1,664/AF is implied for peak use periods, from the Gore Creek study, provided in appendix I (Stratus Consulting 2002).

Another consideration is that most instream flow studies cited here considered rivers in the West. Seasonal water shortages are of greater concern in arid regions and may heighten concern over flows, thereby influencing WTP estimates obtained in CV studies. Furthermore, most of the study rivers are in relatively undeveloped areas. Eastern rivers may be affected by other factors (e.g., shoreline development) that influence the recreational experience and willingness to pay, but are used (for recreation) and valued (for ecological services and protection of endangered species) by possibly more people.

Also, it is important to understand the policy context of the instream flow studies. Most of the studies consider competition between downstream recreational uses and consumptive uses. Consumptive uses such as irrigation and drinking water involve withdrawals that, to a great degree, are never returned to the river. In contrast, hydropower facilities generally affect flow by diverting water around bypass reaches; the water ultimately is returned to the river. However, the point at which the water is returned to the river may be several miles downstream, affecting the ecology and uses of the bypass reach. Also, in some cases, upstream flows do not have to compete with downstream uses. It is not known how these various scenarios may affect values.

CULTURAL WATER VALUES

"Cultural water" refers to water that has an emotional or aesthetic appeal and is used for cultural purposes such as spiritual uses, ceremonial uses, aesthetic water fountains (for both visual and sonic effects), and other art exhibits. Some water sources such as waterways also have historical values that are considered "cultural" as well.

In some of the literature, cultural water use refers to the use of waters to support and maintain the way of life of Native Americans, including but not limited to use for instream flow, habitat for fisheries and wildlife, and preservation of habitat for berries, roots, medicines, and other vegetation (U.S. EPA 2003). However, in this section, Native American water use includes water used for various cultural purposes such as religious applications, and not for uses such as drinking or supporting culturally significant wildlife.

Water Fountains

Humans have appreciated water fountains for centuries. Fountains are a distinguished feature in Spanish architecture, derived in part from Moslem influences (Gelt 2003). As Spain built its colonial empire in the New World, fountains were installed in plazas and courtyards as early as the 16th century (Gelt 2003).

Water fountains today are designed and installed for a variety of purposes, including to create a pleasing visual and auditory effect, to attract attention, to cool an area, and for noise mitigation (Gelt 2003). Water fountains are found in public plazas, in malls, in front of government and office buildings and industrial plants, and in and outside private residences and communities. Water use in fountains varies from a slight trickle to a gushing flow.

Some decorative fountains provide useful services in advertising, because moving or splashing water attracts attention. More extravagant fountains often attract more attention, such as large fountains often used to attract attention to residential developments. A fountain also can be used to cool an area. When a fountain sprays water into the air, the ensuing evaporation has a cooling effect. By creating micro-climates, fountains can make a cooling difference in a relatively small area or zone.

Fountains also mitigate noise. The sounds of splashing, flowing, or moving water can mask bothersome noises. A fountain's ability to mask noise is in direct proportion to the loudness of its water sounds. A fountain with a high-pressured flow dropping or falling from a significant height into a pool or onto a hard surface creates clamorous water sounds, which can mask relatively high volume noises.

Along with serving practical purposes, fountains also provide aesthetic benefits. All fountains, even those designed for a specific use, are generally aesthetically pleasing. To cool air or mitigate noise, water must move, and moving water attracts and appeals. Aesthetics therefore is built into almost all fountains.

One of the most prominent fountains in the United States is the Bellagio Fountain in Las Vegas. At one-fifth of a mile long, with 1,200 nozzles, 4,500 lights, and a $40 million price tag, the Bellagio fountain show is one of the most expensive and ambitious water features ever created. In addition to being Vegas' first fountain of note, the Bellagio Fountain garners national attention. In an article on fountains in *Smithsonian* magazine, design coordinator Jim Garland raved, "It's going to be on a U.S. postage stamp—it's that kind of a fountain" (Stewart 1998).

Spiritual and Ceremonial Water

Many cultures have spiritual relationships with the natural environment and the use of natural resources, especially water. Water soothes and relaxes, inspires reflection, is a source of beauty, and has spiritual qualities recognized in religion and ritual, from baptism to death (Gelt 2003).

Many Native Americans honor rain and other forms of water as a gift bestowed to benefit humans. Their ceremonies pay respect to water and use water. For example, the people of the Pueblo of Isleta, in New Mexico, ingest water or immerse in it for ceremonial purposes (Southwest Regional Assessment Group 2000). This tribe was the first in the nation to include these religious uses of water in their reasons for requiring an upstream water system (the City of Albuquerque) to comply with stricter water quality standards. The Pueblo's contention was that tribal religious ceremonies, or the right of tribal members to practice their religion as it had since time immemorial, was adversely impacted by the contamination of river water by toxic discharges (Southwest Regional Assessment Group 2000).

Summary

It is especially difficult to quantify the aesthetic, spiritual, and historical values of water. These water uses do not generally do not have dollar values readily associated with them because these water values are not bought or sold in the marketplace. They are not relatively large volume uses of water (compared to such extractive sectors as power and irrigation) and thus have not been studied much. Treatment and replacement costs can be examined (e.g., the cost of replacing the water used in these various practices); however, costs are typically underestimates of true value. For example, one may examine the extra costs incurred by the City of Albuquerque to treat the water to an acceptable standard to the Pueblo, but the Pueblo might value their water used for religious purposes at amounts far larger. While it is straightforward to identify the ways in which water values are incorporated into our cultures, unfortunately, there is very little available empirical research on the monetary value of water for cultural uses.

WATER VALUES AND COMMUNITY ECONOMIC DEVELOPMENT

Changes in water use may affect specific sectors of the economy (e.g., tourism, agriculture), which will have secondary effects on the regional economy. Initial impacts are directly experienced through changes in expenditures. Subsequent, indirect impacts are experienced as the other industries that serve as inputs to production experience changes in demand for goods and services. This creates a multiplier effect, because there are multiple rounds of industries purchasing from other industries. Also, there are changes in money put into the local economy in the form of wages and spending by households. All of these economy-wide impacts can be estimated from the industry-specific economic changes by use of a number called an economic multiplier.

The multiplier is a factor that when multiplied by new or increased expenditures (or reductions in expenditures) yields the benefits (or reductions in benefits) to the region. In this context, the term "benefits" usually refers to levels of economic activity, such as total wages or income or employment in the region. Different industries have different multipliers—some industries tend to produce greater regional, or even national, impacts than others.

Michigan State University's "Multiplier" webpage states that the size of a multiplier depends on four basic factors: (1) the overall size and economic diversity of a region's economy, (2) the geographic extent of the region and its role within the broader region, (3) the nature of the economic sectors under consideration, and (4) the year (MSU 2003). Regions with large, diversified economies producing many higher order goods and services will have high multipliers, whereas regions that are not able to buy and hire locally will have low multipliers. Large geographic regions will have higher multipliers, all other things equal, than small areas because transportation costs will tend to inhibit imports. Multipliers vary across different sectors of the economy based on the mix of labor and other inputs and the propensity of each sector to buy goods and services from within the region. Tourism-related businesses tend to be labor intensive and thus tend to have larger induced effects rather than indirect effects. A multiplier also depends on the characteristics of the economy at a single point in time. Multipliers for a given region may change over time in response to changes in the economic structure as well as price changes.

The Minnesota IMPLAN Group (2003) has estimated multipliers for over 500 industrial sectors at the county, state, and national levels for estimating regional, statewide, and nationwide economic projections. Their "Multiplier Report Package" is available for $500 and contains multipliers for 528 industrial sectors, either for the nation, state, or county.

The industrial sectors identified here as potentially affected by water use changes are recreation, tourism, hydropower, agriculture, and municipal water for home use. The literature was examined to determine if general multiplier estimates were available for these sectors, and the results are summarized below.

Different multipliers have been estimated for different types of recreational activities (e.g., fishing, kayaking). Cordell et al. (1990) estimated regional economic multipliers of 2.00 and 2.03 for the total economic effects of water-based recreation expenditures on local economies. That is, for every $1 spent on water-based recreation, the total regional economic impact would be $2.00 to $2.03. Norton, Smith, and Strand (1981) estimated a range of multipliers from 2.03 to 2.88 in an analysis of the total economic value of recreational fishing. Therefore, a range of literature-based multipliers from 2.00 (the low end of the range of Cordell et al.) to 2.88 (the high end from Norton, Smith, and Strand) might reflect expected regional impacts from changes to water-based recreation expenditures.

Changes to recreation use of resources often affect recreation-based tourism. Small communities or communities with simple, narrow-based economies may have tourism multipliers of 1.0, and large communities with broad economic bases may have multipliers approaching 3.0 or 4.0 (TAMU 2003). However, most tourism-related multipliers appear to be under 2.0. Michigan State University operates a "Michigan Tourism Economic Impact Calculator" webpage, which can translate visitor spending in Michigan into economic impacts (MSU 2003). The calculator uses a set of multipliers from 1996 Michigan IMPLAN data; a sales multiplier for a typical tourist spending pattern is 1.6 for statewide impacts, 1.45 for metro regions, and 1.3 for rural regions. A similar multiplier of 1.5 was used for ski-tourism impacts in Colorado (Goldsmith, Seidl, and Weiler 2001), and 1.68 was used in a study that examined the impact of fall tourism on the Vermont economy (UVM 2001). One study that examined the economic impacts of reduced recreational opportunities in northern Tampa Bay lakes used IMPLAN multipliers for the general merchandise retail business. The specific multipliers used were not given in the analysis, but the authors do mention that a more precise multiplier would be derived by modifying the IMPLAN model to reflect the specific types of business being evaluated.

Reductions in water flows available for hydropower could primarily affect the economy through power prices because an alternative might increase the cost of power, which in turn could affect the viability of local industries. This case was illustrated in the Federal Energy Regulatory Commission's (FERC's) analysis of the Ripogenus and Penobscot Mills projects in Maine (FERC 1996). In relicensing proceedings, the applicant (Great Northern Paper) claimed that changes in streamflows would add incremental costs to its production of pulp and paper products that might threaten the viability of some of its mills. FERC largely agreed with the applicant's analysis, concluding that increased power costs could have significant repercussions for the Maine economy. This study used a multiplier of 2.61 to estimate total employment impacts in Maine.

Changes in the amount of water available for agriculture may have economic impacts to the agricultural sector and related industries. The Oklahoma Cooperative Extension Service (Song and Doeksen 2003) estimated two multipliers for the agriculture industry: one for income effects (change in personal income resulting throughout the economy from a $1.00 change in income in a sector) and another for employment effects (change in employment due to a one unit change in the labor force in a specific sector). The income multipliers estimated for the livestock and crop sectors were 2.72 and 2.51, respectively. The employment multipliers estimated for the livestock and crop sectors were 2.02 and 1.69, respectively.

Reductions in water flows available for municipal use may increase the prices businesses and homes pay for water use. These fee increases are often considered "burdens" on small businesses and individuals with low or fixed incomes such as the elderly. No specific municipal government multiplier has been located, but such a multiplier would be similar to tax multipliers: the ratio of the change in aggregate output (or gross domestic product) to an autonomous change in a taxes. A tax cut increases disposable income, which is likely to lead to added consumption spending; therefore, income will increase by a multiple of the decrease in taxes. The reverse is true for tax increases, or in this case, an increase in fees. The formula for calculating the tax multiplier relies on the level of the marginal propensity to consume. No general tax multipliers were stated in the literature examined, but are likely to lie between -1 and -4 (tax multipliers are negative since an increase in taxes causes a decrease in expenditures).

In addition, there may be secondary effects stemming from water flows enhancing local property values and improving or renewing community quality of life and ability to attract tourists and businesses, but multiplier impacts for these aspects were not found in the literature.

LITERATURE REVIEW SUMMARY

This section summarizes the literature available on the value of water in specific sectors of the United States. As demonstrated in the various sections and summarized in Table 5.18, the value of water can vary dramatically, even within a specific sector. These ranges of values are not necessarily due to imprecise measurement of the value of water but in fact represent the true situation that the value of water is site, time, and use specific, such that one value does not fit all. While this table helps develop an appreciation for the relative values of water by sector, it is important to understand the methods used to develop these values for each sector. The specific value of water may vary widely, but there is still much to be gleaned from an overarching view of the literature.

Agriculture is a major user of water in the United States, accounting for 80% of the nation's consumptive water use and over 90% in many western states (Solley, Pierce, and Perlman 1998). Most of the major water markets in the United States are based on transfers either within the agricultural sector or between the agricultural and municipal sectors. Because agriculture plays such a significant role in total water use of the United States, understanding the value of water to this sector of the economy is especially important.

Estimating the value for water in the agriculture sector is based on the concept of farm production functions and the incremental increase in yields resulting from the application of additional quantities of water. Table 5.18 shows the range of estimated values for water when applied to various crops across the nation. In the agricultural sector, one major factor that controls the overall value is the relative amount of purchased irrigation water versus rainfall or natural irrigation. Hops are mainly grown in the Pacific Northwest region, where rainfall is often plentiful, whereas potatoes are typically grown in more arid climates. While these are not the only factors influencing the value of water in the respective crop production, they do shed light on possible reasons for the wide difference in the value of water when applied to each crop.

The total quantity of water used for domestic purposes is minor when compared to agriculture, but clearly residential water use is one of the driving forces in water policy and allocation issues in the United States today. Total U.S. residential water use is approximately 10% of all national consumptive uses of water.

Table 5.18
Relative value of water by sector (2003 USD)

Sector	Low $/kgal ($/AF)	High $/kgal ($/AF)
Agriculture	0.065 (hops) ($21/AF)	4.44 (potatoes) ($1,447/AF)
Residential	4.30 ($1,400/AF)	12.28+ ($4,000+/AF)
CII	0.086 ($28/AF)	2.47 ($805/AF)
Instream	0.03 (recreation) ($10/AF)	2.36 + (recreation) ($770+/AF)
Cultural	Unknown	Unknown

Estimating a single "value" of water in the residential sector is difficult, as in agriculture. Most of the residential water valuation studies do not look at water as in input to household production, though clearly one could. The studies focus on the value of changing current or future water use levels, or changes in the quality of water used for specific purposes. A main approach to estimating the value of water is through the relative elasticity of water demand for given price changes. Another approach is to estimate what individuals would be willing to pay to forego a potential reduction in water consumption. Studies of the WTP of residential customers are limited, but suggest values may be on the order of $1,400 per AF. Interpreting results of surveys that estimate WTP to avoid potential shortages (increase reliability) suggests values may be $4,000/AF or higher.

These illustrative numbers support the intuition that residential water users are willing to pay a relatively high price for water, but there are limitations to the application of these estimates to other geographical areas. First, these estimates are for avoiding water shortages in areas that have rather limited supplies to begin with (e.g., California). It is likely that the residential users' willingness to pay to increase the quantity or reliability of water available to them may be lower than their WTP to avoid a loss of water. Thus utility managers trying to identify the benefits of increasing the supply of water to their consumers would be cautioned against applying these full values.

Nonresidential water use accounts for 53% of total water use in U.S. CWSs, and of that, over 70% is delivered to commercial, industrial, and institutional water users. The value of water to the commercial, industrial, and institutional sectors has not been studied as extensively as the value of water used in the agriculture and residential sectors. Thus, while it is commonly believed that commercial, industrial, and institutional use is a relatively high value application of water, relatively little empirical evidence supports this belief.

A review of seven industrial water valuation studies (the most recent of which is 1982) found the median value of industrial water intake to be $165/AF ($0.51/kgal), with a range of values from $35/AF ($0.11/kgal) to $1,002/AF ($3.07/kgal). Many industrial facilities self-supply their water, and pay little or nothing for their raw water input.

Values of the water used in the CII sector are difficult to estimate. Information on CII withdrawals is limited, and many industries supply a larger portion of their water independent of municipal systems. Values among the sectors vary significantly by region, yet some relative values may be consistent across regions. Another factor affecting value is time. Future market values depend on a particular industry's sustainability with regard to its inputs (water, in particular) and the fate of each industry

Water use and value for instream recreational or ecological purposes are derived from the use of water in producing some other good, say a recreational experience or a quality ecosystem. The concept of instream flows has a variety of meanings, including stream flows required for swimming, floating a canoe, supporting fisheries, sustaining waterfowl habitat, providing downstream water supplies, assimilating wastes, and generating hydroelectric power. Thus water in this sector, as with the agricultural or CII sectors, can be thought to be an input to the production of a good or service. These uses require different quantities of water at different times and at different levels of water quality. The values that individuals place on or derive from instream water can be considered either use or nonuse values.

Studies examining the economic values of instream flows have estimated that most recreational values are between $15 and $60/AF for rivers that offer high-quality recreational opportunities or are used extensively for recreation, and up to $790/AF for reservoirs in the

California Central Valley. Combining recreational values with associated community economic stimulus benefits, instream flows for a whitewater boating course in Colorado generate values of over $1,600/AF in the peak use season.

For environmental values, annual household WTP to protect instream flows specifically for the silvery minnow on the middle Rio Grande was estimated at $28, and $61 to protect minimum instream flows on all major New Mexico rivers and to protect 11 listed fish species. These estimates of annual household WTP imply a range in values of $95/AF to $190/AF to protect instream flows for the silvery minnow.

Estimates of WTP for instream flow must be interpreted carefully. The estimates are sensitive to differences in use levels and the underlying value per day for recreation at the site. Comparing the marginal values requires correcting for key differences in activity levels, value per day of activity, and the baseline flows on which the marginal values are based.

Another consideration is that most instream flow studies cited here were conducted for rivers in the American West. Seasonal water shortages are of greater concern in arid regions and may heighten concern over flows, thereby influencing WTP estimates obtained in CV studies. Furthermore, most of the study rivers are in relatively undeveloped areas. Eastern rivers may be affected by other factors (e.g., shoreline development) that influence the recreational experience and willingness to pay, but are used (for recreation) and valued (for ecological services and protection of endangered species) by more people.

Also, it is important to understand the policy context of the instream flow studies. Most of the studies consider competition between downstream recreational uses and consumptive uses. Consumptive uses such as irrigation and drinking water involve withdrawals that, to a great degree, are never returned to the river. In contrast, hydropower facilities generally affect flow by diverting water around bypass reaches; the water ultimately is returned to the river.

Many cultures have spiritual relationships with the natural environment and the use of water. Water soothes and relaxes, inspires reflection, is a source of beauty, and has spiritual qualities recognized in religion and ritual, from baptism to death. It is difficult to estimate these aesthetic, spiritual, and historical values of water. They do not have observable dollar values because they are not traded in the marketplace. They are not relatively large uses in terms of volume of water (compared to such sectors as drinking and irrigation) and thus have not been well studied. While it is straightforward to identify the ways in which water are incorporated into our culture, unfortunately, currently, there is very little available empirical research on the monetary value of water for cultural uses.

CHAPTER 6
IMPACTS OF FUTURE TRENDS AFFECTING THE VALUE OF WATER

"If there is magic on this planet, it is contained in water. Its substance reaches everywhere; it touches the past and prepares the future." (Loren Eiseley, *The Immense Journey*, 1957)

The previous chapter summarized the existing empirical evidence in the literature for various types of water-related values. It is important to recognize that those literature-based values reflect estimates specific to certain locations, circumstances, and time periods. For various reasons that are discussed below—including the increased scarcity caused by demand growth occurring in concert with dwindling or limited supplies—the value of water is expected to increase over time, perhaps significantly.

Several emerging trends and probable future scenarios will directly shape the public's real and perceived value of water in the future. This chapter briefly explores these trends and outline potential impacts. Trends and related issues that are likely to affect water values in the future include:

- Population growth
- Total watershed management (including tighter restrictions on pollutant loads and instream flows)
- Climate variability and global climate change
- Public environmental attitudes (including economic growth)
- Improvements in water supply technology
- Water efficiency and productivity
- Water rights and emerging water markets
- Rising costs of water provision (including infrastructure renewal)
- Public versus private management of water
- Development of alternative water delivery models.

POPULATION GROWTH

Of the several trends that will directly shape the public's value of water in the future, perhaps the most important will be population growth. The current population of the United States is approximately 292 million. Within 100 years, the population is projected to nearly double to over 550 million. And this estimate represents growth with moderate levels of fertility, life expectancy, and net immigration. Using comparatively high values for the above three factors produces a population level of 1.1 billion people in the United States by 2100—almost quadruple the current population. Under these higher growth assumptions, population levels in 2050 would be double the current level (Means 2001). A significant amount of this growth in U.S. population is expected to occur in the South and Southwest (unless growth control measures, infrastructure limitations, or resource restrictions shift the growth pattern to other parts of the United States).

Many issues and factors could (and will) affect this projection, including immigration rates, changes in fertility rates, medical breakthroughs that prolong life, potential epidemics, population control measures, and so forth. Nonetheless, it is clear that water utilities will need to be in a position to accommodate significantly higher water demands in the future (not withstanding the reduced per capita water usage that increased conservation may produce).

As water demands rise, so too will the potential for shortages. Indeed, the U.S. Department of Interior has identified numerous water "hot spots" in their report entitled *Water 2025: Preventing Crises and Conflict in the West* (U.S. Bureau of Reclamation 2003). These locations include Las Vegas, Reno, Albuquerque, Denver, Houston, Salt Lake City, Flagstaff, Los Angeles, Sacramento, San Diego, Phoenix, and San Antonio.

Should shortages develop, public awareness of water limitations will grow along with media coverage, possibly affecting the water use ethic in these communities. These shortages could affect land use decisions by further injecting water supply reliability into debates about growth and development.

TOTAL WATERSHED MANAGEMENT

Water suppliers need to consider the wide range of regulatory or other mechanisms that affect the quality or availability of their source waters. Total watershed management refers to all the regulatory or other mandates that may alter the future of source waters. Water utility managers need to consider how changes in flows, contaminant concentrations, and allowable extraction volumes may change over time.

Environmental regulations will most likely increase in stringency to manage the pollutant loading of increased population and to protect the environment. The collective mandates of the Endangered Species Act (ESA), Clean Water Act (CWA), and Safe Drinking Water Act (SDWA) will increase the cost of treatment and water supply development. Increasing the diversion of water from existing watercourses is no longer an option for many communities. Sustainable population growth strategies may force difficult land use and lifestyle decisions. Land use, watershed protection, aquatic habitat preservation, and drinking water may converge as hot button issues, with value and use tradeoffs and heightened conflicts becoming increasingly likely and widespread.

The increased pollutant load and water needs associated with population growth will have widespread watershed-level impacts. Providing the water infrastructure and source water supplies to meet this growth will require navigating numerous legislative and regulatory hurdles, as well as growing public debate. Isolated policy decisions in one area of a watershed can have implications for compliance in other parts of the watershed. This may be particularly relevant where pathogen loading, largely unregulated under the current CWA, increases to the point that water utilities must install more advanced technology to comply with the SDWA.

The implications of recovery planning for threatened or endangered aquatic species on water supply development will require a more holistic view of the water cycle within the watershed. Legislative changes under the SDWA amendments in 1996, as well as provisions in the ESA and CWA, have spawned numerous nongovernmental organizations (NGOs) that are now active on a watershed-wide basis. These NGOs are raising the public's awareness of the impact of human activities on water quality. The public may grow more sensitive to the value of these aquatic resources and, in the process, place a higher value on water.

CLIMATE VARIABILITY AND GLOBAL CLIMATE CHANGE

Although there are still scientific uncertainties and political debates, there is growing consensus among credible researchers that global climate is changing. Mean global temperatures increased approximately 1°F in the last century. While there are many factors that can contribute to such a warming of the climate, the mainstream scientific conclusion is that anthropogenic factors (e.g., increased greenhouse gas emissions, changes in land cover) caused the majority of increase in temperatures in the latter half of the century (IPCC 2001). Even without these anthropogenic forcings, our expanding knowledge of climates of past centuries shows that those climates were quite different from that of today (see www.ncdc.noaa.gov). Thus, the recent climate (e.g., the last 30 or 50 years) is not as reliable a guide to future climates as once thought. The combination of our knowledge about climate variability and the likelihood that climate change caused by human factors will continue and could accelerate in the future means that prudent planning in water resources should account for the possibility of changes in mean and extreme climate conditions.

Numerous models have been developed to predict global climate change. While there are significant variances associated with these models, especially regionally, there are common elements that are believed to be likely water resource outcomes of global climate change. These include the following:

- Rising temperatures will affect the extent and amount of winter snowpack in the mountains, as well as the timing of thaw-related releases of snowpack waters to downstream users and storage facilities. This will affect instream flows and temperatures, with widespread implications for reservoir operations, storage, and extractions.
- Within the next century, sea levels will rise by a likely minimum of 0.7 feet. Water supply intakes in estuarine environments could be affected by the movement of tidal zones and saline levels to more upstream portions of coastal surface watersheds. Saltwater intrusion of coastal aquifers also is likely to occur. This will result in the degradation of water quality and source water availability.
- Some increase in the intensity of extreme precipitation and floods is likely. Rainfall events are likely to be more intense than they are today, meaning that it is likely that surface waters will receive higher loads of sediment, nutrients, and pathogens from nonpoint sources, stormwater overflows, and animal feedlots.
- As temperatures rise, irrigation requirements for crops and landscaping could increase. Demands are likely to increase for human uses as well, for more evaporative cooling in commercial and residential buildings, and for more frequent bathing and swimming to cool off during elongated and hotter summers. These demand-side implications may result in increases in per capita water usage (at a minimum, they will limit how much per capita water savings may be realized by additional conservation efforts).
- River and estuary water temperatures will rise, posing increasing problems for cold water fisheries. This will increase the instream flow requirements to moderate water temperature impacts, thereby reducing supplies available for extractive uses. In general, as climate-related stresses on natural environments increase, there are likely

to be continued additions to the list of threatened and endangered species, with associated restrictions and constraints on water extractions and other uses.
- Increases in water temperatures are likely to result in several water quality and treatment challenges for public water supply systems. For example, higher water temperatures may lead to increased formation of disinfection byproducts, or result in the increased likelihood or severity of taste and odor problems (as may be associated with algal blooms).

The effects of global climate change will vary geographically. Media coverage of the impacts will underscore the water supply planning difficulties faced by the water supply community, and the public may become more willing to invest in the necessary infrastructure to "hedge" against global warming effects on water resources.

PUBLIC ENVIRONMENTAL ATTITUDES

As societies and economies mature and become wealthier, demands for environmental amenities increase, as does the value placed on natural systems and environmental quality. Relatively rich societies can afford to look beyond immediate subsistence needs to those elements that are widely considered to add to the quality of life. Additionally, older members of the population have higher voting rates than younger individuals. This economic growth and increased population maturity, if coupled with political activism and environmental ethic, could drive public policy decisions toward more environmentally friendly water use and extraction decisions. Cost may not be as much of an overriding decision factor as it has been at times in the present and recent past. Utilities will need to be increasingly attuned to other water values and tradeoffs.

In addition, should water quality deteriorate or widespread use of degraded source supplies occur, there is some potential for further erosion of public trust and confidence in water utilities as the customer's perception of the safety and aesthetics of tap water is eroded. On the other hand, consumer concern for the environment (particularly aquatic ecosystems) could create public support for increased reliance on nonconventional supplies (e.g., desalination) such that some surface or groundwater sources may be more carefully managed (or left untapped) to preserve water for the environment.

The role of public interest, advocacy, and special interest groups in providing information to utility customers probably will grow in communities where utilities fail to provide needed information. Currently, there is widespread use of the Internet to coordinate advocacy group positions (both against and for water supply development or infrastructure projects). The Internet has become a powerful means of customer communication for utilities, and real-time water quality information available on the web could ultimately become the norm. As customers become better informed, they will most likely engage in water policy issues to a greater degree.

IMPROVEMENTS IN WATER SUPPLY TECHNOLOGY

Water treatment technology is growing in sophistication and ease of operation. The use of desalted water is likely to grow in importance and will accelerate as further breakthroughs in seawater or brackish water desalination are made, or if inexpensive energy sources are developed. Such breakthroughs would provide dramatic and fundamental relief in water-short coastal regions or where significant brackish or wastewater water supplies exist inland. The public is likely to see more adoption of advanced water treatment technologies, including desalination and reuse, as time passes.

Part of this adoption of advanced technologies will happen as water utilities increasingly turn to "degraded" sources of water supply to avoid the conventional water sources (e.g., river diversions, lake intakes) that increasingly face environmental permitting and cost constraints. These new supply sources are likely to include brackish groundwater, seawater, recycled water, and contaminated groundwater.

Providing these highly treated (and typically expensive) water supplies may have two effects on how consumers value water. The first could be that higher treatment levels cause water costs to increase, thereby sending high volume water users an economic signal to reduce usage. The second is that there could be some negative consumer reaction to drinking water from degraded sources (like contaminated groundwater sources). The public's inherent "trust in technology" (or lack thereof) may color their perception of the acceptability of these new supplies.

WATER EFFICIENCY AND PRODUCTIVITY

Important changes have been occurring regarding water efficiency. Improved water use efficiency in its simplest form means lowering the amount of water needed to achieve a unit of production in any given activity. Technological advances in irrigation, cooling, and sanitary applications (e.g., toilet flushing and washing machines) have been reducing the amount of water needed to accomplish these tasks. Behavioral factors related to taste, preference, and resource ethics in water using applications have also been changing.

This is best evidenced by the U.S. Geological Survey (USGS 2003b) report on total trends in water withdrawals by category of usage in the United States. Figure 6.1 depicts the decline in total water withdrawals since 1980, which has been attributed to active efforts to reduce consumption. This decline has occurred during a period when the U.S. population increased from approximately 227 million to 275 million. Note that the decline in water use since 1980 has been mostly in the thermoelectric, irrigation, and industrial sectors. Water use in the public supply sector continues to increase, however, despite increased water use efficiency.

The research team believes that support for restoring aquatic habitats and fisheries, controlling instream water pollution, and reducing contaminant levels in drinking water will continue to grow. Water utilities will be forced to work within formidable constraints. Demand side management probably will be a major feature of water utility practice. These programs may become more actively promoted through rate structures that provide strong financial signals and incentives to users, and through capital investments.

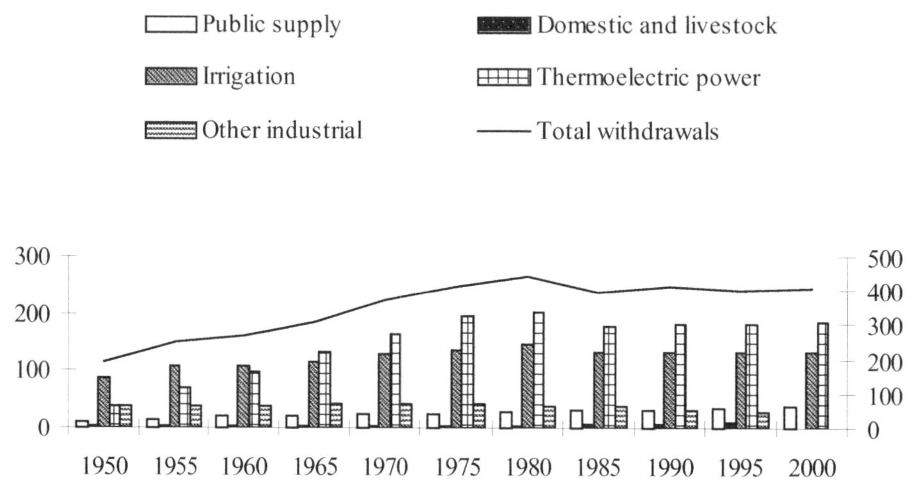

Source: Based on data from USGS 2003b.

Figure 6.1 U.S. water withdrawals, 1950-2000

WATER RIGHTS AND EMERGING WATER MARKETS

Some water is being transferred from willing sellers to willing buyers (frequently agricultural to urban uses), and this transfer will possibly accelerate. There is growing pressure on agriculture to secure a return on their water assets. Significant pressure on agriculture from foreign producers coupled with suburbanization of formerly rural areas and the associated loss of production acreage creates about 1,000,000 acres of farmland loss annually in the U.S. (USDA 2003b). The water supply that serves this production capacity is a logical candidate for transfer to other users (including environmental uses). As part of the Cal-Fed solution in California, an Environmental Water Account has been established to purchase water on the market for dedicated use in environmental restoration or to maintain minimum flow levels in rivers to protect critical habitat for aquatic species.

The development of formal wheeling policies by water resource managers (federal, state, and local agencies) will provide a clearer pathway to effecting such transfers. Reconciliation of third party impacts will be critical to successful transfers. As greater experience is gained with the institutional requirements of such transfers, the framework and logistics of water transfers will become more clear and more easily navigated.

The private sector is involved in water transfers, but, to date, there has been significant public institutional resistance to private sector involvement in water markets. And, while the public is generally supportive of agriculture, water shortages could affect how they view the relative merits of agricultural production (particularly "nonessential crops") versus water supply reliability for cities.

Another water rights factor to consider is the potential emergence of increased claims by Native American tribes. In several parts of the western United States, tribal governments are asserting historical claims to water rights that may have been unrecognized or inappropriately taken in the past. The possible increase in such claims—or of court rulings that more formally recognize such claims—could complicate the future use by municipalities or other users of water rights that they currently hold or hope to acquire.

RISING COSTS OF WATER PROVISION

Water utilities in the United States will face significant infrastructure replacement requirements over the next two decades. The Water Infrastructure Network (WIN) report, *Clean and Safe Water for the 21st Century: A Renewed Commitment to Water and Wastewater Infrastructure,* has indicated that the cost over the next 20 years for expanding, replacing, and maintaining this infrastructure could be up to $1 trillion for water and wastewater infrastructure (WIN 2000).

To accommodate this investment, the magnitude of the required rate increases will be a function of the age of local community infrastructure and how well the facilities have been maintained. The rate implications of replacing this infrastructure are potentially enormous. Water is about to become significantly more expensive (on a per household basis). To support the above investment, household water rates will need to double or triple, *excluding inflation*, over the next 20 years. This will create huge political upheaval for those systems that are not addressing the issue (and for many that are), and could have substantial impacts on how the public values water.

The average cost of water per household in the United States in 1998 was approximately $15.70 per month (Raftelis 1998). Figure 6.2 provides a rough comparison of typical large cities in the United States. The average is about $250 per year currently (assuming a 5% annual increase in water rates from 1999 to 2004 inclusive). Some household water costs are significantly higher. If the actual infrastructure need were closer to $1 trillion, the average annual household cost would increase by almost $1,200 per year, to almost $100 per month.

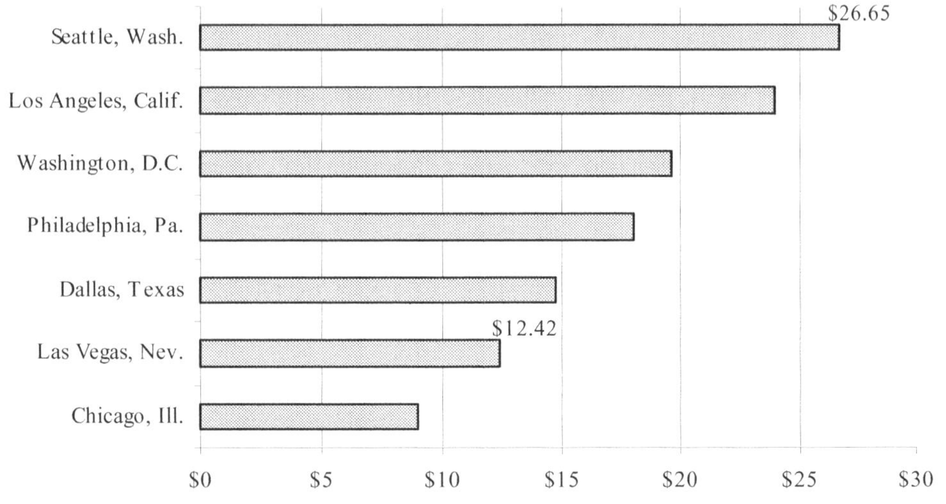

Source: Based on Raftelis 1998.

Figure 6.2 Monthly per household water costs in select cities (1998 USD)

There is a great deal of imprecision in these cost estimates, caused in part by the ongoing debate about the level of expenditure required for water infrastructure renewal. What is not debatable is that water rates (or some other revenue source) will rise to pay for this, or the infrastructure will not be replaced and water service interruptions, emergency repair costs, and cross connections will become far more frequent in the nation's water systems. These outcomes will affect public perceptions and the value of water.

PUBLIC VERSUS PRIVATE MANAGEMENT OF WATER

A sizable increase in the privatization of water supply utilities was expected in the past decade across the United States and the world. While private ownership and pubic/private partnerships have increased, there also has been an emergent debate over ownership of (and who makes decisions about) water resources and water service provision (including pricing). The debate that is emerging is whether water is an essential public good that belongs in the hands of public interests (e.g., municipal governments) or a precious resource best managed by the invisible hand of economic markets (e.g., private enterprise). For example, a current debate among public water utilities in Southern California regards the involvement of private water companies in the development of coastal desalination plants. Staff of the California Coastal Commission has raised these issues in the environmental permitting process for the Huntington Beach desalination plant (and elsewhere).

The general public is not currently well informed about the distinction between public and private operation except where vocal campaigns have been waged either for or against privatization. These campaigns are not always characterized by objective analysis. It is not clear what the general public believes about private versus public involvement in water service provision. Perhaps most citizens are more concerned with adequate, high quality water being available at a "reasonable" price. To further complicate the issue, U.S. experience with privatized water services have not always resulted in the cost savings or gains in efficiency that are often touted as reasons to consider private sector involvement.

A key observation from the AwwaRF Strategic Assessment of the Future of Water Utilities (Means 2001) was the apparent distrust and fear in public agencies regarding the intentions and objectives of private water utilities. The debate over whether water is a public resource or a commodity (or perhaps a bit of both) will continue. Nonetheless, private water utilities will be poised to assist municipalities that either do not have the capacity to raise water rates or do not want the political exposure. Any change in the current federal tax laws that allow tax exempt financing for private water utility infrastructure will provide greater incentives to privatize water utilities.

DEVELOPMENT OF ALTERNATIVE WATER DELIVERY MODELS

The conventional utility approach to providing water is to perform all water treatment at a central plant and distribute treated water through a single distribution system. Many customers, however, prefer to purchase bottled water or install point-of-use (POU) or point-of-entry (POE) devices—at significantly higher cost—because of taste, aesthetic, and perceived health advantages. Also, more utilities are using or considering reclamation and dual distribution system approaches that provide recycled water for at least some of their customers' water needs (e.g., irrigation).

It is clear that many so-called unconventional supply arrangements and agreements are being made and that they are likely to be a continuing trend and influence in the future. This rapidly rising use of bottled water and home treatment devices reveals that the general public is willing to pay considerably more than current tap water prices for water that they perceive to be safer, more convenient, or more aesthetically pleasing.

In addition, continued tightening of drinking water regulations will drive decisions regarding the feasibility of maintaining high quality water from sophisticated, centralized treatment plants through existing distribution systems. Regulatory drivers, including concerns over disinfection byproducts and lead levels, both of which may increase in concentration in the distribution system or through service lines, may make centralized treatment approaches for some water quality regulations very difficult technically and quite expensive. Decentralized approaches to treatment and delivery of premium quality water for human ingestion may become more of a norm in some instances (with centrally treated and distributed waters still being very high quality and applied to most uses, even if not used for drinking per se).

CONCLUSIONS

While it is difficult to predict the future with any high degree of precision or confidence, emerging trends and likely future scenarios provide a strong indication that water values and costs will change in the years ahead, perhaps significantly. Overall, future utility managers are likely to witness increased demands for water in all of its uses (instream and extractive) as populations and incomes grow and other factors push demands upward. At the same time, it is likely that the availability of high quality source waters will be increasingly constrained due to factors such as climate change and increased competition between various users and types of use. As a consequence of demand growth concurrent with supply limitations, the cost of water is likely to increase significantly. This will add pressure to direct water to its most highly valued uses, and the cost and political pressures will reveal just how valuable this resource will become as it becomes more precious.

Other factors are also likely to come into play and affect water costs and values in the future. Infrastructure renewal and regulatory requirements are likely to add significant additional upward pressure on costs. The need to rely more extensively on lesser quality source waters is also likely to add to rising costs, although this may be somewhat counterbalanced if technological advances in treatment render desalination, reuse, and other currently expensive approaches more cost-effective.

Regardless of which projections and scenarios come to pass, it is likely that the future is likely to witness increased competition and conflict regarding the fate and use of the world's finite water resources. Water utility managers and other water professionals inevitably will be drawn into more and more circumstances where their decisions and mandates will need to take a more complete accounting of the values of water in its many varied potential uses.

CHAPTER 7
TOOLKIT FOR WATER MANAGERS

"Perfection of tools and confusion of aims are the products of our time." (Albert Einstein)

Chapter 6 examined trends in the characteristics and management of water resources, demonstrating some of the future issues to consider and the growing complexity facing water managers. This chapter provides an overall framework and set of templates (forms) to assist water managers in working through and tracking each step of a benefit-cost or similar type of analysis in which a management decision can be enhanced by considering the values for water.

The objective is to guide users through the process of developing and interpreting water value information, so that each critical step is identified and implemented according to sound practices. The templates also will help users organize the information that they develop and apply in their assessments.

Perhaps most important, the templates provide a useful mechanism for documenting and communicating key assumptions and findings. This will ensure greater transparency and replicability, and thereby enhance the defensibility and credibility of the analyses performed. The templates also will facilitate reviews of the analysis by other utility professionals, governing bodies, and stakeholders.

Once the water supply management problem has been identified, the following steps make up the framework and are discussed in the sections that follow:

- Establish the baseline (usually the outcomes associated with the "no action" status quo).
- Identify water utility options that could help solve that problem.
- Identify the full range of benefits and costs applicable to each option.
- Quantify the benefits and costs to the degree feasible, and record the underlying assumptions, data, and results.
- Qualitatively describe key benefits and costs that cannot be quantified.
- List all values in one summary, listing the quantified values along with the qualitatively described values.
- List all omissions, biases, and uncertainties, and describe their potential impact on the analysis.
- Conduct sensitivity analysis on key values or variables, and record the results.

ECONOMIC VERSUS FINANCIAL ANALYSIS

The example provided in the discussion that follows illustrates benefit-cost analysis. The same logic and templates can be used for other types of economic assessments as well (e.g., for cost-effectiveness analysis). It is intended to facilitate qualitative analysis as well as (or in lieu of) quantitative assessments.

It is important to note that an economic analysis (such as developed using these templates) is different and more inclusive than a financial analysis. Financial analysis and economic analysis are both used in the planning and development of water projects. However, there is an important distinction between the two analyses. Financial analysis is typically used by the utility or agency as a cash flow analysis. The purpose of the financial analysis is to determine if a water project can be financed and what the debt service, O&M, and other recurring costs will be over time. A financial analysis answers questions such as, Will projected revenues and other funding sources be sufficient to pay for project capital and operating costs? Will water rates need to be raised to cover those costs (and if so, by how much)? Only direct costs to the agency and project revenues are considered in this type of analysis.

In contrast, an economic analysis is a more comprehensive investigation of potential water projects and management decisions. Economic analysis generally takes into account costs and benefits not only from the perspective of the water agency, but also from the perspective of society (i.e., the broader community) as a whole. Economic analysis takes into account not only the financial costs and revenues accounted under the financial analysis, but also the wider range of benefits and costs of a project. These can include direct benefits such as avoided cost of the development of a new water supply as well as nonmarket benefits and costs (e.g., environmental impacts from the new recycled water project). Economic analysis answers questions such as, Is the value of the all of the benefits of a project greater than the value of all the costs? How does the cost of (or values generated by) one option (e.g., to supply a given amount of water) compare to the cost (or values) of other projects or options?

Economic analysis allows for a comparison of the full range of costs associated with the project to the full range of benefits. In choosing between project alternatives, the alternatives can be ranked according to their net present values, which represent the present value of net output that will be generated over the life of the project. The project with the highest net present value (assuming the same discount rate) is more desirable. In addition, economic analysis can also be applied to determine the allocation of costs and funding responsibility on a equitable basis, by identifying the proportion of total project benefits a stakeholder is expected to enjoy (i.e., it can help identify who bears the costs versus who receives the benefits).

PROJECT IDENTIFICATION: DEFINE THE POLICY OPTIONS (INCLUDING THE STATUS QUO BASELINE)

The first step in performing a benefit-cost or similar type of analysis is to define the options to be evaluated. This includes defining the baseline (status quo) to clearly articulate how "doing nothing" would affect the applicable outcomes and impacts. In some instances, the status quo may include undertaking some already planned actions, and under such circumstances these should be considered as a part of the baseline. These baseline impacts and outcomes become the threshold from which the impacts of other options under consideration are identified and measured. The baseline may already include conservation programs that are not projected to change in the future, regardless of which other water supply options are pursued.

For the illustration carried through in the templates in this chapter, the management issue being evaluated is consideration of new water supply options, in a setting where the utility recognizes that its current supply sources may be inadequate to handle anticipated demands in the not-too-distant future. Thus, the options defined and evaluated are water supply alternatives intended to help meet growing demands.

In this context, part of defining the baseline entails projecting how large a water supply shortfall would be expected under the "do nothing" alternative. This in turn is useful for assessing the types, scale, and timing of the impacts of maintaining the status quo (e.g., how much water the community will be short, how often, and what the implications may be in terms of water use restrictions and economic activity). This baseline becomes the benchmark against which the alternative supply enhancement options should be assessed.

Issues in Defining the Baseline

Defining the appropriate baseline is in many cases one of the most challenging aspects of an economic analysis. In many instances, the challenge is not analytical, but rather a matter of perceptions and motives for impacted stakeholders, customers, and those with local governance responsibilities. This is because the manner in which the baseline is established and characterized will, in effect, define the problem that the water resource projects or policy options are intended to address. Since the baseline determination defines the problem, it creates the framework and context within which the potential solutions are evaluated.

For example, if a water agency is investigating options to increase the quantity of water it can deliver in the future, the first step is to predict and clearly characterize what will happen if new water sources are not brought on line (i.e., the status quo). This baseline definition therefore will entail projections of water demand growth over the coming years and decades, so that the magnitude and timing of future projected shortfalls in supply can be described. However, the assumptions underlying the demand forecast can easily become the basis for fundamental disagreement with managers, public officials, customers, and stakeholders. This is because these demand forecasts have embedded in them some assumptions on "hot button" topics such as the rate of growth in local populations and economic activity, the extent and effectiveness of conservation programs, and so forth. No-growth advocates may, for example, want to see limited water supplies in the future as a way of creating a bottleneck that will limit the number of new residents in the community.

One way to help alleviate some of the problems with defining the baseline is to be explicit about what assumptions and estimates are embedded in the analysis of what happens if no action is taken (i.e., if the water supply continues as it does now, status quo). Stakeholders and other impacted parties will then be able to discuss the explicit assumptions, scenarios, and estimates in a direct manner. This is more productive than having an obscure basis for the baseline, which would contribute to having the debate occur on more veiled issues and hidden (or not so hidden) agendas such as attitudes about growth. And, if some interested parties want the growth debate to be a central part of the analysis of water supply policies, then the pros and cons of growth can be added to the evaluation of future supply enhancement options (where it belongs) rather than having it be part of the premise for why the baseline is wrong and, therefore, why the economic analysis should not be done in the first place.

Evaluating Options

Once the baseline is established, the next step is to develop and evaluate the options that will be compared to the baseline (and compared to each other). When evaluating a range of alternative supply options, it may be useful to frame the comparison based on each option delivering the minimum amount of water needed, and also factoring in the time period within which the water would need to be supplied. That is, it typically is useful to scale the options to match the projected minimum delivery requirements. Projects that are scaled to meet these minimum requirements can then considered to be the relevant water supply options.

Alternatively, project options can be evaluated at any scale, so long as the different quantities of water delivered by each option are factored into the analysis. This may be necessary when different supply options have different levels of capacity or production (i.e., the estimated volume of water that can be saved via a conservation program may be different than the amount produced by a potential reclamation facility, and this may differ from the amount of new surface water that might be tapped via another option). When options tend to be available at different scales, it may be useful to consider some options in combinations (e.g., a mix of source water options that combine to produce a target yield). This will help planners see which combination of options is the best alternative for meeting the projected shortfall between supply and demand.

IDENTIFICATION OF PROJECT IMPACTS RELATIVE TO BASELINE

The next step is to ensure all relevant costs or benefits are identified for each water supply option. One way to help identify all the applicable benefits and costs is to use a simple checklist that embodies all the types of impacts that are typical of the projects under consideration, and then comparing what is anticipated under each option to the items on the list. One possible characterization of this checklist is provided in Table 7.1.

This first stage of the evaluation process does not entail any actual quantification or monetization of costs and benefits. Instead, it is an effort to develop a thorough inventory of the cost and benefit categories that apply to a particular project. It is important to identify all impacts no matter whether the analyst anticipates that they can be monetized or described qualitatively. It is important to survey stakeholders in this process to understand viewpoints and identify benefits and costs that are important to them.

The list of possible benefits and costs might be divided into six main categories, as shown in Table 7.1. Applicable parts of this list will vary with the project considered. For instance, some projects such as water recycling, desalination, and water conservation may have avoided costs compared to a baseline, whereas conventional new water supply projects such as dam construction or groundwater importation may not have avoided costs. More detail for some of the categories may be helpful, depending on the issues at hand. Site- and option-specific subcategorization may be worth developing, if circumstances merit.

Table 7.1 is developed here in the context of the water supply illustration, providing an overview for tracking the costs and benefits associated with the water supply options to be compared. The costs to the water agency are the direct financial costs associated with the proposed water supply option. These can include capital equipment or construction costs, water supply acquisition costs, land costs associated with water right acquisition, treatment and distribution costs, administration fees paid to a water supplier (e.g., water district), and the agency's additional administration costs.

The proposed project may also mean that water supply and wastewater costs compared to the baseline can be avoided. The project may avoid water supply acquisition and capital costs, avoid treatment plant or wastewater plant expansion capital costs, or avoid the variable costs associated with the expanded water supply treatment plant or wastewater plant.

Table 7.1
Overview of benefits and cost categories across water supply options

Category	Import groundwater	Water recycling	Option 3	Option 4	Option 5
	Costs to water agency (internal financial costs)				
Capital		✓			
Water	✓				
Land	✓				
Treatment	✓	✓			
Distribution		✓			
Admin fees	✓				
Admin costs		✓			
	Avoided water supply and wastewater costs				
Supply costs		✓			
Treatment capacity		✓			
Wastewater capacity		✓			
Treatment variable costs		✓			
	Water reliability and quality				
Water quality (aesthetics)	✓	✓			
Water quality (regulatory compliance)					
Quality reliability	✓	✓			
Supply reliability	✓	✓			
	Public health and safety				
Change in risk of illness (morbidity)	✓				
Change in risk of premature fatality (mortality)					
	Environmental and recreational impacts				
Sourcewater protection		✓			
Downstream habitats		✓			
Environ. restoration					
Groundwater	✓				
Coastal ecosystems					
T&E species		✓			
Terrestrial ecosystems					
Recreation		✓			
	Economic, social and equity impacts				
Economic development/growth	✓	✓			
Resource access		✓			
Resource location					
Aesthetics					
Cultural values	✓	✓			

✓ indicates that the category of impact exists for the given supply option.

A third major category of impacts is water quality and reliability. The project may have an impact on drought resistance or overall system reliability. Changes in the level or consistency of water quality may be associated with some options, with possible engineering and cost implications related to blending or addressing seasonal variations in quality. Changes in water quality may have public health or aesthetic impacts, or both, implying that costs or benefits may arise with respect to meeting health standards or addressing taste and odor problems.

The fourth major category is public health and safety. This reflects any change in the risks of illness (morbidity) or of premature death (mortality) due to changes in the quality of water delivered by one option as compared to an alternative. For example, some source water options or treatment plant process alternatives may pose greater or lesser degrees of exposure to microbial agents, chemical contaminants, disinfection byproducts, and so forth. To the extent that options are associated with changes in potential exposures (and, hence, risks), these should be articulated and included in the analysis. Quantifying and valuing changes in risks to public health are challenging exercises, but at a minimum the potential changes need to be described in an informative qualitative manner. For a further discussion of challenges and options available to quantify or value changes in health risks, the AwwaRF report *Quantifying Public Health Risk Reduction Benefits* provides useful summaries and examples (Raucher et al. 2002).

A fifth major category is environmental and recreation impacts. Aquatic ecological impacts can be divided into three subcategories: sourcewater protection, downstream habitats, and environmental restoration. Other environmental impacts may include those associated with groundwater resources (such as subsidence or salt water intrusion) or loss of wetlands. Coastal ecosystems, threatened or endangered species, and terrestrial ecosystems may also be affected by some options in some settings. Recreation impacts may occur instream (e.g., rafting, swimming) or near-stream (e.g., bird watching), and can include impacts on flatwaters (i.e., in lakes or reservoirs) as well as instream in flowing waters.

The sixth major category in our checklist is economic, social, and equity impacts. These environmental justice issues may include consideration of the location of impacts, resource access issues, aesthetics, or cultural values. It also includes the impact of the project on the local economy, whether considered in a positive light such as helping to build or sustain community economic development, or considering the negative impacts from growth. Overall, the objective is to identify who the likely beneficiaries are, who is likely to be affected by the impact, and whether they will be impacted in a manner they will consider negative or positive. This enables the utility to consider stakeholder outreach efforts (e.g., to enlist beneficiaries) in recognition of positive outcomes and possible ways to compensate or offset potential negative outcomes.

SCREENING ANALYSIS

The next step in the process is to screen the list of potential costs and benefits for a project to determine which impacts are so small (or mitigated) that they can be dropped from the analysis, which must be qualitatively described (because quantification is not generally feasible), and which impacts can and should be quantified. The three screening criteria used in this step are described below. Figure 7.1 shows the screening analysis process as a flow chart.

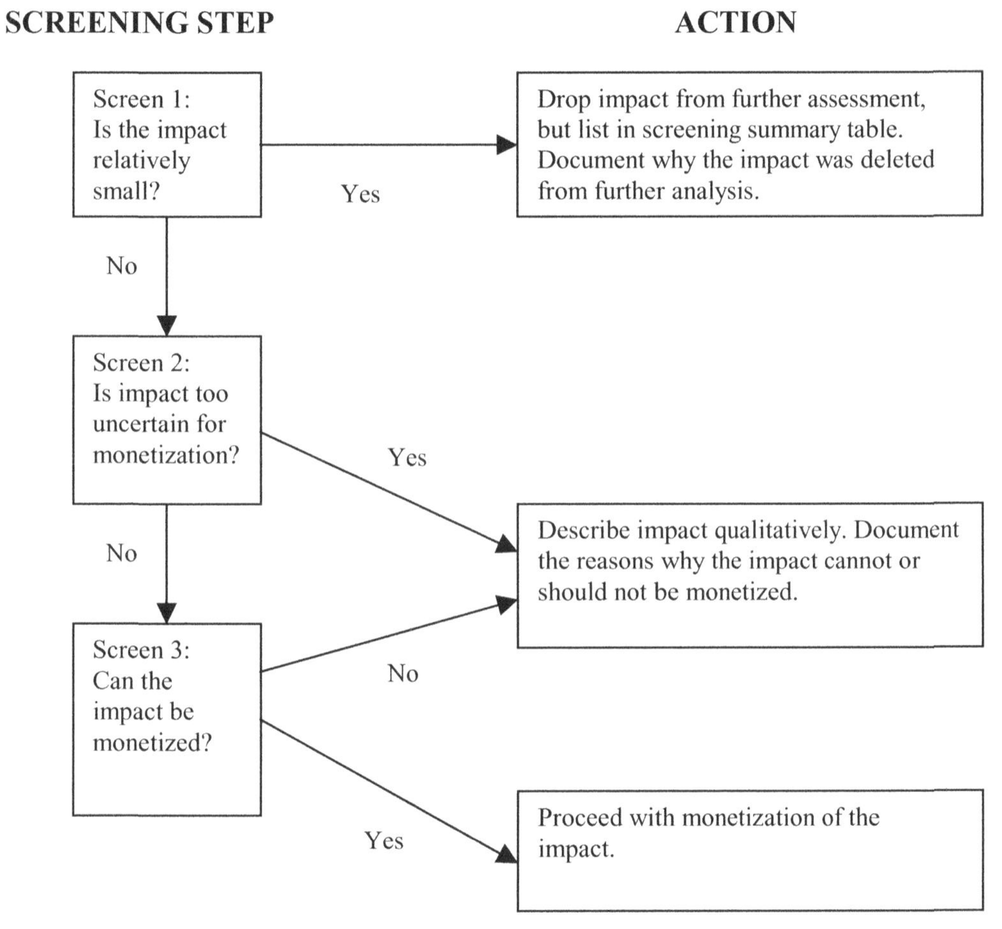

Figure 7.1 Screening analysis flow chart

Screen 1: Is the Impact Relatively Small?

This screen considers whether the cost or benefit will be very small in absolute terms, or in relative terms compared to the other impacts. If the impact is so small as to be insignificant, then perhaps it can be eliminated from the analysis in an effort to save or focus resources. This is a matter of judgment, and it is important to document the reasons behind such a decision.

Some impacts can be dropped from the assessment, even if they are not small, when the impacts are going to be mitigated through some activity embedded in the project design and cost. For example, if a supply option may adversely impact some wetlands, but as part of the project there will be restoration or creation of other nearby wetlands, then the wetland impact will be mitigated and does not belong in the analysis. However, it will be important to carefully assess and document that the impacts are indeed mitigated (e.g., that the newly restored wetlands will provide the same or better ecological services than those that may be adversely impacted).

Screen 2: Is the Impact so Uncertain or Changing so as to Resist Economic Quantification?

This screen considers the situation where the impact is so changing or uncertain (e.g., due to scientific uncertainty or time lags in natural processes) or sensitive (e.g., due to political considerations, legal uncertainties, or cultural sensitivity) that any attempt at economic assessment would be impossible or not useful. In this case, it is important to explicitly recognize that economic valuation may not be possible or useful, but that continued recognition of the impact is important through qualitative characterization.

Screen 3: Can the Impact be Quantified in Economic Terms?

This screen considers whether available data and methods are sufficient for monetization of the impact. If the data and methods are available, then the analyst can proceed with quantifying the impact and then converting the impact into monetary terms. If monetization is not possible, it is important to continue recognition of the impact through qualitative characterization.

In some instances, it will be feasible to quantify the impacts in physical terms (e.g., acres of wetlands, numbers of fish, acre-feet of water), but not in monetary terms. In these instances, the analyst should provide results in quantified physical units, even though monetization may not be feasible.

The results of the screening process for each project can be recorded in a screening summary table. Table 7.2 shows an example of the screening process applied to a water recycling project. The affirmative answers to Screen 3 are recorded first under "benefits and costs receiving full or partial economic valuation." Affirmative answers to Screen 2 are recorded next under "benefits and costs requiring qualitative assessment." These are benefits (denoted by "+") or costs (denoted by "-") that cannot be quantified within the analysis, but that may be important and therefore warrant informative qualitative discussion. Affirmative answers to Screen 1 are recorded last under "impacts deleted from further analysis: impacts that are relatively small or mitigated."

BENEFIT AND COST QUANTIFICATION

Establishing a Baseline

As noted above, a clear baseline should be established before proceeding with the quantification of benefits and costs. In the illustration used here, this means identifying the water supply situation and costs that exist or will exist in the absence of the proposed project or the options being evaluated. The situation in the absence of the project could be the existing water supply and wastewater treatment situation (i.e., a do nothing scenario), or could include the most likely future water supply option that would be built in the absence of the options to be evaluated. Analysts can define the baseline in a way that is most useful for their application, but should be careful to assess the impacts of each option against whatever the impacts would be for the baseline they have defined.

Table 7.2
Summary screening analysis (water recycling project)

Benefits and costs receiving full or partial economic valuation
- ❍ Capital equipment costs
- ❍ Treatment operations and maintenance costs
- ❍ Distribution system capacity costs
- ❍ Distribution system variable costs
- ❍ Avoided supply acquisition costs
- ❍ Avoided wastewater treatment capacity
- ❍ Avoided wastewater treatment variable cost
- ❍ Improved water supply reliability
- ❍ Improved in-stream recreation

Benefits and costs requiring qualitative assessment*
- (+) Impact on fisheries
- (+) Value of source water protection
- (+) Impact on threatened or endangered species
- (-) Differential access to recycled water according to geographic location
- (+) Community economic development impacts

Impacts deleted from further analysis: Impacts that are relatively small or mitigated
- ❍ Reduced air pollution from reduced pumping costs†

*(+) denotes benefits, and (-) denotes costs (or negative benefits).
†In some areas, reduced air pollution would be an important benefit that would more appropriately be placed under one of the preceding categories.

In a situation where a water recycling, desalination, or long-term water conservation project is being considered, the most likely conventional water supply option that meets the projected water needs may be considered the baseline. For example, that option may be a dam construction project. The environmental and recreational impacts of this project then can be assumed to be part of the baseline.

In defining and measuring impacts of an option relative to this baseline, one might find that the indirect potable water reuse project option may reduce variable treatment costs that would otherwise have been incurred to treat the same water for wastewater discharge. After establishing baseline costs, the estimate of the variable cost savings for wastewater treatment can be applied in the analysis as a benefit of the reuse option (or, it can be shown as a cost for the status quo baseline option). Likewise, the recycling project may delay or eliminate the need to expand the wastewater treatment facility, such that capital outlays avoided can be applied as a benefit of the reuse option.

As another example, if a water reclamation option is being considered relative to another option that would acquire additional surface water rights from irrigators, then it is important to consider all the cost elements associated with each option so that they can be compared fairly on a true apples-to-apples basis. For example, water reclamation may cost $350/AF of water delivered. The acquisition of agricultural water rights might be obtained at $180/AF. At first glance, it appears to make sense to acquire the surface water rights rather than pay nearly twice as much for reclamation. However, the analyst here needs to recognize that the cost of a water right covers only the expense of the raw water. The newly acquired water also would need to be conveyed to a water supply treatment plant, treated to meet applicable standards, and then pumped for delivery to the utility customers (the added supply would also add to wastewater loads, thereby adding more cost). If the cost to convey, treat, and deliver the new surface water amounted to \$270/AF, then the new surface water option would in fact cost a total of \$450/AF of potable water delivered (\$180 + \$270 = \$450). Thus, when the full lifecycle costs of delivering water under each option is considered, the new source water option may no longer be the most cost-effective on a dollar per acre foot basis.

Quantification Methods

After establishing a clear picture of which costs and benefits can be quantified and establishing the baseline, the quantification of benefits and costs can follow. Direct costs to the water supply agency are often established through engineering estimates. It is important to explicitly recognize any uncertainty in these estimates by recording the range of likely values. Other costs and the benefits can be monetized using a variety of methods discussed in more detail in previous chapters. Primary valuation methods such as hedonic estimation, contingent valuation surveys, or travel cost surveys are considered more robust than secondary methods such as benefits transfer. However, cost and time restrictions often prohibit use of primary research methods, and careful application of the benefits transfer method can produce benefits estimates that are likely to be reasonably accurate for the purposes of comparing water supply options. However, if overall project evaluation outcomes are very sensitive to assumptions and benefits transfer results (as can be ascertained via sensitivity analyses, as discussed below), then a utility may need to consider a more refined analysis, perhaps using some primary methods.

No matter what quantification method is used, the first step is to make annual estimates of the cost or benefit, and to project the years that they apply to in the future. The present value of the benefits stream should then be calculated to facilitate comparison on an apples-to-apples basis to benefits or costs incurred with other starting or ending dates, or different periods of time. There are many debates about the appropriate choice of discount rate for net present value calculations. In considering new water supplies, the average cost of capital to public agencies is often used, as is the average cost of capital to the public (or ratepayers), which is generally higher than the cost of capital to public agencies. In nominal terms (i.e., including anticipated inflation), the latter rate currently is in the range of 8 to 10%, given today's interest rates on credit card debt. In contrast, in some instances a real rate of social time preference may be a more suitable discount rate, and that is about 3%.

Whatever discount rate is selected, analysts need to check for consistency in how inflation is handled. The research team's recommendation is to conduct the analyses in real terms (i.e., excluding any adjustment for inflation). However, if cost or other estimates are not available in inflation-free terms, then the discount rate applied should be based on a nominal inflation-adjusted rate as well. For a more complete discussion of the temporal aspects of economic analysis, the reader is referred back to chapter 4.

Recording the Inputs and Results of Benefit-Cost Quantification Efforts

Table 7.3 offers an example template for recording the key inputs used for monetizing the benefits of a project option. Comparable templates should be used for key cost elements of each option as well.

Table 7.3
Detail on benefit value derivation for water recycling project

Benefit category	Annual quantity	Unit value used	Comments
Improved water supply reliability	20,000 AF of use by Memory Chip Corp.	$140-$200/AF of added value to industrial user	Industrial use value based on interview with plant manager, based on avoiding production cutbacks due to periodic drought year water use restrictions.
Improved instream recreation	3,000 user days added for anglers in the county, based on 10% increase over baseline levels.	Mean user day value of $36.90 for fishing (Loomis 1998a)	User day value likely to be underestimate because of high quality recreational experience for cold water gamefish in the affected stream segment.
Nonuse values for sustaining habitat for threatened or endangered species	200,000 households in impacted counties assumed to hold nonuse values related to the species and habitat impacted.	A low of $6.88 (2002 USD) per household per year, to a high of $10.24 (2002 USD) per year per taxpayer	The low-end value based on a benefits transfer from the research published by Boyle and Bishop (1987) for shiners, and high end from Cummings, Ganderton, and McGuckin (1994) for Colorado squawfish.

In the template, note that the unit value and the annual quantity used to develop estimates are recorded, along with relevant comments on the valuation method and data used. This table will provide a record of the assumptions and inputs used in the benefit-cost calculations, for each relevant benefit (or cost) component or category. This documents the basis of the estimates (so that in the future, the analyst or others can determine where the results came from) and also promotes transparency. This template also enables easier reviews of the analysis and thus can bolster the confidence others have in the results generated.

The template illustrated in Table 7.4 can be used to record all costs and benefits for a project, with results depicted in present value form. Ranges of values should be recorded, if applicable, to reflect any likely variability or uncertainty in the estimates. Benefits requiring qualitative assessment also are explicitly recorded on this form, to help ensure that the full range of costs and benefits—including those that could not be quantified—are kept in mind through each stage of the analysis.

It is often important to understand and record the distribution of costs and benefits across project stakeholders. Some benefits may accrue to the water supply agency while others may accrue to the wastewater agency or some key stakeholder group. External costs or benefits may affect customers or public interest entities. Understanding the incidence of costs and benefits for each project stakeholder can help illuminate the interests motivating project stakeholders. This understanding may point to an equitable cost sharing arrangement for the project that can be justified to the stakeholders. Table 7.5 adds a column to Table 7.4 that allows recording of the entity to which project benefits or costs mainly accrue.

Table 7.4
Costs and benefits of water recycling project
(present values, 10% discount rate, thousands of 2002 dollars)

	Dollar amount
Costs	
Capital construction costs	5,000
Treatment operations and maintenance costs	900
Distribution system capacity costs	9,000
Distribution system operation and maintenance costs	1,100
Total costs	16,000
Benefits (monetized)	
Avoided supply acquisition costs	4,000-6,000
Avoided wastewater treatment capacity expansion	8,000-10,000
Avoided wastewater treatment variable costs	300
Improved water supply reliability	3,000-4,000
Improved in-stream recreation	500-800
Improved non-use value for threatened or endangered species	150-200
Total monetized benefits	15,950-21,300
Benefits requiring qualitative assessment	
Impact on sourcewater ecosystem (fisheries impact)	+
Differential access to recycled water according to geographic location	-
Community economic development	+
Monetized net benefits (monetized benefits minus costs)	>(50) to 5,300+

+ indicates positive benefits anticipated, but not monetizable with readily available data.
- indicates costs anticipated, but not monetizable with readily available data.

IMPRECISION IN DERIVED ECONOMIC VALUES

Omissions, Biases, and Uncertainties

All estimates of economic values, no matter what method was used to derive them, will have some factors that cause them to deviate from their "true" value. If the reason for this deviation from the true value is known, it should be noted. The reasons for imprecision in estimates can typically be divided into three categories: omissions, biases, and uncertainties.

Omissions are information gaps that affect project valuation. It is important to identify them, and indicate whether they would affect net valuation results positively or negatively, if possible.

Biases refer to any factor causing benefit or cost estimates to be larger or smaller than their actual value. For example, if a benefit value depending on flow was estimated in a year with higher than normal flow, and could not be adjusted for the average value, then the estimate is biased, and this should be noted.

Uncertainties result from the fact that quantification involves estimating or predicting changes in the physical world or in socioeconomic relationships that may not be completely understood. This knowledge gap can cause estimated values to be different from the true value. Key uncertainties should be acknowledged in the analysis, and addressed through sensitivity analyses.

Table 7.5
Costs and benefits of water recycling project
(present values, 10% discount rate, thousands of 2002 dollars)

	Dollar amount	Stakeholder accruing cost or benefit
Costs		
Capital construction costs	5,000	Water utility
Treatment operations and maintenance costs	900	Water utility
Distribution system capacity costs	9,000	Water utility
Distribution system operation and maintenance costs	1,100	Water utility
Total costs	16,000	
Benefits (monetized)		
Avoided supply acquisition costs	4,000-6,000	Water utility
Avoided wastewater treatment capacity expansion	8,000-10,000	Wastewater district
Avoided wastewater treatment variable costs	300	Wastewater district
Improved water supply reliability	3,000-4,000	Memory Chip Corp.
Improved in-stream recreation	500-800	General public, whitewater boaters
Improved non-use value for threatened or endangered species	150-200	General public, Sierra Club
Total monetized benefits	15,950-21,300	
Benefits requiring qualitative assessment		
Impact on sourcewater ecosystem (fisheries impact)	+	General public, Trout Unlimited members
Differential access to recycled water according to geographic location	-	General public
Community economic development	+	Irrigation district
Monetized net benefits (monetized benefits minus costs)	> (50)-5,300+	

+ indicates positive benefits anticipated, but not monetizable with readily available data.
- indicates costs anticipated, but not monetizable with readily available data.

It is important to note that variability and uncertainty are often identified together but have distinct meanings. Variability reflects differences that exist as a basic state of the world—conditions may differ from time to time and from place to place (e.g., the amount of rainfall varies from year to year). Variability can often be described as a range of values, differing according to measurement in different time periods or locations (or other changes). It is recommended that this range of values be recorded explicitly in cost or benefit estimates.

Table 7.6 offers a template for listing and describing key omissions, biases, and uncertainties. The legend at the bottom can be used as a guide for indicating the degree that the omissions, biases, and uncertainties will influence the estimate in either the positive or negative direction. This type of table is useful for presenting key limitations in the empirical analysis, and thereby helping decision-makers and stakeholders recognize elements that might alter outcomes.

Table 7.6
Omissions, biases, and uncertainties and their effect on the water recycling project

Benefit or cost category	Likely impact on net benefits*	Comment
Instream recreation	+	User day adjustment for instream recreation value is likely to result in an underestimate because the high quality of the recreation available consistently draws from a wider region.
Water supply reliability	++	Value per acre foot estimate for value to industrial water use is likely an underestimate—alternative for providing high quality water with the same reliability through desalination is about $1,000/AF.
Impact on sourcewater ecosystems	+	Compared to the baseline, the recycling project would preserve quality habitat for the existing high quality trout fishery.
Equity—differential access	-	Only certain areas of town would have access to reclaimed water until additional distribution system infrastructure is in place.
Community economic development	+	Project would preserve flows available for agricultural diversions compared to the baseline, and help refine/attract CII customers.

*Direction and magnitude of effect on net benefits:
+ Likely to increase net benefits relative to quantified estimates.
++ Likely to increase net benefits significantly.
- Likely to decrease benefits.
-- Likely to decrease net benefits significantly.
U Uncertain, could be + or -.

Table 7.7
Sensitivity analysis applied to discount rate for the water recycling project (NPV, in thousands of dollars)

Discount rate	Monetized benefit	Cost	Monetized net benefit
0%	49,000-51,500	30,000	19,000-21,500
3%	39,500-41,700	26,000	13,500-15,700
6%	29,500-34,000	22,000	7,500-12,000
9%	15,950-21,300	16,000	(50)-5,300
12%	8,500-14,000	11,000	(3,500)-3,000
15%	2,500-8,000	8,000	(5,500)-0

Sensitivity Analysis

In many cases it will be useful to explore the impact of uncertainties or key assumptions such as discount rates through the use of sensitivity analysis. Sensitivity analysis involves systematically changing some key input variable, and then seeing how it affects the outcome of the analysis. The change in results with the change in inputs can illuminate how important the impact of uncertainty in a particular variable is to the outcome. Sensitivity analysis is often performed by varying a particular input by equal amounts greater to and less than the current value. For instance, if a discount rate of 9% has been chosen for the main analysis, that value might be varied in increments of 3 percentage points from 0 to 15% for the sensitivity analysis. Table 7.7 shows sensitivity analysis for the discount rate applied in this fashion to the water recycling example. In this example, benefits are particularly sensitive to changes in the discount rate, because many benefits are realized further into the future than are the corresponding costs.

FLOW DIAGRAM AND GENERAL GUIDANCE

To help guide readers through the steps laid out in the discussion and templates above, the research team developed a flow diagram (Figure 7.2) that is an overview of the key steps and shows how they link to one another. In addition, step-specific guidance and cross references are provided in Table 7.8. The combination of the flow diagram and associated summary guidance table gives utility professionals a concise roadmap to help them get started in conducting economic analyses of key options facing their water agency.

In addition, extra (blank) versions of the templates of the key tables that go along with each step are at the end of this chapter. Further, the case studies in chapter 8, and in associated appendices, provide further illustrations of the use of the economic analysis process and the templates, and also reveal some of the empirical values that may be applicable for a utility staffer's application of the approach. While economic analyses of the type described in this report are not amenable to cookbook application, the combination of materials in this chapter and other parts of this report should provide key assistance to a utility looking to apply value-based economic analyses to pending water supply, capital improvement, or other options under consideration.

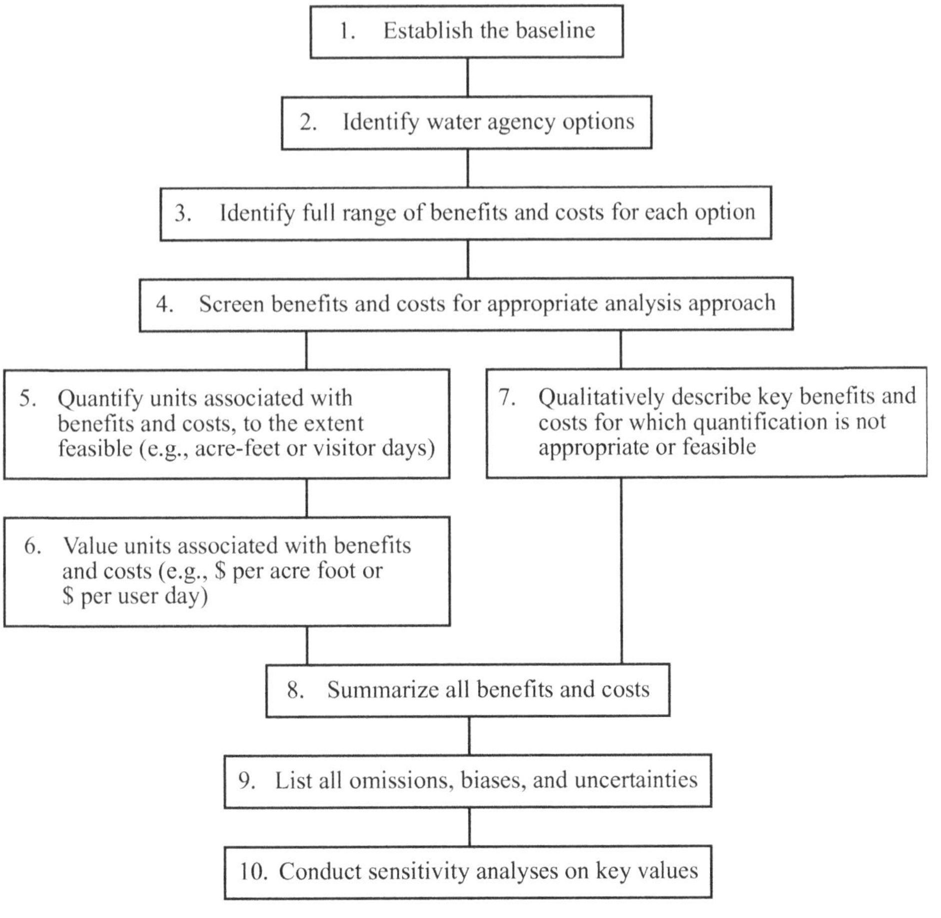

Figure 7.2 Overview flow diagram of steps for economic analysis

Table 7.8
Guidance and cross references for steps of economic analysis

Step	Guidance and key cross references
Step 1: Establish the baseline	Define the outcomes associated with the "no action" status quo. This base case may entail doing nothing or undertaking already planned actions. The baseline is the mark against which changes resulting from the project alternative are measured. It is important to define scale and timing of the impacts of the baseline, be explicit about assumptions, and engage stakeholders about their perspective of what happens under a no action, status quo baseline. *Cross reference*: Discussion early in chapter 7.
Step 2: Identify relevant water agency options	Identify and develop all the relevant utility options that will be compared to the baseline, and to each other. It is useful to scale project options to a common size or objective (e.g., to meet projected minimum water delivery requirements). For options available at different scales, it is helpful to consider staging or combinations of options. *Cross reference*: Discussion early in chapter 7.
Step 3: Identify full range of relevant benefits and costs for each option	Develop a thorough inventory of all likely costs and benefits associated with each of the project alternatives (options). Include costs and benefits beyond those faced by the utility alone or customers alone. *Cross reference*: Checklist to aid identification of benefits and costs is in Table 7.1 (Template 1). The case studies in chapter 8, and associated appendices, offer some illustrations.
Step 4: Screen benefits and costs, to determine type of analysis	In the screening step, the analyst determines which costs and benefits can and should be analyzed quantitatively, which should be described only qualitatively, and which are insignificantly small and can be eliminated from further analysis. *Cross reference*: The screening process is diagrammed in Figure 7.1, and results are illustrated in Table 7.2 (Template 2) and some case studies (chapter 8).
Step 5: Quantify benefits and costs, to extent feasible, so that they can be assigned monetary values (if feasible)	In the first step of valuing a benefit or cost, the amount (quantity) of the outcome (e.g., water or resource use) should be established. These quantity outcomes may be a volume of water delivered (e.g., AF), number of recreational user outings (e.g., recreational angling days per year), or in whatever units the outcomes are most readily and meaningfully measured. It is important to match the quantity units of measurement to whatever metric is available for the corresponding dollar values (e.g., if the valuation in step 6 uses a $/household measure, then the quantification in step 5 should be aimed at estimating the number of households affected). Ranges of quantity estimates (rather than a single point estimate) may be used to better represent variability or uncertainty associated with resource use estimates. *Cross reference*: Annual quantities for resource use and explanations are shown in Table 7.3, and can be entered in Template 3.
Step 6: Value the benefits and costs in monetary terms	Once the quantity of resource use has been estimated, a per unit dollar value often can be assigned to the benefit or cost, to reach a total value (quantity times per unit value). The per unit values can be expressed as dollars per unit of water (e.g., dollars per acre-foot) or dollars per unit of resource use (e.g., dollars per visitor day). Ranges of values may be used to better represent per unit resource valuations. Annual benefit or cost values should be projected over the project life (either annualized, or as a net present value, as per step 8). *Cross reference:* Unit values used and comments can be entered in Template 3, as shown in Table 7.3. Ranges of values from the literature are provided in chapter 5, and the case studies in chapter 8 (and associated appendices) reveal how these and other value estimates may be obtained and applied.

(continued)

Table 7.8 (Continued)

Step	Guidance and key cross references
Step 7: Qualitatively describe key benefits and costs for which quantification is not appropriate or feasible	It may not be feasible or desirable to express some types of benefits or costs in quantitative or monetary terms (as per screening in step 3). However, it is always important to describe these nonquantified benefits and costs in a meaningful, qualitative manner. These benefits and costs may be described qualitatively, in part, by using a simple scale indicating likely impact on net project benefits. Impacts can be qualitatively rated on a 5-point scale ranging from very positive to very negative. Qualitative ratings should be accompanied by descriptions of the impact, and should be explicitly carried through the analysis. *Cross reference*: The qualitative impacts should be listed along with the quantitative estimates in Template 4, as shown in Table 7.4.
Step 8: Summarize all present value (or annualized) costs and benefits	Quantitative benefit or cost projections over time (from step 6) should be discounted to present value at an appropriate discount rate. These net present value of monetized benefits and costs should be summarized in one location, along with the listing and ranking of those benefits described only qualitatively (from step 7). *Cross reference*: Results can be assembled and displayed in a summary table as in Template 4, shown in Table 7.4. Table 7.5 and Template 5 include identification of the prime stakeholder to which the benefit or cost accrues.
Step 9: List and assess all omissions, biases and uncertainties	All omissions, biases, and uncertainties associated with the estimated benefits and costs, should be explicitly documented. The impact that these may have on the final outcome of the analysis (e.g., in terms of their likelihood of increasing or decreasing net benefits, or an uncertain direction of change in net benefits) should be noted. *Cross reference:* An example of a useful approach to organize and present this information is provided in Template 6, as shown in Table 7.6.
Step 10: Conduct sensitivity analyses of key values	Sensitivity analyses should be conducted on key variables or benefit and cost estimates, to explore and communicate the impact of assumptions, uncertainty, or natural variability. Use sensitivity analyses to identify which assumptions or uncertainties have the largest impact on the outcome of the analysis (e.g., identify which assumptions might change the net benefits of an option from positive to negative, or alter the ranking of options in terms of their relative net benefits). *Cross reference:* The results of these sensitivity analyses can be recorded in a manner similar to Template 7, as shown in Table 7.7.

CONCLUSIONS

Whenever an analysis is being conducted that involves weighing the impacts of various water management options, it is important to conduct the analysis is an orderly and systematic manner, following best practices. The flow diagram and templates provided as a toolkit in this chapter are intended to help water managers systematically and properly assess the benefits and costs of water management options they may be considering, or better develop and document water value considerations that are applicable to the water management decisions they are making.

EXTRA, BLANK TEMPLATES

To assist water professionals in applying the framework described here, blank templates of each is provided in the pages that follow.

Blank Templates

Template 1
Overview of benefits and cost categories across water supply options

Category	Option 1	Option 2	Option 3	Option 4	Option 5
Costs to water agency (internal financial costs)					
Capital					
Water					
Land					
Treatment					
Distribution					
Admin fees					
Admin costs					
Avoided water supply and wastewater costs					
Supply costs					
Treatment capacity					
Wastewater capacity					
Treatment variable costs					
Water reliability and quality					
Water quality (aesthetics)					
Water quality (regulatory compliance)					
Quality reliability					
Supply reliability					
Public health and safety					
Change in risk of illness (morbidity)					
Change in risk of premature fatality (mortality)					
Environmental and recreational impacts					
Sourcewater protection					
Downstream habitats					
Environ. restoration					
Groundwater					
Coastal ecosystems					
T&E species					
Terrestrial ecosystems					
Recreation					
Economic, social and equity impacts					
Economic development/growth					
Resource access					
Resource location					
Aesthetics					
Cultural values					

✓ indicates that the category of impact exists for the given supply option.

Template 2
Summary screening analysis

Benefits and costs receiving full or partial economic valuation

❍
❍
❍
❍
❍
❍
❍
❍
❍

Benefits and costs requiring qualitative assessment

( )
( )
( )
( )
( )

Impacts deleted from further analysis: Impacts that are relatively small or mitigated

❍

Template 3
Detail on benefit value derivation for water recycling project

Benefit category	Annual quantity	Unit value used	Comments

Template 4
Costs and benefits of water recycling project
(present values, x% discount rate, <enter year> dollars)

	Dollar amount
Cost components	
Total costs	
Benefit components	
Total monetized benefits	
Benefits requiring qualitative assessment*	
Monetized net benefits (benefits minus costs)	

* (+) indicates positive benefits anticipated, but not monetizable with readily available data; and (-) indicates costs anticipated, but not monetizable with readily available data.

Template 5
Costs and benefits of project
(present values, x% discount rate, <enter year> dollars)

	Dollar amount	Stakeholder accruing cost or benefit
Cost components		
Total costs		
Benefit components		
Total monetized benefits		
Benefits requiring qualitative assessment*		
Monetized net benefits (monetized benefits minus costs)		

* (+) indicates positive benefits anticipated, but not monetizable with readily available data; and (-) indicates costs anticipated, but not monetizable with readily available data.

Template 6
Omissions, biases, and uncertainties and their effect on the project

Benefit or cost category	Likely impact on net benefits*	Comment

*Direction and magnitude of effect on net benefits:
+ = Likely to increase net benefits relative to quantified estimates.
++ = Likely to increase net benefits significantly.
- = Likely to decrease benefits.
— = Likely to decrease net benefits significantly.
U = Uncertain, could be + or -.

Template 7
Sensitivity analysis applied to discount rate (thousands of dollars)

Discount rate	Monetized benefit	Cost	Monetized net benefit
0%			
3%			
6%			
9%			
12%			
15%			

CHAPTER 8
CASE STUDIES ON THE VALUE OF WATER

"Whiskey is for drinking; water is for fighting over." (Mark Twain, attributed)

The research team has developed, in concert with participating utilities, a series of case studies that reveal various issues associated with the value of water, and how these values (or value-based concepts) can be used to help inform water management decisions. These case studies are provided in appendices C through I, and cover a broad range of water management issues and locations from across the United States.

The case studies span the nation from Massachusetts (low and no flow conditions in the Ipswich River due to excessive withdrawals) to California (the value of recycled water in San Diego). They also stretch from Florida (developing high cost regional water supplies, including desalination, in the Tampa region) to Arizona (the value of instream and Phoenix-area extractive water uses for the Verde River). Additional case studies include the "triple bottom line" approach to measuring utility performance in San Francisco, the value of water for instream recreational purposes in the Colorado Rocky Mountains, and the cost of alternative water supply options for the water-constrained city of El Paso.

This chapter provides a summary overview of these case studies. As condensed in Table 8.1, each case study summary provides abbreviated discussions of case-specific background and the water management issues that are (or could be) better informed by the inclusion of value-related concepts or measures. Interested readers can find the full-length case studies in the appendices.

THE TRIPLE BOTTOM LINE APPROACH (SAN FRANCISCO, CALIF.)

Water utilities provide services that have value beyond the simple financial balance sheet of utility-incurred costs and revenues received from customers. A broader economic perspective of the total value of water service provision includes social as well as environmental benefits. One approach to addressing this issue that has evolved from Australian applications is called the triple bottom line (TBL).

TBL starts with the conventional financial bottom line that a utility develops to record its costs incurred and revenues received. This is the standard financial bottom line that accountants develop and auditors review. However, this bottom line does not include other valuable contributions that a water agency provides to its community on other levels. Hence, TBL adds a second bottom line to include social values, and a third bottom line to include environmental issues.

The case study (provided in full in appendix C) examines the history and concept behind the TBL approach, and provides examples of how it has been applied in four utilities spanning Australia, New Zealand, and the United Kingdom. Tables 8.2 and 8.3 provide examples of the environmental and social bottom lines, respectively, that these overseas utilities have developed.

Table 8.1
Case study illustrations for the value of water

Case study	Water management issues	Types of values addressed	Values derived for key water uses (and methods used to estimate value)
San Francisco, Calif. (appendix C)	The mission of water systems goes beyond the financial bottom line of the utilities' internal costs and revenues. Water service provision has broader social and other values that need to be recognized.	Recognizing the nonfinancial components of the value of services provided by a water supplier. Triple bottom line concept described as a way to more systematically reflect the full social and environmental values of water service provision.	Empirical aspects have not been developed, but the research team examined the implied social value assigned to reuse and diversification of water supply options, based on voter support for these options (despite their comparatively higher costs) through a recent voter referendum.
Ipswich River, Mass. (appendix D)	Regional growth has led to over-extraction of surface and groundwaters, leaving the river dry in some summer months. Water management options to restrict extractions and provide for sustainable water use include importing water and increased conservation.	Environmental and economic impacts of low flow and no flow conditions on (1) ecosystems, (2) commercial and recreational fisheries, (3) near-water property values, and (4) supply reliability of drinking/municipal water. Possible solutions include importing Massachusetts Water Resources Authority (MWRA) water, reducing wastewater flows sent out of basin, and improving conservation measures.	The value of leaving enough water instream for various uses and purposes has not been accounted for, leading to over-extraction. Empirical evidence reveals local residents' and officials' rankings of various uses (e.g., habitat, drinking water) and their WTP to restore flows to sustain healthy fish populations (an average of $31 per household per year).
Tampa Bay Water Regional Supply, Fla. (appendix E)	Development and use of desalination (and enlarged surface water capacity) mandated by regional water management district to protect wetlands, estuaries, lakes, and properties from the adverse impacts of over-extraction of groundwater.	Environmental sustainability and increased reliability of water supply; avoided damages to wetlands and other resources from former rates of groundwater mining; value of reliable water for regional economy.	The lower bound estimate of the social value of having a more environmentally sustainable and reliable water source may be deduced by examining the difference in cost between former reliance on local groundwater and more expensive use of regional supply.

(continued)

Table 8.1 (Continued)

Case study	Water management issues	Types of values addressed	Values derived for key water uses (and methods used to estimate value)
Verde River, Ariz. (appendix F)	Potential extraction of water upstream likely to adversely affect both instream and municipal use values downstream.	Critical aquatic habitat for threatened and endangered species; instream and near-stream recreational use values; and value of water to Phoenix metro area municipal utilities.	$16.3 million per year for threatened and endangered (T&E) species preservation ($900/AF per year); based on benefits transfer (BT) of stated preference surveys. $35 million per year for recreational use ($2,000/AF per year); based on BT of user day value estimates. $3.5 million per year for municipal utility use ($200/AF per year); based on cost of replacement water rights.
San Diego, Calif. (appendix G)	Reuse water is becoming increasing valuable in this community where source waters are scarce, expensive, and subject to drought and other limits.	Value of reuse water to a high value-added commercial enterprise, reflecting the value of having a more reliable source of water (subject to less supply fluctuation due to drought, etc.) than potable supplies.	Empirical work limited, but implies a value of reuse water to pharmaceutical complex (especially the reliability premium). Also suggests a value of reuse from broader community-wide and utility perspective (given scarcity and cost of alternative raw and potable supply options).
El Paso, Texas (appendix H)	What mix of new water supply options are most economical in a region where historical use of groundwater supply must be curtailed, surface supply options are very limited, and rapid economic and population growth is increasing demands.	Full cost accounting and proper baseline approaches to provide a true apples-to-apples comparison of a range of water supply options (e.g., water rights acquisitions, desal, conservation, reuse, water ranching).	Ranges of values for water in various uses are developed (agricultural, residential, commercial). Comparative costs are developed for broad array of supply options.
Vail and Golden, Colo. —whitewater parks (appendix I)	Value of instream recreational river flows in a changing state economy (i.e., transition from traditional agriculture and mining economic base to a more sustainable path driven by tourism and recreation).	Value of instream waters for recreation and tourism (to participants and for local economies), and instream value of water for hydropower production. Opponents concerned about potential limits on future upstream extractive uses.	Recreational value to users based on benefits transfer of user day values; local economic impact estimated by use of existing multipliers. Instream use of the waters for recreation and tourism creates appreciable value (e.g., over $1 million per year for a whitewater course), and preserves quality and quantity of water for downstream users. Values comparable to the value of electricity generated by nearby large hydropower facility.

Table 8.2
Examples of environmental bottom lines as reported by four water utilities

Severn Trent, UK	City West Water, Australia	Sydney Water, Australia	Water Care, New Zealand
29% of river lengths improved from 1990 to 1998	Upgrade of wastewater treatment plant	Conserve water supplies and prevent need for new dams	Construction of ultraviolet (UV) light wastewater treatment plant
Water saved from 1994 to 1998 by reducing leakage from Severn Trent Water pipes is enough to fill a quarter of a million swimming pools	Launched Community Environment Awards and committed to broadening program to Sustainability Awards in 2004	Provide beneficial and environmentally acceptable management of effluent	Progression of restoration of Manukau Harbor, resulting in the return of aquatic and bird life
Nitrogen and sulphur dioxide emissions from sewage sludge incineration have been reduced by 87% and 99%, respectively, in between 1994 and 1998	Volume of water consumed per capita reduced from 249 to 229 liters per day	Maximize beneficial reuse of treatment by-products	Continuation of the drive for eco-efficiency by maximizing biogas and hydroelectric generation
Almost a million trees have been planted on Severn Trent Water land between 1994 and 1998	Majority of targets in Environmental Improvement Plan implemented	Contribute to collaborative research to improve environmental performance of assets	
We currently generate around 7% of our electricity from totally renewable sources	Enviro Partnerships Initiative launched	Responsibly manage land and water assets	
Around 37,000 tonnes of industrial waste were recycled in 1998/1999		Encourage waste minimization	
Over 3 million visitors to Severn Trent Water Conservation and Recreation sites each year		Minimize environmental impact of energy use	

Table 8.3
Examples of social bottom lines as reported by four water utilities

Severn Trent, UK	City West Water, Australia	Sydney Water, Australia	Water Care, New Zealand
89,702 lead pipes have been replaced by Severn Trent Water between 1994 and 1998.	High drinking water quality in spite of low rainfall	Protect public health	Maintenance of Ministry of Health "A" grade quality of drinking water
Our employees and customers raised over £4 million for sustainable water-related projects in Africa and Asia, and the work continues to encourage donations.	Ranked 10th out of 195 companies in the 2002 Managing Work/Life Balance Benchmarking Study	Look after the health, safety, and welfare of all employees, contractors, and others that may be affected by our sites	Delivery of a new water source and treatment plant to the Auckland region to meet its growth and water quality demands
Active support of WaterAid since it was first established, and in 2000 alone, over £745,000 was raised by staff, customers, suppliers, and community groups working together.	Hardship assistance provided to 215 customers at a value of $96,000	Ensure all customers, especially those with special needs, maintain fair and affordable access to essential services	Continuation of the employment equity programme, including multicultural awareness education
At Creswell Crags in Nottinghamshire, Severn Trent Water is helping to re-create the way our ancestors survived at this internationally important Stone Age site.	Improved customer service	Eliminate discrimination in employment	
Public education through website.	Established a Business Women's Network	Enjoy the trust and credibility of our customers we serve, and have open, honest relationships with stakeholders	
	Community consultation on development of TBL report	Support communities in which we operate through a variety of community investment opportunities	

The case study examines how TBL could be applied to the San Francisco Public Utilities Commission (SFPUC), which provides water for the City of San Francisco and 29 other Bay Area water agencies. While SFPUC does not currently apply a TBL framework, the case study reveals how this could be developed and applied in the context of key elements of the agency's capital improvement program. A sample "report card" using TBL in this manner is provided in Table 8.4.

Table 8.4
Selected capital improvement program projects in a triple bottom line framework: The SFPUC "Report Card"

Value category	Value objective	Rating	Selected capital improvement program projects
Social	Reliability of water delivery	++	Irvington Tunnel alternatives: a new facility will be constructed to provide a back-up for the Irvington Tunnel.
		++	Summit Reservoir rehabilitation and seismic upgrade: the reservoir will be rehabilitated and seismically upgraded to achieve current seismic standards and to meet all State Health requirements.
	Reliability of water supply	++	Calaveras Dam replacement: The dam replacement project will evaluate, plan, and then construct a new dam that will have increased terminal storage and will be seismically reliable.
		++	New northwest reservoir: A new reservoir will be constructed to alleviate the water supply problems of the northeastern zone of San Francisco.
		++	Recycled water project: Implement recommendations of Recycled Water Master Plan.
	High quality water	++	Sunol Valley Water Treatment Plant improvements: The operational and storage capacity of the plant will be increased through construction of a new reservoir and the addition of new filters and basins.
		++	Sunset Zone circulation improvement: The project will reduce the number of outlying zones dependent on Sunset Zone water, thus increasing circulation and reducing water contamination in the Sunset Reservoir.
Environmental	Restore Chinook salmon	+	Alameda Creek fishery enhancement: Project will release water from Calavaras Dam and recapture the water in the upper Sunol.
	Protection of Environmental Value of Watersheds	+	Valley for the purposes of establishing and/or enhancing fish populations.
Economic	Efficient capital expenditure program	++	Efforts are being made to ensure that the projects of the Capital Improvement Program are being planned and implemented in an efficient manner. Efficiency achievements are currently unknown, as the Program is in its first year of implementation.
	Provide cost effective products and services	++	The Capital Improvement Program was influential in determining the need for an SFPUC-wide performance assessment program, which began in July 2003.

Key: + represents positive improvement.
++ represents very positive improvement.

OVER-EXTRACTION OF ZERO-PRICED WATER (IPSWICH RIVER, MASS.)

When scarce and valuable natural capital assets such as water can be used without charge and with no institutional limits to govern who can extract or how much they can take, then invariably the resource becomes overused. This is what economists refer to as "the tragedy of the commons," wherein common property (shared) resources are not used or managed in a sustainable way to maximize the value generated.

Even though Massachusetts is considered to be a relatively water-abundant and modestly growing state (compared to more rapidly growing and arid regions such as the American Southwest and California), it still has sizeable challenges posed by increasing demands for water by municipal water supply systems. The Ipswich River, for example, has been impacted by increasing groundwater withdrawals and direct river water extraction associated with residential development in the watershed. The problem has grown to the point that stretches of the river now face extremely low flow and no flow conditions in summer months. The absence of water from the river has several adverse consequences in terms of losses in values for ecologic, recreational, aesthetic, commercial, and other services that would otherwise be provided by retaining higher summertime instream flows.

The state and communities in the region find the low or no flow situation unacceptable, and are attempting to resolve the problems. The case study describes several options that are under consideration. These include encouraging conservation, importing water from the Massachusetts Water Resources Agency (MWRA), and providing more in-basin return flows from treated wastewaters. The costs and benefits of these various options have not been fully estimated (though estimates exist for some components). However, the general perception is that considerable value can be gained from programs and policies that ensure that suitable flows remain in the river year-round. The benefits will consist of restoring lost or reduced services (e.g., avoiding the loss of recreational, aesthetic, and other services due to currently inadequate flows).

In addition to articulating some of the losses associated with overuse of water resources, this case study also reveals a more fundamental problem that arises when there are neither economic signals nor institutional (i.e., political, regulatory) controls to manage use levels for a common property resource. Local groundwater and river water in the Ipswich basin have, until recently, been seen as free goods that are available to anyone with properties that provide land-based access. If a price or fee were assessed on all users based on the volume of water extracted (i.e., the water resource itself had to be paid for by users), then the associated economic signals would help regulate use levels.

DEVELOPING HIGH COST REGIONAL SUPPLIES TO REPLACE UNSUSTAINABLE USE OF LOCAL GROUNDWATER (TAMPA BAY REGION, FLA.)

The greater Tampa Bay region, including Pinellas, Pasco, and Hillsborough counties, has relied almost exclusively on groundwater from the Floridian aquifer to meet its water supply needs. However, as populations have grown and groundwater extraction increased to meet rising demands, a series of adverse environmental and economic consequences became apparent. These included appreciable declines in the levels of lakes and wetlands, land subsidence (including damages to many houses and properties), and reduced flows (and increased salinity) in streams, rivers, and estuaries.

The economic value of the environmental and other consequences of overpumping have not been estimated, but were significant. The adverse impacts led regulators, local public officials, and local water agencies to explore ways to resolve the problem. Because of restrictions placed on groundwater pumping permits, a region-wide approach to developing alternative water sources was pursued. Tampa Bay Water was formed as a wholesale agency to supply water to meet region-wide needs, using a combination of new desalination facilities and increased storage of surface waters.

The water developed by Tampa Bay Water will be expensive, especially relatively to the low cost of the previous reliance on local groundwaters. However, the added cost of having desalinated water (projected to be $2.60/kgal, but which may end up being considerably higher) is nonetheless considered a good investment for the region. The value of this water is probably well over the cost when one factors in (1) the damages (negative values) that would otherwise be incurred associated with continued reliance on groundwater (i.e., the avoided cost of overusing groundwater); (2) the increased reliability of the regional water supply due to the drought-insensitive nature of desalinated water and the increased surface water storage capacity; and (3) the ability of the region to continue to follow its planned population and economic growth trajectory (which would have been severely constrained absent the development of regional alternatives to groundwater).

COMPLEMENTARY USES OF INSTREAM AND EXTRACTIVE USES GENERATE CONSIDERABLE TOTAL VALUE FOR LOCAL RIVER WATER (VERDE RIVER, ARIZ.)

The Verde River is an important resource in north-central Arizona. The Verde River is Arizona's only designated Wild and Scenic River. In the designation, the river was found to contain outstanding scenic, fish and wildlife, and historical and cultural values. These and other nonconsumptive values such as recreation (fishing, boating, hiking) and protection of threatened and endangered species are generally linked to the upper to middle reaches of the Verde River. The river also supports consumptive use values such as agricultural, municipal, and industrial use. The river is a major supply source for several cities in the Phoenix metropolitan area, at the downstream edge of the Verde watershed. Although some small municipal and agricultural uses are supplied in the upper Verde, the majority of the current consumptive use is located downstream of the nonconsumptive uses. As a result, instream and extractive uses generally do not currently compete for water.

A key issue for the Verde watershed is growth of extractive uses in the upper Verde watershed. The population of major cities and towns within the Verde watershed has more than doubled in the last 20 years and is projected to more than double again within 50 years. The region's population is projected to continue to increase in part due to its attractiveness as a retirement location. This includes growth of the cities of Prescott and Prescott Valley, which currently have no significant source of surface water, relying on groundwater pumping for their supply and facing declining groundwater levels locally. Without another water source, increasing demand might force Prescott to look to the groundwater supplies that provide the baseflow for the Verde River headwaters, as well as perhaps the Verde River itself for supply.

As a result of this continuing and projected growth, management of the headwaters of the Verde watershed is of particular concern. Groundwater in the Little Chino Valley is the source of springs that constitute the Verde's headwaters. Current extraction of groundwater has resulted in groundwater declines in some areas and reduced flow at the source of the Verde River headwaters. Future extraction of these groundwaters may cause significant impacts to the baseflow of the Verde River itself. Clouding the water management picture is the fact that little is known about how much groundwater is actually in storage in many areas of the Verde watershed, or about how much water use in the Upper Verde may affect continued availability of water for instream and extractive uses below the headwaters in the Verde Valley.

This case study evaluated the value derived from flows in the Verde that originate at its headwaters (i.e., the springs fed by the groundwater in the Little Chino Basin). Results show that the combined non-consumptive instream values calculated for the Verde River sum to the range of $48.1 million to $54.5 million per year. As shown in Table 8.5, this value range includes greater than $35 million in instream or near stream use values for recreational fishing, hunting, and wildlife watching, and $13.1 million to $19.5 million in nonuse values such as the value of preserving threatened and endangered species. A potential complete loss of baseflow in the upper Verde was estimated to reduce downstream municipal use values by $3.5 million per year. These values sum for a total ranging from $51.6 million to $58.0 million per year. This is an underestimate of total water values on the Verde River, and does not include other recreation values (for swimming, float boating, camping, picnicking, and sightseeing), historical and cultural values, and other nonuse values (e.g., existence and bequest motives) aside from preserving T&E species, and additional values from the Wild and Scenic River designation. This valuation also did not attempt to value the water used as it flows down the Verde River before it reaches Phoenix for irrigation, drinking water, power generation, and other industrial/commercial uses such as dairy, golf, sand and gravel, water bottling, and fish hatcheries.

Table 8.5
Benefits of Verde River water
(annualized values, millions 2002 dollars)

	Dollar amount
Benefits	
Upstream instream values	
Use values	> $35.0
Nonuse values	$13.1 - $19.5
Downstream municipal values	$3.5
Total monetized benefits	> $51.6 - > $58.0
Benefits requiring qualitative assessment	
Use values—recreational values from swimming, boating, camping, picnicking, and sightseeing	+
Nonuse values—added value from designation of Verde River as a Wild and Scenic River	+
Nonuse values—historical and cultural values	+

+ indicates positive benefits anticipated, but not monetizable with readily available data.

This analysis shows that upper Verde instream uses are highly valuable, estimated to be worth over $50 million a year. This value provides a benchmark for considering whether extraction of water from the source groundwater basin and upper river by Prescott may impose costs (foregone values) that outweigh the value that may be realized by new users in Prescott. The study did not attempt to value the water if used by Prescott, or assess the extent that Prescott extractions might adversely affect instream or extractive vales downstream, but it is evident that if groundwater or upper river withdrawals do reduce instream flows, then the impact on downstream values would be sizeable.

RECLAIMED WATER PROVIDES VALUE AS A RELIABLE SUPPLY FOR LOCAL BUSINESSES (SAN DIEGO, CALIF.)

Recycled water is becoming an increasingly valuable water source in San Diego as water users, the San Diego County Water Authority, and other regional water management agencies (such as the MWD of Southern California) continue to discover the benefits of water recycling.

In response to the need for dependable water and the desire to retain and recruit high-valued industry, the City of San Diego developed its Guaranteed Water for Industry Program. The program exempts research and development or industrial manufacturing firms from mandatory water restrictions in times of drought. To qualify, the business must be located in an area where reclaimed water is available, use reclaimed water on its premises to the fullest extent possible, and participate in all applicable city water conservation programs.

The San Diego area has a relatively strong, high-value biotechnology employment base, and two of the first participants in this program are the R.W. Johnson Pharmaceutical Research Institute and the PharMingen biotechnology company. Both companies use reclaimed water for irrigation and cooling, and one PharMingen site is dual-plumbed for reclaimed water use. These are two of the companies located on "Biotech Hill" in Torrey Pines that are drawing water from the North City Water Reclamation Plant in San Diego.

These companies are realizing multiple benefits from the Guaranteed Water for Industry Program. First, they are attracted to the higher drought reliability offered by the program. Although water is not a large factor in the production chain, water cutbacks due to drought can be disruptive to business operations. Recycled water also is less expensive to the customer. This is in part because the city pays the capital costs for retrofitting existing facilities to receive recycled water, and in part because recycled water is priced at less than half the cost per acre-foot of potable water offered to similar customers. This relatively low price ($350/AF) is achieved through financial incentives from MWD and the San Diego County Water Authority to stimulate the use of recycled water in their service areas, and in part because wastewater customers pick up all of the treatment costs since wastewater flows already need to be treated to a high level before outfall to the ocean. Recycling customers pay only distribution costs from the plant to the end customers.

The city also receives substantial multiple benefits from water recycling. One benefit of water recycling is that it diversifies the mix of water supplies the city can offer to customers. Some of these sources are imported from the Colorado River or Northern California and can dramatically fluctuate with weather and possible disruptions in distribution. Water recycling also increases local control. Recycling projects also help reduce wastewater flows, helping the city reduce wastewater discharges to the Pacific Ocean as part of a solution to violations of EPA rules. Recycling also can be a cost-effective solution to meeting increased water demands, especially when increased reliability is considered.

At the regional level, the San Diego County Water Authority is strategically identifying alternative and supplemental supply sources, in part due to increasing uncertainty regarding the availability of imported supplies. Supplemental sources will extend limited potable water supplies to meet the growing demands within San Diego County. As part of the overall Water Resources Plan adopted by the Authority, there is a need to identify local supply sources and increase water use efficiency. Reclaimed water is expected to expand further in importance as a local supply source, growing from 13,700 AF of recycled water use in the Authority's service area currently to a projected 53,000 AF per year by 2020.

Beyond the direct regional benefits, large-scale recycling also can provide significant improvements to aquatic habitats impacted by waters imported by San Diego. For example, enhanced instream flows in the Sacramento-San Joaquin Delta region can improve and increase aquatic and terrestrial habitats, and improve ecological functions to support sustainable populations of diverse and valuable plant and animal species. These enhancements generate value for individuals assured that specific plant or animal species or the whole ecosystem will be healthier and knowing that these ecosystems will be available for future generations to enjoy.

DEVELOPING AN APPLES-TO-APPLES COMPARISON OF WATER SUPPLY OPTIONS FOR A WATER-SHORT REGION (EL PASO, TEXAS)

El Paso Water Utilities (EPWU) and its neighboring communities along the U.S.-Mexican border have signed an international compact for significantly reducing their reliance on groundwater sources. The compact is the result of historical over-reliance on groundwater at depletion rates that have far exceeded natural recharge. In response, EPWU's pumping rates from its local groundwater aquifers have slowed, since their peak in 1989. At the same time, the El Paso area has experienced population growth rates among the highest in North America. This combination of factors has forced El Paso to consider a wide range of water supply options to meet continued growth in demand in a sustainable fashion. Those options include additional surface water from the Rio Grande, desalination of brackish groundwater locally, importation and desalination of groundwater from "water ranches," recycling of municipal wastewater for nonpotable use, and an aggressive water conservation program.

El Paso's largest source of future surface water is through the El Paso County Water Improvement District #1 (EPCWID#1, or the District) and the Rio Grande Project. The project is run by the U.S. Bureau of Reclamation, and almost all water has historically been used for irrigation purposes. A recent option for obtaining surface water is an adjudication of a 1949 permit. EPWU is in the process of reaching a Memorandum of Understanding (MOU) with the District for a settlement of 11,000 AF of water. There is no charge for water under this agreement, but there are administrative charges of $15/AF from the District, and a $5/AF from the Bureau of Reclamation, so that the total cost to EPWU for the water under this agreement would be $20/AF.

Another option for surface water would be a large-scale acquisition of irrigation surface water rights from the Rio Grande through the District. A March 1998 contract between the USBR and the District allows the District to enter into long-term agreements with third parties such as EPWU. EPWU has signed such an agreement with the District that expires in 2010-2011. The price of water under this agreement is $208/AF, with scheduled price escalations each year.

This water is subject to $15/AF from the District and $5/AF from the Bureau, for a total cost of $228/AF.

For either surface water option, the water supply acquisition costs are only a portion of the total cost to provide this water to EPWU customers. Treatment of this water is expected to be accomplished at the Johnathan Rogers and Canal Street treatment plants, with a majority of treatment expected at the Johnathan Rogers plant. Total treatment costs for an additional 15,000 AF of water, based on a recent expansion, are $251/AF. Transmission facilities required to connect the expanded facility with the distribution system add another $58/AF.

Thus, the combined cost of treatment and distribution for surface water from the District through the MOU is approximately $309/AF ($251 + $58). Total cost for water through the March 1998 contract—including acquisition cost, fees, treatment, and distribution—would be approximately $537/AF ($309 + $228).

Desalination is another source of potable water that will come on-line in the near future (by 2006). A joint desalination plant is being built with Fort Bliss (a local military base), using reverse osmosis. The plant is designed to deliver 27 mgd (or approximately 30,250 AF per year). Based on annualizing the capital costs over a 20 year period at 5%, the cost of this desalinized water, after accounting for possible uncertainties, will range from $414 to $622/AF (assuming brine wastes can be disposed via relatively low cost deep well injection).

Groundwater importation involves the cost for acquiring ranches in west Texas, where the ranch ownership also conveys to EPWU the rights to underlying groundwaters. The total cost of this "water ranching" also includes desalination treatment costs and conveyance costs. For the groundwater importation scenario analyzed, acquisition cost for 38,800 AF per year of groundwater is approximately $246/AF. The cost of treating and conveying the water approximately 70 miles to El Paso is estimated to be $2,054/AF. After applying an uncertainty factor to the combined acquisition, treatment, and conveyance costs, groundwater importation is estimated to cost between $1,840 and $2,760/AF.

Reclamation is another source of water considered. Each acre-foot of reclaimed water produced and used represents an acre-foot of water that will not need to be acquired, conveyed, and treated to potable standards. The cost of reclamation was estimated based on the Northwest reclamation plant, which is the least efficient of El Paso's reclamation facilities, but is still considered to reasonably reflect reclamation costs. Reclaimed systems serving peak demand customers allow EPWU to avoid expansions of the potable supply network. This avoided cost for peak capacity expansion is a benefit of reclamation and should be netted out of reclaimed water costs to maintain an apples-to-apples comparison of water supply option costs. After netting out the cost of potable supply distribution upgrades that otherwise would have been required in the area, the cost to meet firm demand (670 AF per year) was calculated to be $2,337/AF. For planned demand (1,750 AF per year) the cost is $981/AF, while for demand at capacity, the cost is $282/AF.

Water conservation programs present some of the same issues for establishing the correct baseline for comparing water supply options on an apples-to-apples basis. Conserved water has already been treated and uses existing distribution systems. Therefore additional treatment or other costs do not need to be added to this option. A range of water conservation options were analyzed. The low flow showerhead option was found to be the cheapest option, with a best-guess cost of $53/AF of water saved. Water costs saved for the refrigerated air rebate program in which EPWU teams with El Paso Electric to offer an incentive to replace evaporative coolers with refrigerated air units were found to range from $362 to $486/AF. The turf replacement program, which pays established customers to replace turf areas in their front lawns with low water-using landscapes, was found to have a best estimate return of $605/AF. The horizontal-axis (front-loading) clothes washer rebate program to replace existing top-loading machines was found to have a most likely return of $666/AF. And the single-family ultra-low-flush toilet rebate program to replace toilets using 3.5 gallons or more with those using 1.6 gallons or less was found to have a most likely return of $1,266/AF.

This case study thus provides a good illustration of how a water utility can develop a suitable economic basic for evaluating the relative costs of a wide range of options for meeting growing demands. In the El Paso setting, many of the supply options are relatively expensive. Large-scale water producing options, such as desalination and the acquisition and treatment of irrigation rights to Rio Grande waters, will cost from $400 to $600/AF (and perhaps more). This raises the question of whether the value of those waters to EPWU customers, and the community as a whole, is high enough to justify the expense. Indications are that the benefits exceed the costs.

INSTREAM FLOWS GENERATE SIZABLE ECONOMIC VALUES FOR REGIONAL ECONOMIES BASED ON OUTDOOR RECREATION AND TOURISM (VAIL, COLO.)

As water uses in the western United States shift from being dominated by extractive uses such as agriculture and mining to a more diversified base that factors in newer industries such as recreation and tourism, water law and water managers face challenges in adapting. Recreation and tourism based uses often leave water in the stream, such as in the case of a whitewater boating or kayaking course. Traditional prior appropriation water law was not designed to recognize nonconsumptive uses, and, in many western states, instream flow maintenance was not allowed as a beneficial use. Several water cases in Colorado involving whitewater boating parks have defined the courses as controlling and changing the flow of the river, qualifying as water diverted for beneficial use. Another legal requirement is that the use demonstrate economic value. This case study examines the economic value generated from a whitewater boating course on Gore Creek in Vail, Colo.

The Vail Whitewater Park provides value to boaters using the course, spectators (particularly during special events), and the community as a whole. The total recreational value to the boaters at the Vail Whitewater Park has two components. The first is what people actually pay to kayak through expenditures such as equipment and travel costs, and the second is the component they would be willing to pay over and above what they currently pay. This second component is based on the WTP of boaters for use of the park, and is known as consumer surplus.

In this study, the expenditures calculated for non-event boater use were the cost of kayak equipment, automobile expenses, and the value of travel time. Lodging cost also is included for non-local kayakers. The total cost to kayak is estimated to range from $36.35 to $61.35 per visit assuming an average roundtrip travel distance of 30 miles. Non-local boaters, who make up approximately 25% of park users, are assumed to have additional expenses for lodging, meals, and other expenses. Out of town boaters are estimated to spend an additional $324 per day.

Willingness-to-pay values for kayaking were derived using the benefits transfer method to approximate user day values for consumer surplus, obtaining values per unit of use for similar types of activities from studies conducted in similar locations. Those unit values are multiplied by the amount of use. For purposes of this study, a range of $35.62 to $39.91 per person, per outing, is applied as the value of kayaking, based on the most appropriate studies of similar uses from the peer-reviewed published empirical literature.

Total user value per day was derived by adding the boating expenditures and the consumer surplus values. This resulted in a range of total use values per day for local boaters of $71.83 to $101.26.

Typical daily (nonevent) use was estimated through interviews with two local boat shop owners who are also frequent users of the park to range from 1,044 to 2,345 boater-days per season. For comparison purposes, a linear regression correlating average boater days with flow was estimated, resulting in a total number of boaters for the season equal to 3,474. Assuming total non-event boater days per season of 2,345 (the upper end of the estimated range from the interviews), the total value of non-event boater use totals approximately $394,000.

Use of the whitewater park for special events is another aspect of the park's value. In 2001, Vail hosted its annual whitewater festival over Memorial Day weekend at its new whitewater park. The event was sponsored by Teva, and attracted top kayakers from the United States and around the world. In part because of the new park and the high level of competition, the 2001 event attracted the largest crowds in the event's 7-year history, with approximately 2,300 spectators on Saturday, and attendance for the remaining portions of the weekend around "several hundred." A survey of spectators on behalf of the local tourism and convention bureau revealed the average Saturday spectator spent a total of $133 for food, lodging, and shopping. Other events hosted at Vail's whitewater park include its whitewater rodeo series, which consists of 11 weekend competitions. These events also attracted sponsors. Total non-event value was conservatively estimated to be at least $305,900 (2,300 spectators x $133 per spectator) in spending by spectators at the Saturday competition alone and only for the Teva Whitewater Festival.

Increased expenditures by boaters and spectators generate economic stimulus for the community, which creates a larger economic effect as it ripples through the regional economy. Precise estimates of the multiplier for kayaking expenditures in Vail have not been estimated, but multipliers estimated for water-based recreation expenditures range from 2.00 to 2.88 or higher in the existing literature. This study used 2.88 as the multiplier (actually 1.88, after subtracting 1.0 to adjust for existing value), and multiplied it by the kayaking expenditures (minus the travel time calculation), the additional expenditures from non-locals, and the expenditures by spectators to obtain an economic stimulus estimate of approximately $1.1 million per year.

The value of the benefits provided by the Vail Whitewater Park in 2001, including kayaker use values, spectator expenditures, and economic stimulus, totaled at least $1.8 million annually (see Table 8.6). This value is consistent with a 2002 estimate of the value of a new whitewater park in Breckenridge, Colo. (a lower bound total value estimate of $1.4 million per year), and a 2000 study of a whitewater park in Golden, Colo. (which is more easily assessable from the Denver metro area) at a range of $1.4 to $2.1 million annually. The total flow claimed by Vail for the whitewater park is approximately 11,000 acre-feet from March through October, implying a value of $16.36/AF for nonconsumptive instream use of water. However, when values per AF are estimated for the peak use period of high flows (May, June, and July), then the value is much higher (i.e., $1,664/AF).

Table 8.6
Benefits of Vail Whitewater Park
(annualized values except where noted, 2001 dollars)

	Dollar amount
Monetized benefits	
Boaters	
Expenditures	$114,553
Additional expenditures nonlocals	$189,945
Consumer surplus	$88,559
Spectators at competitive events	
Expenditures	>$305,900
Economic stimulus for community	$1,121,097
Total monetized benefits	$1,820,054
Capitalized value over 20 years, discounted at 7%	$20.6 million
Benefits requiring qualitative assessment	
Spectators at competitive events consumer surplus	+
Spectators and near-stream recreators (non-event)	+
Benefits to event sponsors and vendors	+
Beneficial use of water downstream of kayak course (heavy use corridor)	+
Community identity, quality of life	+
Option value	+

+ indicates positive benefits anticipated, but not monetizable with readily available data.

This estimate of value is considered a conservative underestimate, in part because many benefits were left out of this analysis. Omitted benefits include, but are not limited to, increases in property values for landowners near the park, the consumer surplus of spectators and other non-boating recreators, and improved or renewed community identity and quality of life.

CHAPTER 9
A SUGGESTED AGENDA FOR FUTURE RESEARCH

"If we knew what it was we were doing, we would not call it research, would we?" (Albert Einstein)

The main conclusions and observations from this research project have been summarized in the Executive Summary. In this chapter, some items for future research are suggested, based on the findings of the work performed in this project, including the findings of an expert workshop (using the Nominal Group Technique) that was organized and facilitated by the National Water Research Institute as part of the research team.

1. Perform additional research to directly estimate the WTP of households for residential water service from water supply agencies. Although providing water for residential customers is the core focus of virtually all water utilities, remarkably little is known about how much households value their residential supplies, especially for different types of uses (e.g., essential uses such as for drinking, cooking, and bathing, as contrasted to discretionary uses such as lawn watering). The extensive available literature on the own-price elasticity of demand offers some glimpses of the price sensitivity of quantity demanded, but does not reveal very much about values (WTP). In addition, the elasticity estimates reveal consumer data for only very small (marginal) potential changes in prices and quantities demanded relative to current rates and use levels. The suggested future research also should examine how various attributes (e.g., aesthetic qualities, reliability) of the water affect household values, relative to various types of use. This will help water utilities understand what aspects of water service are most valued by their customers, and in turn it will be useful to have value estimates to compare to costs for future utility options to alter reliability, quality, and (or) quantities available.
2. Estimate the monetary worth (dollar value) of water for various uses. As noted in this report and described under item 1 above, there is surprisingly little available empirical information that has been estimated for the value of water in different possible uses. The lack of WTP estimates for residential water customers is a key data gap (as noted above), but additional value data for other water uses and attributes are needed as well. This might include the value of irrigation water, corrected for the distortions created by subsidies that reduce the costs of many farm inputs and prop up prices of many farm outputs. Additional insights on the value of reliable and secure water for community water system customers also would be useful. In chapters 4 and 5 of this report, the research team has attempted to assemble and interpret the available empirical information on values, but the valuation data are sparse and often of questionable reliability or transferability. Some additional research to empirically measure values in key use categories would be helpful, so that decision-makers have a better general idea about the comparative worth of competing water uses and values. This will help utilities and water resource managers set more informed priorities and help direct water to its highest and best sustainable uses.
3. Develop a common parlance (a clear set of terms) that will describe the types of values that people derive from water. As noted in chapter 2, people with different

backgrounds, belief systems, and interests value water in different ways. We need to make sure technicians, decision-makers, and the public do not speak past each other. There is a risk that when discipline-specific or other technical terms are used in some circles to describe values or categories of benefits and costs, lay persons (customers, stakeholders, the media, and some public officials) who have different ways of conceiving of or expressing water-related values may feel that their interests have not been considered (e.g., that the benefits they care about do not appear to be accounted for). Ultimately, making good public policy decisions about water requires that the value terminology be clear and reflect what the public, decision-makers, and water professionals perceive. The taxonomy should be broad and all-inclusive, reflecting "values" regardless of whether or not they are amenable to quantification or monetization.

4. Make all water values count in decision-making. Current water planning and decision-making processes rely on engineering and financial data, and limited economic information, to help improve decision-making. However some factors (values) do not readily lend themselves to financial and engineering-based evaluation processes. Yet these values may be considered very important to some community members, and failing to take these into account can result in unanticipated and unintended negative social, cultural, environmental, or economic outcomes (and perhaps lead to strong opposition to water planning initiatives). In this report, the research team has developed templates to help practitioners identify and keep track of often elusive values. To help promote broader recognition and consideration of diverse and elusive values, it may be useful to develop a research project that entails a small number of detailed case studies in which the templates are applied systematically, and within which stakeholders and public officials are enlisted in focus groups to review the efforts at various stages. Refined and expanded templates and associated guidance could then be developed, and training for utility professionals could be provided based on the experiences gained.
5. Develop pilot applications of the triple bottom line approach for selected North American utilities. The TBL approach has considerable intuitive appeal because it enables utilities to describe and track their progress along a suite of meaningful measures beyond the limited financial accounting of incurred costs and collected revenues. The pilot applications can be used to help refine and expand the TBL approach to best suit U.S. and Canadian circumstances, and to help promote its broader application. A workshop would be useful to present and foster discussion of the pilot applications, and to help define improvements and resolve emerging problems.

APPENDIX A
ISSUES AND METHODS FOR ESTIMATING AGRICULTURAL WATER DEMANDS AND VALUES

This appendix provides a discussion of some of the issues and the methods available for estimating water demands, and the values associated with those uses. The focus in this appendix is on the agricultural sector. Agriculture is used here for several reasons. First, it is the largest consumer of water in many parts of the United States (see Table A.1). Therefore, it is important to understand what drives the demand for water in agriculture, how the values for irrigation water can be estimated, and why it is complicated to estimate those values. Second, agricultural demand and values for water are well studied and ultimately tied to market-based prices for farm inputs and outputs. This enables a relatively in-depth assessment of the issues and the methods available for deriving empirical values.

FACTORS OF AGRICULTURAL WATER DEMAND

Agricultural water demand is the product of two, relatively independent processes. The first is the biophysical production activity that delivers water to the field and ultimately to the plant that transforms the water and other constituents into the harvestable yield. It is within this process that the *marginal physical productivity* of water can be modeled and estimated (i.e., the response of yield to changes in water applications). Often the subject of conventional crop studies and experimentation, analysis of water-production relationships provides a fundamental piece of information on which the agricultural demand for water is based.

Table A.1
Irrigation consumptive use in selected river basin/states

	Irrigation's consumptive use (MAF/year)	Irrigation's share of state consumptive use (%)
Columbia River Basin		
Oregon	2.7	96
Washington	2.5	91
Idaho	3.8	99
Montana	1.6	93
Colorado River Basin		
Wyoming	2.4	95
Colorado	4.4	94
Utah	1.7	88
New Mexico	1.5	85
Arizona	2.8	83
Nevada	0.9	79

Source: Based on data from Bonnie Colby, University of Arizona, presentation on Climate Change and Water Utilities, National Center for Atmospheric Research, Boulder, Colo., March 2004.

The second process involves the often-complex mix of market and government actions that ultimately determine the output and factor prices the farmer faces. This process is only weakly understood and very difficult to systematically model or describe. Therefore, water demand analyses typically make strong assumptions regarding output and factor prices, using, for example, historical averages, current trends, or a range. Water demand depends fundamentally on the demands and market values of the crops that are produced. Therefore, changes and shifts in consumer food demand, commodity and trade policies, and income levels can affect long-run demands for water in agriculture—these are discussed in a later section.

The product of the marginal physical product of water and the output price gives rise to what economists call the *value of marginal product* or, when aggregated across activities, farms, etc., the demand for water. In other words, it is the maximum that could be paid for the water, given its contribution to production and net income (i.e., covering the costs of all other inputs).

Basic Farm-Level Problem

The valuation methods and approaches will be more clearly understood if the basic farm-level management problem is considered first.[1] This basic premise of this problem, given in Young (1996), is that a farmer seeks to maximize profits—or net returns to fixed factors of production—adhering to the following constraints[2]:

- Natural or environmental endowments (climate, weather, soil type, and quality)
- Production technology (seed varieties and genetics, biological response to managed and unmanaged factors)
- Farm resource endowment (land, labor, capital, water)
- Input and product prices
- Political, legal, and institutional constraints and incentives.

In this problem, the profit-maximizing farmer allocates the farm resources, including water, to achieve the greatest return. The farmer continues to use more of the resource as long as the gain in revenue (i.e., marginal revenue) exceeds the cost. The gain in revenue is determined by the physical productivity of, in this case, water, multiplied by the price of the product. The physical productivity of water is most clearly evident in regions that are naturally water-short; and where irrigation is helpful to production, the addition of water leads to additional gains in yield. However, as with most other variable inputs, the yield gains from adding more water begin to lessen, and may at some point even become negative (for example, when the soil becomes water-logged or flooded over an extended period of time). Thus is derived the basic property of a

1. A thorough presentation of this problem is given by Boggess, Lacewell, and Zilberman (1993). Additional insight into the farm-level problem is given by Young (1996), who also clearly examines the methods of estimating farm-level water demands. A more technical treatment is found in Moore (1999), in which he develops a model for estimating shadow prices for water based on water's contribution to irrigator profit and ability to pay. Moore, Gollehon, and Carey (1994) examine irrigation water demand using a multicrop framework.

2. This objective can be modified, as discussed in the later section on risk, to a more general form that maximizes utility. This enables production risk, variability, and farmer risk preferences to be accounted for in the problem.

downward sloping demand for water. In the case of an individual farmer, where product price is essentially constant—either through market forces or by government action—the demand for water essentially mirrors the marginal physical productivity.

When water supply conditions change, a number of factors become important to the farm-level water-use problem and give rise to possible adjustments and adaptive responses. Gibbons (1986) identifies the following factors that affect irrigation demand and the assessment of the value of water to which the demand for irrigation water is sensitive to:

Management

- Availability of alternative, water-saving crop varieties
- Crop mix
- Irrigation scheduling and land use decisions, e.g., which land to irrigate versus which to dry up, when to irrigate.

Factors Affecting Crop Mix and Land Use

- Crop price
- Input costs
- Risk characteristics (plant risk, price and market risk, policy risk)
- Management effort
- Physical conditions (soil and climate, water quality, terrain)
- Government programs and policies.

Technology and Infrastructure

- Alternative irrigation methods
- Water delivery systems, e.g., changes in water-use efficiency.

METHODS AND APPROACHES TO VALUE AGRICULTURAL WATER USE

Estimating agricultural water value and water demand is complicated by several important realities. First and foremost, irrigation water demands and price elasticities cannot be estimated directly in the conventional manner, based on observed quantities of water used and various price levels. Variation in water prices faced by farmers in a given region is, generally, insufficient to permit conventional statistical analyses (Young and Haveman 1985, Colby 1989b). Second, each of the alternative approaches, including those described below, has significant limitations that can lead to problems of bias, particularly when value estimates are needed for long-run public policy purposes.

Approaches to estimating agricultural water demands include the following: (1) production function, (2) residual value and variants, and (3) property transactions, including land sales, water right sales, and water leases. Each of these is described below and, where available, appropriate examples are given.

Production Function

A direct approach to estimating the value of water for a specific crop uses experimental data to estimate a production function. As described above, this approach estimates water demand by valuing the marginal contribution of water to crop yields based on assumed crop prices and the marginal physical product. This approach has been widely used to measure the value of water at the field level. The approach, however, is limited in its capacity to adequately account for the many and varied farm-level responses that are often possible when water costs change. For example, this approach is unable to account for changes in crop mix, management, or technology that would be expected as part of the normal adjustment process.

As shown in Tables 5.2 and 5.3 in the main body of the report, based on values originally reported by Gibbons (1986), the value of water varies considerably across regions and across crops. These estimates derive from controlled experiments conducted by USDA that estimated crop water production functions. Values are based on the value of production lost when the crop water is reduced by 10% from its yield-maximizing level. Highest values are associated with vegetable crops, and, by region, the highest values occur in California and in the pacific Northwest.

Residual Valuation

To better account for farm-level responses, researchers developed methods based on farm-level budgets. These budgets have been used directly and in linear programming models to impute the value of water based on the residual value approach. These methods allocate the total value of output among each resource in the production process. After assigning appropriate prices to all inputs, save for one—water for example, the residual of the total value is imputed to the remaining resource. This approach assumes that (1) market prices for all the resources, except the one being valued, are equal to their value of the marginal product; and (2) the total output value can be shared among each resource according to marginal productivity, and total value is completely exhausted as a result (Young and Haveman 1985, Gibbons 1986, Colby 1989b, Young 1996).

The residual valuation method is the foundation of three variant approaches used in practice: (1) farm budget analysis, (2) change in net income (CINI), and (3) programming methods.

Farm Budget Analysis

Farm budget analysis extends the production function approach to better account for the differences between actual farm conditions and controlled experimental conditions. This approach takes into account all the farm-level costs and permits an estimate of the maximum outlay possible for water. It is important to note that this approach, as well as the others using residual value methods, rely on a particular estimate of the expected change in yields associated with a change in water use. Thus, there remains a direct linkage to the concept of crop-water productivity, which after all is the principal source driving the economic valuation of farm water use.

Change in Net Income

A key limitation in the use of farm budgets alone is that it is readily recognized that when the amount of one input changes, water for example, the optimal mix of all inputs will most likely change. Adjustments and adaptations in the production process will take advantage of technical and economical substitution possibilities. These adjustments can be accounted for—with additional information—in the CINI approach.

The CINI approach is a variant of the residual method that has been adopted and highlighted for use by the U.S. Water Resources Council (1979, 1983). The approach recognizes that the value of irrigation water is a measure of the net economic contribution of water to the value of agricultural production and that willingness to pay for an incremental increase in water is bounded by the net producer income associated with that increment (Young 1996). Essentially, this is the multicrop extension of farm budgets.

In principle, the approach is transparent to apply. Net income is measured in two states, one with and one without the change in water. The difference in the net income of the two states divided by the change in the amount of water used provides an estimate of the water value implicit in the process. By expanding the number of crops and activities in the analysis, the approach is more holistic of the farm response and adaptation, particularly in accounting for changes in crop mix, land use, management, and technology (factors that are instrumental in determining the underlying value of water). However, to accurately account for these factors, more information is needed or assumptions must be made about the farm-level response between the *with* and *without* states. Critical among these assumptions are (1) crop types and acreages, (2) crop response to water changes (i.e., crop-water productivity), and (3) the effectiveness and cost of alternative technologies and the management requirements.

Programming Methods

Programming models are a further extension of the residual valuation method and the CINI approach, in particular. In this approach, the farm-level planning problem is framed and solved as an optimization problem such as the one described earlier in which the farmer desires to maximize net returns subject to resource and process constraints. This extends the CINI approach by making many of the exogenous assumptions in CINI internal to the programming model. Whereas in CINI assumptions had to be made about several other management decisions and their effect on yields and costs, programming models can incorporate these decisions implicitly in the optimization process, for example, optimizing the mix of crops or the choice of technology. The result is a more continuous and transparent accounting of the decision making process, potentially improving the accuracy of the resulting estimate of water value. As in the previous residual approaches, programming relies on explicit farm budgets—including the crop-water production effects.

Programming approaches have enhanced the capability of analyzing water use decisions in several different directions. For example, sequential or multistage processes, common in agriculture, can be explicitly modeled where irrigation is sequenced throughout a growing season and conditioned by the stage of crop growth and soil-moisture conditions (see, for example, Yaron and Dinar 1982; Taylor and Young 1995). Programming models have also been used for regional water resource planning and intersectoral allocation (e.g., Vaux and Howitt 1984, Booker and Young 1991, Hurd et al. 1999). These models typically model water demands by sectors and by regions and, hence, rely on previous efforts and water demand models as inputs, further increasing the complexity of analysis. Such integrated approaches, however, are a useful platform for exploring water allocation decisions and consequences from a variety of perspectives.

Property Transaction Valuation

Market transactions are a potentially attractive source for information regarding the market value of water, including water for irrigation. There is a measure of credibility in observing actual transactions between buyers and sellers; the valuation process is relatively transparent, i.e., not unduly clouded by analysts assumptions; and such measures—assuming fully informed traders—can reflect a forward-looking perspective on value trends. Transactions that can be observed and analyzed for their value of water include permanent transfer of water rights, leasing contracts, contingency contracts (including water banking arrangements), and land value comparisons with and without water. With regard to the last, statistical regression can be used to infer the contribution of water to land values, a technique that is sometimes referred to as *hedonic analysis*.

Use of this method is limited by the scarcity of studies, due in part to physical, economic, and institutional characteristics that constrain activity. Furthermore, regarding some of the potentially most common sources of data, Colby (1989a) suggests that though it is common for cities that have acquired rights in excess of their current needs (often in anticipation of future growth) to lease the water back to farmers, the rates that are commonly charged are generally intended to cover the city's distribution costs rather than reflect agricultural economic potential.

Water Right Valuation Approaches

Colby (1989a) discusses the appraisal and valuation approaches for assessing the value of water rights, both in general and in agriculture. It is important to note, however, that valuation that arises from water right markets is not necessarily reflective of a specific type of use, such as agriculture. Appraisal values should properly reflect the "highest and best" use—defined as "the reasonable and probable use of the property that generates the highest present value" (Colby 1989a, p. 182). Determining the highest and best use for water can be more complicated than in land appraisals. Not only are new uses possible, but also water can be transported to new locations. For estimating agricultural water value, the use of water right transfers should be limited to those that are intended primarily for agricultural use and thus net of speculative values. This can be difficult to control for since the speculative value of water for other uses, in part, is an important driver in many markets for water.

Colby (1989a) identifies the following four approaches:

- *Sales Comparison Approach.* Market data are assembled on comparable transactions. With the emphasis on comparability—data should be reflective of the current market conditions vis-à-vis, for example, location, quality, and transactions costs—especially with regard to third party impacts.
- *Income Capitalization Approach.* This approach, much like the residual valuation methods described above, analyzes the stream of net returns that can be generated by

the asset, in this case, water. This is likely to be a common approach to valuing water in agricultural uses because of the productive role of water in agricultural production.

- *Analysis of Land Value Differentials.* Land sales with and without irrigation water are analyzed in this approach and compared to net the principal effect of water on the price difference. More sophisticated statistical methods based on this premise, often termed hedonic price models, aim to tease out the specific contribution of water rights to the overall land value. Presumably, there should be a close affinity with this approach and the income capitalization approach.
- *Development Cost Approach or Least-Cost Alternative.* This is a "cost-based" approach that does not really bear any necessary resemblance to the demand or value of water. It is primarily used as a last resort technique to estimate the costs of alternative water supplies. Since there may be no buyers potentially willing to pay these alternative costs—the use of this technique, especially for agricultural water valuation, is most likely inappropriate but is included here for completeness.

In particular when water right values are available, care must be used to select an appropriate discount rate to annualize the values so that they are comparable to values in other uses.

ADDITIONAL CONSIDERATIONS IN THE VALUATION OF AGRICULTURAL WATER USE

Estimating appropriate and useful values for agricultural water is generally a complex process involving many issues and considerations. This section highlights several issues that have particular significance. Some of these have been touched on above, but in this section they receive emphasis and greater attention. Topics addressed here include:

- Risk and uncertainty
- Water supply reliability
- Market distortions and their consequences on water demand
- Technology, water use efficiency, and substitution possibilities
- Indirect benefits and costs of agricultural water use.

Risk and Uncertainty

Farm-level irrigation decisions are commonly made in the presence of three sources of uncertainty: variability in biological relationships and seed-stock quality, variability in resource conditions and states of nature, and variability in market conditions and prices (Boggess, Lacewell, and Zilberman 1993). In addition, some farmers may face uncertainty with respect to water supply, because periodic drought and relative seniority of water rights can create uncertainty.

Agriculture is an inherently risky process and input decisions such as irrigation scheduling can influence the nature and magnitude of some of these risks. For many irrigated crops, farmers use water not only to contribute to expected yields but also to effectively reduce yield and quality variability. For example, a farmer might over-irrigate in early stages of plant growth to help ensure that the crop becomes well established and avoids any water stress in this critical stage of development. In this case, additional water acts as a level of insurance that guards against risks, for example, weather conditions that might become less favorable and cause plants to become water-stressed. Particularly when faced with relatively cheap and abundant water supplies, farmers will use water in sufficient amounts to avoid possible limiting effects (Gibbons 1986). This suggests that in estimating seasonal irrigation water demand there can be an insurance component that exceeds the minimum efficient water requirements. For example, Taylor and Young (1995) estimate that irrigation demand is raised by about 16% (raising the value of the marginal product for 1 AF of water from $37 to $43) in eastern Colorado by the risk-reducing effects of additional water.

Antle (1983) demonstrates that risk and the risk characteristics of inputs are important to agricultural producers independent of their risk preferences (e.g., attitudes toward risk such as *risk aversion*). He shows that risk matters even to producers who are otherwise characterized as *risk neutral*. He does this by contrasting the conventional, static model of behavior—in which case risk matters only if decision makers have risk preferences, i.e., are risk averse—with a model more typical of agricultural in which production and input decisions occur through time, in a multistage, sequential manner (e.g., planting, maturation, and harvesting). In this case, input decisions that affect outcome variability in the early or intermediate stages of production can, in fact, have a determining effect not only on the variability of returns but also on the level of expected returns. For example, consider the effects of pre-planting irrigation on the outcome of establishing young healthy plants. Good yields and quality (and hence revenues) depend not only on water applied during the growth and maturation of the crop but also directly on how well the crop is initially established. High variability in the establishment stage, Antle shows, diminishes expected outcomes in subsequent stages such as maturation and harvest.

This is, in fact, a very important finding with direct relevance to the issue of irrigation demand. No longer is the assumption of risk aversion required to establish the importance of risk in decision making. Dynamic production processes alone are sufficient for risk to matter. Antle further notes, "Dynamic, risk-neutral models may be a better first-order approximation to the decision problems farmers actually face."

These results suggest that the risk-reducing potential of irrigation—and the increase in irrigation demand that results from it—is not necessarily an artifact of risk aversion and the loss of personal well-being attributable to variability (as many studies suggest) but can have a material and productive role in enhancing profitability.

Water Supply Reliability

Reliability of water supplies is a cornerstone of economic development and settlement in the western United States. Reliability is central to the water laws and institutions, known as prior appropriation, that have evolved in the West. To provide a measure of "reliability"—at least with respect to who had the "right" to use what nature made available—prior appropriation doctrine was adopted as the institutional basis for water rights in the West. Under this approach, which applies in large measure to all the Western states, the right of use is given in order of who first diverted the water and put it to "beneficial use." This first in time, first in right approach gives rise to the distinction of seniority, i.e., senior and junior water rights. The more senior the water right (i.e., the older the claim), the more secure and, hence, reliable the right to access and use the available supply. Without respect for the geographic position of the user, either upstream or downstream, distance to the water source, or type of use (e.g., municipal and agricultural), the right of use of the senior holder supercedes that of all junior claims. In times of shortages, the water that is available follows the seniority of rights and goes to those who were first to risk and invest the capital and labor to develop an economic foundation in the arid regions of the West (Ciriacy-Wantrup 1964, Getches 1997, Sax et al. 2000).

Buoyed by capabilities in engineering and construction, western settlers banded together to finance water development projects, ushering in the technological fixes that would provide water for the ever-growing populations enticed into settling the West. When and where private financing was insufficient to carry projects forward, the U.S. government lent a hand to reclaiming desert. The U.S. Bureau of Reclamation brought a vision that expanded and released settlements from its former constraints (Wahl 1989). Such investments might not otherwise have been made if there were no protection and security against subsequent developers who might otherwise install diversions farther upstream and thus erode or undermine the value of those investments.

Therefore, reliability provides measurable economic benefits that can readily be measured by the marginal value of seniority (or priority date) in water right sales and transfers. More senior water rights have a measure of security against drought years and can be expected to provide a more reliable or firm level of water yield.

Agriculture is often but not always a holder of senior water rights. For example, as Taylor and Young (1995) describe, some irrigators in eastern Colorado along the Arkansas River, because of the relatively junior status of their water rights, can expect adequate water for an entire irrigation season in only one out of every four years. The result in this case is that much of the water rights in this region have been sold and transferred to municipalities that in turn lease water back to farmers until such time as the water is needed.

APPENDIX B
INSTREAM FLOW VALUATION CITATIONS

A literature search identified 48 studies concerning values of instream flows. Their citations are presented in Table B.1, along with the document type, location of study area, if recreation values were estimated, if nonuse values were estimated, if values for ecological services were estimated, and if any other values attributable to instream flows were estimated.

Table B.1
In-stream flow valuation literature

Citation	Document type	Location	Recreation valued?	Nonuses valued?	Ecological services valued?	Other values?
Amirafathi, P., R. Narayanan, A. Bishop, and D. Larson. 1985. A Methodology for Estimating In-stream Flow Values for Recreation. Water Resources Planning Series (UWRL/P-85/01). Logan, Utah: Utah State University, Utah Water Research Laboratory.	Academic paper	Northern Utah	y			
Bailey, B. Undated. Minimum In-stream Flow Legislation and the Section 404 Permitting Process: The Impact of State Water Quality Laws. Washington, D.C.: Environmental Defense Fund.	Report	General				y
Berrens, R.P., P. Ganderton, and C.L. Silva. 1996. Valuing the Protection of Minimum In-stream Flows in New Mexico. *Journal of Agricultural and Resource Economics*, 21(2):294-309.	Journal article	New Mexico		y	y	
Brown, T.C. 1991. Water for Wilderness Areas: In-stream Flow Needs, Protection, and Economic Value. *Rivers*, 2(4):311-325.	Journal article	General			y	
Brown, T.C., and J.W. Duffield. 1995. Testing Part-Whole Valuation Effects in Contingent Valuation of In-stream Flow Protection. *Water Resources Research*, 31(9):2341-2351.	Journal article	Five Montana rivers	y	y	y	
Brown, T.C., and T.C. Daniel. 1991. Landscape Aesthetics of Riparian Environments: Relationship of Flow Quantity to Scenic Quality Along a Wild and Scenic River. *Water Resources Research*, 27(8):1787-1795.	Journal article	Colorado	y			y

(continued)

Table B.1 (Continued)

Brown, T.C., J. Taylor, and B. Shelby. 1991. Assessing the Direct Effects of Streamflow on Recreation: A Literature Review. *Water Resources Bulletin*, (6):979-989.	Journal article	Nationwide	y			
Butkay, S.M. 1989. The Economic Value of In-stream Flows for Recreation: A Contingent Valuation Approach. Master's thesis. University of Montana, Missoula.	Thesis	Big Hole and Bitterroot rivers, Montana	y			
Christensen, J.D., M.E. Monaco, R.J. Livingston, G. Woodsum, T.A. Battista, C.J. Klein, B. Galeprin, and W. Huang. 1998. Potential Impacts of Reduced Freshwater Inflow on Apalachicola Bay, FL Oyster (*Crassostrea virginica*) Populations: Coupling Hydrologic and Biological Models. Silver Spring, Md.: National Oceanic and Atmospheric Administration.	Report	Florida				y
Colby, B.G., J. Leones, C. Mullahy-Koenig, and L. Ryan. 1994. River Recreation and the Economy of Northern New Mexico. Tucson, AZ: University of Arizona, Department of Agricultural and Resource Economics.	Academic paper	Northern New Mexico	y			
Crandall, K.B., B.G. Colby, and K.A. Rait. 1992. Valuing Riparian Areas: A Southwestern Case Study. *Rivers*, 3(2):88-98, Studies in the Science, Environmental Policy and Law of In-Stream Flow. Tucson, Ariz.: University of Arizona.	Journal article	Hassayampa River Preserve in Arizona	y			
Daubert, J.T., and R.A. Young. 1981. Recreational Demand for Maintaining In-stream Flows: A Contingent Valuation Approach. *American Journal of Agricultural Economics*, 63(4):666-676.	Journal article	The Cache la Poudre River in northern Colorado	y			
Douglas, A.J., and R.L. Johnson. 1993. In-stream Flow Assessment and Economic Valuation: A Survey of Non-market Benefits Research. *International Journal of Environmental Studies*, 43:89-104.	Journal article	General	y	y		
Douglas, A.J., and J.G. Taylor. 1998. Riverine Based Eco-tourism: Trinity River Non-market Benefits Analysis. *International Journal of Sustainable Development and World Ecology*, 5(2):136-148.	Journal article	Trinity River, California	y		y	y
Duffield, J.W., and D.A. Patterson. 1991. Field Testing Existence Values: An In-stream Flow Trust Fund for Montana Rivers. Paper presented at the American Economics Association Annual Meeting, January 4. New Orleans, La.	Report	Upper Missouri River Basin, Montana	y			

(continued)

Table B.1 (Continued)

Duffield, J.W., T.C. Brown, and S.D. Allen. 1994. Economic Value of In-stream Flow in Montana's Big Hole and Bitterroot Rivers. (No. RM-317). Fort Collins, Colo.: Rocky Mountain Forest and Range Experiment Station, USDA Forest Service.	Report	Big Hole and Bitterroot rivers, Montana	y	y		
Duffield, J.W., S. Butkay, and A. Stewart. 1990. Economic Value of Recreation and Preservation Benefits of In-stream Flow.	Report		y			
Duffield, J.W., C.J. Neher, and T.C. Brown. 1992. Recreation Benefits of In-stream Flow: Application to Montana's Big Hole and Bitterroot Rivers. *Water Resources Research*, 28(9):2169-2181.	Conference paper	Study site is two Montana streams	y			
Duffield, J.W., C. Neher, M. Josephson, and R. Josephson. 1991. Market Value of Agricultural Water Leased for In-stream Flows.	Journal article	Big Hole and Bitterroot rivers, Montana	y			
Duffield, J.W., C. Neher, D. Patterson, and S. Allen. 1990. In-stream Flows in the Missouri River Basin: A Recreation Survey and Economic Study. Final report. Helena, Mont.: Montana Department of Natural Resources and Conservation.	Report	Montana	y			
Epp, D.J., and K.S. Al-Ani. 1979. The Effect of Water Quality on Rural Non-farm Residential Property Values. *American Journal of Agriculture and Economics*, 61(3):529-534.	Journal article	Pennsylvania				y
Flug, M., and R.H. Montgomery. 1988. Modeling In-stream Recreational Benefits. *Water Resources Bulletin*, 24(5):1073-1081.	Journal article	New River Gorge, West Virginia	y			
Frymier, L.G., and C.H. Mitchell. 1997. A Comparative Analysis of Value Between Users and Non-users of the White River. Proceedings of the 1996 Northeastern Recreation Research Symposium March 31-April 2, 1996, on Lake George in Bolton Landing, New York (pp. 79-81). Randor, Pa.: USDA Forest Service Publications Distribution.	Conference paper	White River, Vermont	y	y		
Gillilan, D.M., and T.C. Brown. 1997. *In-stream Flow Protection: Seeking a Balance in Western Water Use*. Washington, D.C.: Island Press.	Book	The West	y	y	y	y

(continued)

Table B.1 (Continued)

González-Cabán, A., and J. Loomis. 1999. Measuring the Economic Benefit of Maintaining the Ecological Integrity of the Río Mameyes in Puerto Rico. Research paper PSW-RP-240. Albany, Calif: Pacific Southwest Research Station, U.S. Forest Service, U.S. Department of Agriculture.	Report	Puerto Rico	y		
Johnson, N.S., and R.M. Adams. 1988. Benefits of Increased Streamflow: The Case of the John Day River Steelhead Fishery. *Water Resources Research,* 24(11):1839-1846.	Journal article	John Day River in Oregon	y		
Koberstein, P. 1997. What's a River Worth? River Values, 8-12. American Rivers.	Journal article	Missouri, Columbia, and Blackfoot rivers			y
Kyler, D. 2001. Letter of 9 February 2001 from David Kyler, Executive Director, Center for a Sustainable Coast to the Honorable Roy Barnes, Governor, State of Georgia.	Letter	Georgia			y
Leefers, L., and D.M. Jones. 1996. Assessing Changes in Private Property Values Along Designated Natural Rivers in Michigan. Lansing, Mich.: Michigan State University, Department of Forestry.	Academic paper	Michigan			y
Leones, J. 1997. Measuring Regional Economic Impacts of Streamflow Depletions. *Water Resources Research,* 33(4):831-838.	Journal article	Rio Grande, New Mexico	y		
Loomis, J. 1987b. The Economic Value of In-stream Flow: Methodology and Benefit Estimates for Optimum Flows. *Journal of Environmental Management,* 24:169-179.	Journal article	Colorado, New Mexico, Utah	y	y	
Loomis, J. 1987a. Balancing Public Trust Resources of MonoLake and Los Angeles' Water Right: An Economic Approach. *Water Resources Research,* 23:1449-1456.	Journal article	Mono Lake and Los Angeles		y	
Loomis, J. 1999b. Recreation Meta-Analysis Database. Colorado State University, Ft. Collins, Colo.	Database	Nationwide	y		
Loomis, J., and J. Cooper. 1990. Economic Benefits of In-stream Flow to Fisheries: A Case Study of California's Feather River. *Rivers,* 1(1):23-30.	Journal article	Feather River in Northern California	y		

(continued)

Table B.1 (Continued)

Loomis, J., P. Kent, L. Strange, K. Fausch, and A. Covich. 1999. Measuring the Total Economic Value of Restoring Ecosystem Services in an Impaired River Basin: Results From a Contingent Valuation Survey. *Ecological Economics*, 33(2000):103-117.	Journal article	South Platte River, Colorado	y	y	y	y
Loomis, J.B. 1998a. Estimating the Public's Values for In-stream Flow: Economic Techniques and Dollar Values. *Journal of the American Water Resources Association*, 34(5):1007-1014.	Journal article	Nationwide	y	y		
Loomis, J.B., and M. Creel. 1992. Recreation Benefits of Increased Flows in California's San Joaquin and Stanislaus Rivers. *Rivers*, 3(1):1-13.	Journal article	San Joaquin and Stanislaus rivers, California	y			
Narayanan, R. 1986. Evaluation of Recreational Benefits in In-stream Flows. *Journal of Leisure Research*, 18 (2):116-128.	Journal article	Northern Utah	y			
Narayanan, R., D.T. Larson, A.B. Bishop, and P. Amirfathi. 1983. An Economic Evaluation of Benefits and Costs of Maintaining In-stream Flows. Washington, D.C.: U.S. Department of the Interior.	Report	Bear River watershed, northern Utah	y	y	y	
Neher, C.J. 1989. The Economic Value of In-stream Flows in Montana: A Travel Cost Model Approach. Master's dissertation. University of Montana, Missoula.	Dissertation (Masters)	19 rivers and streams in Montana	y			
Shelby, B., T.C. Brown, and R. Baumgartner. 1992. Effects of Streamflows on River Trips on the Colorado River in Grand Canyon, Arizona. *Rivers*, 3(3):191-201.	Journal article	Colorado River, Arizona	y			
Day, L.R. 2001. Testimony of Laura Rose Day, Watershed Project Director, Natural Resources Council of Maine in Support of LD 12488, An Act of Require Public Information Regarding Water Use by Major Water Users, before the Natural Resources Committee (4 April 2001).	Testimony	Maine	y			
TCPS (Texas Center for Policy Studies). 2002. Community and Economic Benefits of Texas Rivers, Springs and Bays. Austin, Texas: Texas Center for Policy Studies.	White paper	Nationwide	y			
USDA (U.S. Department of Agriculture). 1991. Economic Value of Recreation and Preservation Benefits of In-stream Flow in Montana.	Report	Montana	y			

(continued)

Table B.1 (Continued)

Wade, W.W., G.M. Mccollister, R.J. McCann, and G.M. Johns. 1988. Estimating Recreation Benefits for In-stream and Diverted Users of Waterflows of the Sacramento-San Joaquin Rivers Watershed.	Report	Sacramento-San Joaquin rivers, California	y
Walsh, R.G., D.M. Johnson, and J.R. McKean. 1988. Review of Outdoor Recreation Economic Demand Studies With Non-market Benefit Estimates, 1968-1988. Fort Collins, Colo.: U.S. Forest Service, Rocky Mountain Forest and Range Experiment Station.	Report	Nationwide	y
Walsh, R.G., R. Ericson, D. Arosteguy, and M. Hansen. 1980. An Empirical Application of a Model for Estimating the Recreation Value of Instream Flow. Fort Collins, Colo.: Colorado Water Resources Research Institute, Colorado State University.	Report	Walsh	y
Ward, F.A. 1987. Economics of Water Allocation to In-stream Uses in a Fully Appropriated River Basin: Evidence From a New Mexico Wild River. *Water Resource Research*, 23(3):381-392.	Journal article	New Mexico	y

APPENDIX C
THE VALUE OF WATER IN THE SAN FRANCISCO PUBLIC UTILITIES COMMISSION SYSTEM: A VIEW THROUGH THE "TRIPLE BOTTOM LINE" LENS

INTRODUCTION

Public water utilities have historically operated with a focus beyond that of financial bottom lines to one of providing the broad social good of safe and adequate water to a wide range of users. More recently, an additional emphasis has been placed on utilities to operate in ways that preserve and protect ecosystems. In effect, utilities operate in ways that achieve three sets of values: social (e.g., quality of life), financial, and environmental. The need to take multiple values into account is reflected in a concept established in the 1990s called Triple Bottom Line (TBL), which identifies explicitly how an entity performs according to these three objectives. While the SFPUC does not describe its operations using TBL terminology, it clearly has been operating in a triple bottom line context, and this case study highlights the extent to which the nontraditional bottom lines—social and environmental—influence the SFPUC's operations, as well as indicate its larger contribution to society. The triple bottom line offers the advantage of more clearly articulating to the public the role of these values in managing public water supplies.

We begin below by introducing the relationship—actual and potential—between water utilities and the triple bottom line approach, with examples drawn from Australia and England. The next section gives an overview of the SFPUC system, its characteristics, trends, and key actors, and presents a potential TBL framework with associated *values of water* that may be useful for the SFPUC. This case study closes with a summary of how the SFPUC'S Capital Improvement Program (CIP) measures against these values, organized within a triple bottom line framework—with the revelation that *values outside of the traditional financial analysis to society at large are almost an order of magnitude greater than the costs of the improvements*. Finally, relevant lessons and conclusions are drawn for the larger water utility community.

WATER UTILITIES IN THE TRIPLE BOTTOM LINE CONTEXT

Broad Concepts of the Triple Bottom Line

Since its first introduction in the late 1990s by author and green business consultant John Elkington, the triple bottom line has steadily gained traction as an important tool for sustainable development (Elkington 1997). TBL is a way to view how an entity (e.g., a business, government agency, civil society group, public utility) functions in richer and fuller terms—including not just financial measures but also social and environmental measures—so that conclusions can be drawn about that entity's broader role in and impact on sustainable development as well as influence behavior toward sustainability. In other words, the TBL enables users to determine how well—or poorly—key, relevant sustainability concerns have been internalized in an organization's day-to-day functions and decision-making. In practice, the term triple bottom line is most often used to describe an approach to planning or reporting. Applied to these processes, TBL can offer a framework for carrying out a performance assessment that hones in on key economic, social, and environmental objectives. It is important to note that a primary goal of preparing a TBL report is to contribute to an ongoing stakeholder dialogue. Reports alone provide little value if they fail to inform stakeholders or support a dialogue that influences the decisions and behavior of both the reporting organization and its stakeholders, and therefore they are best designed in a format accessible and understandable to a broad audience.

The most widely accepted definition of TBL is that developed by the Global Reporting Initiative (GRI; www.globalreporting.org), which defines each of the three components as follows:

- The *economic dimension* of sustainability concerns an organization's impacts on the economic circumstances of its stakeholders and on economic systems at all levels. Economic impacts can be divided into direct and indirect impacts. Broadly speaking, economic performance encompasses all aspects of the organization's economic interactions, including the conventional measures used in financial accounting as well as intangible assets that do not systematically appear in financial statements. However, reporting on economic performance, as articulated in the Guidelines (GRI 2003), has a scope and purpose that extends beyond that of traditional financial reporting. Financial indicators focus primarily on the profitability of an organization for the purpose of informing its management and shareholders. By contrast, economic indicators in the sustainability reporting context focus more on the manner in which an organization affects the stakeholders with whom it has direct and indirect economic interactions. While financial performance indicators are well developed, indicators of organization-level economic performance are still evolving.
- The *social dimension* of sustainability concerns an organization's impacts on the social systems within which it operates. Social performance can be gauged through an analysis of the organization's impacts on stakeholders at the local, national, and global levels. In some cases, social indicators influence the organization's intangible assets, such as its human capital and reputation. Through its consultative process, GRI has selected social indicators by identifying key performance aspects surrounding labor practices, human rights, and broader issues affecting consumers, community, and other stakeholders in society. Not all social indicators are quantitative—several are qualitative measures of the organization's systems and operations, including policies, procedures, and management practices. Given the diversity of social situations and issues that confront them, organizations should use stakeholder consultation to ensure that the social impacts on which they report are as complete as possible.
- The *environmental dimension* of sustainability concerns an organization's impacts on living and non-living natural systems, including ecosystems, land, air, and water. The environmental dimension of sustainability has achieved the highest level of consensus among the three dimensions of sustainability reporting. It is particularly important to provide environmental performance information in terms of both absolute figures and normalized measures (e.g., resource use per unit of output). Both measures reflect important, but distinct, aspects of sustainability. Absolute figures provide a sense of scale or magnitude of the use or impact, which allows the user to consider performance in the context of larger systems. Normalized figures illustrate the organization's efficiency and support comparison between organizations of different

sizes. Reporting organizations are encouraged to relate their individual performance to the broader ecological systems within which they operate. For example, organizations could seek to report their pollution output in terms of the ability of the environment (local, regional, or global) to absorb the pollutants.

Each organization shapes objectives along these three dimensions in forms relevant to local stakeholders.

Benefits of TBL Approach

The triple bottom line provides an integrated view of how an organization is performing. A group of Australian organizations have identified the following benefits of using TBL (Australian Government Department of Environment and Heritage 2003):

- Embedding sound corporate governance and ethics systems throughout all levels of an organization. Currently many corporate governance initiatives are focused at the Board level. TBL helps ensure a values-driven culture is integrated at all levels.
- Improving management of risk through enhanced management systems and performance monitoring. This may also lead to more robust resource allocation decisions and business planning, as risks are better understood.
- Formalizing and enhancing communication with key stakeholders such as the finance sector, suppliers, community, and customers. This allows an organization to have a more proactive approach to addressing future needs and concerns.
- Attracting and retaining competent staff by demonstrating that an organization is focused on values and its long-term existence.
- Enhancing the ability to benchmark performance both within industries and across industries. This may lead to a competitive advantage with customer and suppliers, as well as enhanced access to capital as the finance sector continues to consider nonfinancial performance within credit and investment decisions.

The advantages given above are from a wide range of industries. The following section presents how the TBL has been implemented by several water utilities outside the United States.

TBL and Water Utilities

Historically, water utilities have had a strong social focus, placing a high priority on providing safe and reliable water supplies. There have, however, been environmental consequences, both positive and negative, associated with providing water, such as those associated with large water storage and transport projects. But while certain large-scale water management decisions may have significant detrimental consequences, the impact of water utilities in economic, social, and environmental terms is far broader. The management priorities of water utilities are, quite appropriately, water supply, delivery, and quality, yet additional priorities exist—from the enhancement of water revenue to the improvement of water safety to the sustainable management of watershed land holdings. For some water utilities, a view through the TBL lens reveals the full array of existing priorities beyond the financial and long-held social values of water quality and safety, to newer social values such as equity in treatment facility siting or sustaining habitat through releasing water from a reservoir for fishery enhancement (Davis 2001). In other words, many water utilities are increasingly managing their systems with a triple bottom line approach, albeit without adopting or even recognizing TBL as a deliberate guiding principle. However, some in the water sector have explicitly taken a TBL approach. The GRI, as mentioned above, is being applied by hundreds of organizations around the world, including a number of major water utilities. Sydney Water (Australia), City West Water Limited (Australia), Severn Trent (United Kingdom), and Water Care Services (New Zealand), for example, have used the GRI Reporting Guidelines directly for the development of sustainability reports, while others have adapted the GRI approach to suit their unique needs (GRI 2003). The approaches used by each of these water utilities are presented below to highlight how each has defined a triple bottom line most relevant to the stakeholders in their region, while maintaining the concept of presenting their performance across a much broader set of criteria than financial viability.

Sydney Water's operation provides 1.6 billion liters of water to 4 million customers daily and clearly demonstrates the benefit of using a triple bottom line approach to water utility management. Sydney Water has defined three key objectives: to protect public health, to protect the environment, and to be a successful business. Each year, the Sydney Water *Towards Sustainability* report examines the progress the utility has made toward stated objectives in three categories: environmental, social, and economic, as illustrated in Figure C.1, from Sydney Water's "Sustainability Snapshot 2002." To determine progress, measurements are made that are appropriate to the criteria. Progress indicators are not always monetized, which further reflects the broad approach to values. For example, progress toward protecting public health by supplying safe drinking water was evaluated by examining compliance with Australian drinking water guidelines and by reviewing the number of public-health-related complaints about the stormwater system. Using the TBL framework, Sydney Water's program activities are reviewed in terms of their adherence and support of the utility's established objectives (Sydney Water 2004).

The other three utilities, Severn Trent (Severn Trent 2004), Water Care (Water Care 2004), and City West Water (City West Water Limited 2004), as well as Sydney Water, all present their triple bottom line via their websites, and three of the four (all but Severn Trent) have formal TBL reports. All have somewhat similar objectives or values in the three TBL categories presented both in summary form (such as a report card) and in much greater detail (several pages for each objective or value). The number of objectives or values presented varies, as can be seen in Tables C.1, C.2, and C.3, which compare the stated values for each utility for each bottom line categories—environmental, social, and economic—respectively. Clearly there is a broad range of values important to these utilities and their stakeholders. However, there are some common themes, as presented below.

When looking at the environmental bottom lines, shown in Table C.1, the specifics of the values are quite different; however, they can largely be grouped into three major categories:

1. Environmental protection of land or water resources controlled by the utility
2. Reduction of impacts on the broader environment
3. Encouragement of local communities/stakeholders to better consider environmental implications.

SUSTAINABILITY SNAPSHOT 2002

PROGRESS RATING

▸	On track	Evidence demonstrating a positive long-term trend in relevant indicators towards the objective.
•	Inconclusive	No clear long-term trend in relevant indicators towards or away from the objective. Mixed results, positive trends for some indicators and negative trends for others towards the objective.
□	Need more work	Data demonstrating a negative long-term trend in relevant indicators towards the objective.

	OBJECTIVES		COMMENT
Environmental objectives (the 10 objectives of Sydney Water's 2001-2005 Environment Plan)			
Objective 1	Conserve water supplies and prevent the need for new dams	▸	Active demand management being pursued to meet targets and offset population growth and supply availability issues
Objective 2	Provide beneficial and environmentally acceptable management of effluent	▸	Major investment in operations and infrastructure to reduce overflows, improve effluent quality and meet urban growth
Objective 3	Maximize beneficial reuse of by-products from Sydney Water's treatment processes	▸	A very high level of reuse achieved and actions being pursued to diversify markets
Objective 4	Manage stormwater systems under Sydney Water's control to minimise pollution	▸	Stormwater Environmental Improvement Program delivered to schedule and management plans implemented
Objective 5	Encourage waste minimisation and cleaner production in the commercial and Industrial sectors	▸	Trade Waste Policy's effectiveness in reducing industrial pollution strengthened by tailoring agreements that encourage cleaner production and effective risk management
Objective 6	Minimise waste and maximise reuse, recovery and recycling waste products	▸	Highly successful initiatives gaining momentum across the organisation
Objective 7	Responsibly manage all Sydney Water's land and water assets	▸	Significant progress made in managing our heritage and natural resources
Objective 8	Minimise environmental impact of Sydney Water's use of energy	□	Improvements in energy management and green power initiatives outstripped by energy requirements needed for new infrastructure
Objective 9	Continue to improve environmental management practices in Sydney Water	▸	Internationally recognised environmental management systems in place and a high level of compliance with environmental regulation
Objective 10	Contribute to collaborative research to improve the environmental performance of assets	[1]	Sydney Water initiated and contributed to a range of research
Social objectives			
Objective 11	Protect public health by supplying safe drinking water and effective wastewater and stormwater services	▸	Drinking water quality meets Australian guidelines; swimability at all beaches continues to improve
Objective 12	Ensure that all customers, especially those with special needs, maintain fair and affordable access to essential services	▸	Rights to water services and existing programs to meet customers' special needs now explicitly expressed in new Customer Contract
Objective 13	Eliminate discrimination in employment and promote equal employment opportunity	•	Sydney Water benchmarks well against accepted measures but seeks further assurance through additional programs
Objective 14	Look after the health, safety and welfare of all employees, contractors and any other people at, or who may be affected by our sites	▸	Major focus on safety program is entrenching cultural change
Objective 15	Enjoy the trust and credibility of the communities/customers we serve, and have relationships with stakeholders that are open and honest	▸	Trust and credibility continue to grow through active engagement of customers and stakeholders
Objective 16	Support the communities in which we operate through a variety of community investment initiatives	▸	Community investment program continues to grow, and deliver corporate environmental and social objectives
Economic objectives			
Objective 17	Efficiently deliver a capital expenditure program that will provide environmental and public health benefits to the community	[2]	On track delivering major capital program; unable to measure efficiency gains for the current year
Objective 18	Provide products and services that are cost competitive	▸	Ongoing reductions in operating costs do not impede increasing levels of service
Objective 19	Manage risks that can potentially affect the value of Sydney Water and its ability to deliver products	▸	Exposure is minimised through active management of environmental and social risks
Objective 20	Improve shareholder value by generating commercial return from the assets we employ	▸	Good return on the Government's Investment in Sydney Water indicates the business' viability

[1] To be reported in 2003
[2] To be reported in 2004

Source: Based on Sydney Water 2004.

Figure C.1 Sydney Water example of tracking TBL-related objectives

Table C.1
Examples of environmental bottom lines reported among four water utilities

Severn Trent, UK	City West Water, Australia	Sydney Water, Australia	Water Care, New Zealand
29% of river lengths improved from 1990 to 1998	Upgrade of wastewater treatment plant	Conserve water supplies and prevent need for new dams	Construction of UV wastewater treatment plant
Water saved from 1994 to 1998 by reducing leakage from Severn Trent Water pipes is enough to fill a quarter of a million swimming pools	Launched Community Environment Awards and committed to broadening program to Sustainability Awards in 2004	Provide beneficial and environmentally acceptable management of effluent	Progression of restoration of Manukau Harbor, resulting in the return of aquatic and bird life
Nitrogen and sulphur dioxide emissions from sewage sludge incineration have been reduced by 87% and 99%, respectively, in between 1994 and 1998	Volume of water consumed per capita reduced from 249 to 229 liters per day	Maximize beneficial reuse of treatment by-products	Continuation of the drive for eco-efficiency by maximizing biogas and hydroelectric generation
Almost a million trees have been planted on Severn Trent Water land between 1994 and 1998	Majority of targets in Environmental Improvement Plan implemented	Contribute to collaborative research to improve environmental performance of assets	
We currently generate around 7% of our electricity from totally renewable sources	Enviro Partnerships Initiative launched	Responsibly manage land and water assets	
Around 37,000 tonnes of industrial waste were recycled in 1998/1999		Encourage waste minimization	
Over 3 million visitors to Severn Trent Water Conservation and Recreation sites each year		Minimize environmental impact of energy use	

Table C.2
Examples of social bottom lines reported among four water utilities

Severn Trent, UK	City West Water, Australia	Sydney Water, Australia	Water Care, New Zealand
89,702 lead pipes have been replaced by Severn Trent Water between 1994 and 1998	High drinking water quality in spite of low rainfall	Protect public health	Maintenance of Ministry of Health "A" grade quality of drinking water
Our employees and customers raised over £4 million for sustainable water-related projects in Africa and Asia, and the work continues to encourage donations.	Ranked 10th out of 195 companies in the 2002 Managing Work/Life Balance Benchmarking Study	Look after the health, safety, and welfare of all employees, contractors, and others that may be affected by our sites	Delivery of a new water source and treatment plant to the Auckland region to meet its growth and water quality demands
Active support of WaterAid since it was first established, and in 2000 alone, over £745,000 was raised by staff, customers, suppliers, and community groups working together	Hardship assistance provided to 215 customers at a value of $96,000	Ensure all customers, especially those with special needs, maintain fair and affordable access to essential services	Continuation of the employment equity programme, including multicultural awareness education
At Creswell Crags in Nottinghamshire, Severn Trent Water is helping to re-create the way our ancestors survived at this internationally important Stone Age site.	Improved customer service	Eliminate discrimination in employment	
Public education through website	Established a Business Women's Network	Enjoy the trust and credibility of our customers we serve, and have open, honest relationships with stakeholders	
	Community consultation on development of TBL report	Support communities in which we operate through a variety of community investment opportunities	

Table C.3
Examples of economic bottom lines reported among three water utilities

City West Water, Australia	Sydney Water, Australia	Water Care, New Zealand
Pre-tax profit of $100 million (payment of $856,000 to government)	Effectively deliver a capital expenditure program	Commitment to a three-year price path
Infrastructure assets valued at $30 million were completed	Provide products and services that are cost-competitive	Completion of the sonar and closed circuit television inspections of those wastewater assets with a high consequence of failure
AAA credit rating retained	Manage risks that can potentially affect our value	Maintenance of the A+ short-term Standard and Poor's credit ratings
New strategic direction explicitly focuses on sustainability	Improve stakeholder value by generating commercial return from the assets we employ	

Similarly, looking at the social bottom line, shown in Table C.2, the values can be grouped into five broad categories:

1. Protect public health
2. Provide customer service, with special consideration of those with special needs
3. Eliminate discrimation
4. Engage communities in decision-making
5. Support broader social goals.

Economic bottom lines in Table C.3 show a similar pattern, where values can be grouped into four categories:

1. Profit and good return on investments
2. Good credit ratings
3. Reduction of risks
4. Sustainability of operations.

Across these three bottom lines and four utilities, not all categories are covered in each utility, yet for the most part virtually all water utilities touch on the broad categories presented above. There is something to be learned from utility to utility, in having some form of consistency in reporting, and to make clear both internally and externally the full range of considerations utilities face in any decision-making—from short-term, day-to-day operations to long-term capital investments.

A TBL FRAMEWORK AND THE SFPUC

The SFPUC approach to water management is driven by a variety of goals and values. The SFPUC's mission is to serve San Francisco and its Bay Area customers with reliable, high quality, and affordable water and wastewater treatment while maximizing benefits from power operations and responsibly managing the resources—human, physical, and natural—entrusted to its care (SFPUC 2002). While SFPUC actions often display consideration of social and environmental values as well as financial values, the triple bottom line approach has not been established as the Commission's explicit guiding principle for water management. Values of SFPUC staff, regulatory guidelines, the good will benefits of diversifying values, and public pressure all influence the SFPUC decision-making process. The net result, as is explored in this appendix, is a water utility whose activities increasingly incorporate the three value sets of the triple bottom line.

While the SFPUC is not formally using the triple bottom line, there is an implicit and growing recognition of the potential usefulness of using a triple bottom line approach since the SFPUC already factors in financial, environmental, and social values. Consideration of these values is not purely altruistic on the part of SFPUC; decisions that sustainably manage watershed land holdings, for instance, also tend to ensure higher water quality. Similarly, water recycling programs can fulfill supply needs just as they make it possible to avoid additional extraction from surface water sources. Activities such as removal of unnecessary dams may not be cost-effective options in financial terms, but are nevertheless added to the SFPUC portfolio because they address a key social or environmental value. Examples such as these serve on the one hand to highlight the integrated nature of management decisions in terms of their financial, social, and environmental costs and benefits, and on the other to highlight the nonmonetary values that can drive decision-making when social and environmental bottom lines are given some weight. The SFPUC experience provides an instructive example of this growing trend.

System Profile

Physical Characteristics

Water for the SFPUC system comes from both local groundwater and surface water sources, and from the Tuolumne River watershed, located about 160 miles east of San Francisco in the central Sierra Nevada. Three water supply and delivery systems constitute the regional SFPUC system: the Hetch Hetchy system, the Alameda system, and the Peninsula system (SFPUC and BAWUA 2000). The regional system components include "over 280 miles of pipelines, over 60 miles of tunnels, 11 reservoirs, 5 pump stations and 2 water treatment plants" (SFPUC 2003a).

Key Actors and Stakeholders

The SFPUC is a department of the County and City of San Francisco that provides water, wastewater services, and municipal power to the city and surrounding region. SFPUC sells two-thirds of the system water to 29 wholesale customer agencies in the Bay Area. The majority of these wholesale customers rely entirely on the Hetch Hetchy system for their water supplies. Some customers have alternate sources of water, such as groundwater or the California State Water Project (Bay Area Economic Forum 2002).

In 1958, the Bay Area Water Users Association (BAWUA) was formed (and incorporated in 1992) to represent and advocate the interests of member communities receiving wholesale supplies from the SFPUC (SFPUC and BAWUA 2000, CPRC 2001). Although the BAWUA provided suburban water agencies with a forum to "to monitor contracts, interactions, and compliance with the Master Agreement signed with San Francisco in 1984," the association (and its member communities) did not have authority regarding SFPUC management and operation decisions and activities. Recent state legislation established a new regional water agency—the Bay Area Water Supply and Conservation Agency (BAWSCA)—with the authority to purchase water from other sources, implement conservation activities, and monitor and negotiate with the city more effectively on SFPUC issues such as the major capital improvement projects that began in 2003. The agency will "oversee water deliveries to the suburban cities and small water districts." In companion legislation, a financing authority that includes wholesale customers and the city was established to facilitate the funding of repairs and other capital improvements to the SFPUC system (Enge 2003a).

Residential, business, and institutional customers use water from the SFPUC system. The business and institutional sectors that have significant water use include "the computer, semiconductor, biotechnology, automotive, aerospace, electric utility, telecommunications, [and] tourism ... sectors, [and] schools and hospitals, and smaller water-dependent business such as restaurants, glass and metal fabricators, beverage plants, and food processors" (Bay Area Economic Forum 2002). These business and industrial users constitute almost one-third of the San Francisco water demand and 25% of the demand in the BAWUA service region. In addition, regulatory agencies and environmental groups are important stakeholders in the region.

Current System Dynamics—Trends in Supply and Demand

As of 2002, the SFPUC system provided an average of 260 mgd to approximately 2.4 million customers in the Bay Area. Of the total water provided, the City of San Francisco uses approximately one-third, and suburban water agencies and other wholesale customers use the remaining two-thirds. Shifts in land use in the suburban areas of the region from farmland to housing and commercial uses have increased the demand for water owing to the additional residents and jobs. Countering this increase in demand has been an increase in conservation and water use efficiency by the suburban customers, stemming primarily from the response to the 1986-1992 drought. Water purchases by suburban customers are currently below 1985 levels. Not withstanding the efficiency gains, water use is expected to increase in the future. The overall system demand in 2030 is projected to be 303 mgd, and in 2050 may reach 310 mgd (SFPUC and BAWUA 2000).

The maximum supply available from the SFPUC, defined as firm delivery capability (the water delivery the system can sustain over historical hydrologic conditions), is 239 mgd. As described above, local Bay Area watersheds provide approximately 15% of this supply, with the remainder provided by the Hetch Hetchy water system. Current and projected future demands are well above the available supply levels (e.g., by more than 50 mgd by 2030; SFPUC and BAWUA 2000). Suggested activities to narrow the gap between supply and demand were evaluated and presented in the 2000 Water Supply Master Plan, published by the SFPUC and the BAWUA. In 2002, the SFPUC approved a series of capital improvement projects, including projects to increase water supplies, and use demand management tools such as water pricing, to constrain growth in demand. Improvements will be made to both the regional and local water systems.

A Sample TBL Framework for the SFPUC

Following the example of the Sydney report, *Towards Sustainability*, there are a number of potential values that the SFPUC could report on as well as use for decision-making based on the TBL approach. The TBL approach is advantageous, primarily because social and environmental values enter into decisions whether made explicit or not. Stating these values explicitly allows utilities and stakeholders to make decisions in a more transparent setting.

Values for the triple bottom line in the context of the SFPUC are broadly defined as follows. *Financial* values represent the financial aspects of SFPUC operations and management, such as the utility's revenue and expenditures, and might include efficient capital expenditures and provision of cost-effective products and services, comparable with industry best practices. S*ocial* values represent the importance of water to society, including both basic houshold needs and the role of SFPUC's services in local economic activities. From the perspective of social values, three aspects that have historically guided and continue to be important are reliability of water delivery, reliability of water supply, and provision of high quality water. *Environmental* values represent the importance of water to environmental and ecological processes, conditions, and sustainability, with active stewardship by the SFPUC. Important environmental values might include restoration of salmon and protection of the broader set of environmental values of watersheds. This template for a triple bottom line is illustrated in Table C.4.

SFPUC CAPITAL IMPROVEMENT PROGRAM IN A TBL FRAMEWORK

The previous section presented a triple bottom line framework that could potentially measure SFPUC's actions across financial, social, and environmental objectives. Recently, a major CIP was initiated to address concerns about the seismic vulnerability and reliability of the system. While the projects in the program do not address all objectives with equal emphasis, it should be noted that the SFPUC is engaged in numerous other activities that address the range of identified objectives. The CIP is presented below in the context of TBL to illustrate the utility of the framework in examining a given set of decisions. In particular, the framework highlights the effect of the CIP across a broad range of values—beyond simple financial terms. For example, the goals of Water-Related SFPUC Capital Improvement Projects include:

Table C.4
Template for a triple bottom line for SFPUC

Value category	Value objective
Financial	Efficient capital expenditures Provide cost-effective goods and services
Social	Reliability of water delivery Reliability of water supply High quality water Affirmative action goals and priorities to women and minorities in awarding contracts
Environmental	Restoration of salmon Protection of environmental value of watersheds Avoid increased extraction of in-stream flows

- *Regional water system*: Maintain water quality; achieve water supply reliability and delivery during droughts and other catastrophic events.
- *Local water system*: Maintain water quality for in-city water delivery; improve water delivery and supply reliability for in-city customers.
- *Clean water system*: Minimize odors and visual impacts at the Southeast Water Pollution Control Plant; reduce sewer failure and flooding in the streets throughout San Francisco; maintain and enhance the combined sewer collection, storage, and treatment facilities.

Key Economic Objectives in the SFPUC System

The SFPUC clearly strives to protect the long-term financial interests of all ratepayers as well as the investments of the utility. Water sales to the city and wholesale agencies in 2004 are projected to be 121.9 million ccf and to generate $150.3 million in revenue (retail sales revenue is projected to be $72.6 million and suburban wholesale revenue is projected to be $77.7 million) (SFPUC 2003b).

Growing concern over the risk of earthquake-related water delivery disruption—and growing recognition of the value of a secure delivery system as identified in the Bay Area Economic Forum Report (2002, p. 32)—prompted a far-reaching and coordinated response in the Bay Area in the form of the CIP. The total amount needed for capital improvements identified for the SFPUC system is $3.6 billion—the largest funding program ever proposed by the SFPUC. These improvements address needs for the local water system, the regional water system, the wastewater system (referred to as the clean water system), and the power system.

The scale of the projects required to provide increased reliability of both supply and delivery is difficult to justify on financial grounds alone. On the other hand, the Bay Area Economic Forum Report, discussed below in more detail, demonstrated the broader economic and social impact of potential failure of the water system.

The SFPUC presented the CIP to the public through a referendum so that area residents could decide the value of a reliable water supply. A local bond was presented as Proposition A on San Francisco's November 2002 ballot. The measure asked voters to approve a $1.63 billion revenue bond to pay for a portion of the SFPUC's capital improvements to the Hetch Hetchy water system. The measure was supported by most of the business and civic organizations in the city and some environmental groups, but opposed by a coalition of neighborhood groups, homeowners, hotels, and realtors and by the Sierra Club (Enge 2003b). It passed with 59.7% support (KQED 2002). A portion of the bond ($700 million) will be repaid through increases in San Francisco water rates. To address the remaining $2 billion, the state legislature created a regional financing authority that includes representatives from SFPUC wholesale customers and the city. Wholesale customers will directly contribute the remainder of the costs either through this financing authority, which can issue revenue bonds, or through a water rate surcharge (Bay Area Monitor 2003, SFPUC 2003a). The net impact on rates is at least a doubling over the next 15 years.

The CIP directly addresses the economic objectives of the SFPUC system, as follows.

Efficient Capital Expenditure Program

The CIP is a huge capital undertaking, with 77 projects costing $3.6 billion in expenditures that need to be planned and implemented. During planning phases for the CIP, the SFPUC completed a long-range financial plan that will serve as a guide for the funding aspects of the CIP and other SFPUC activities. The plan will be updated annually and, according to SFPUC, is "based on sound financial principles to ensure that project costs remain on track. Cost estimates are based on industry standards for projects of similar scope and size. In addition, engineers will assess the various design alternatives to select the most cost effective way to complete each project." Additional review of the efficiency and success of the capital expenditures will be accomplished through a new Program Controls System that will track the status of capital projects to "generate monthly management reports indicating whether projects are meeting budget and timeline targets," which will facilitate rapid evaluation and action on projects that are failing to meet the targets (SFPUC 2003b).

Provide Products and Services That are Cost-Effective, and Comparable to Industry Best Practices

Fueled in part by the beginning of a large, complex, capital improvement program, an SFPUC-wide performance assessment program was initiated in July 2003 to conduct a management and operations evaluation of all departments and bureaus. The performance assessment will (a) compare selected SFPUC work practices against the industry's best practices, (b) document internal strengths, (c) identify significant opportunities for improvement, (d) coordinate recommendations across departments, and (e) develop a prioritized implementation plan.

Key Social Values in the SFPUC System

The SFPUC manages an extensive network of aging water supply and delivery facilities in a predominately dry region of the country. This network traverses numerous major and minor seismic fault lines, serving in most cases as the sole water supply to a large and growing urban and suburban population. In dry years, water supply has proven to be inadequate to the region's needs, yet securing additional supply can involve costly tradeoffs. Clearly, the SFPUC faces a range of significant challenges. The CIP represents a strong recognition of the social value of water, as highlighted in the following discussion of four social objectives related to water.

Reliability of Water Delivery

Seismic vulnerability and associated costs. With over 2.4 million customers dependent solely on the SFPUC for reliable water, the importance of the security of the water delivery system can scarcely be overemphasized. Within San Francisco County, nearly all of the water consumed—96%—is supplied by the SFPUC (SFPUC and BAWUA 2000). In San Mateo County, for users within the SFPUC service area, that figure is 91%. In Santa Clara and Alameda counties, both of which have significant local and external water sources, the share of SFPUC supply is still quite significant, at 58% and 44%, respectively.

Of the range of water users within the SFPUC system, few if any have a casual dependence on their water supply. Households are of course critically dependent on water to meet basic needs. For the majority of businesses, water plays a key role in some aspect of their operations—from water-reliant manufacturing processes to restaurant and hospitality services. Even those that have no direct need for water, such as retail outlets, are dependent on water for things like fire protection and sanitation. Similarly, high-tech companies, many of which have helped to drive the regional economy, are quite dependent on the secure and reliable provision of high quality water. Two-thirds of the 28 corporations surveyed by the Silicon Valley Manufacturing Group reported that a 30-day water service interruption would shut them down entirely, while over half reported that they would expect multi-million dollar economic losses. In San Francisco, businesses and institutions account for a full 30% of water demand; in the larger BAWUA area, the figure is 25%. The centrality of water to the SFPUC area economy comes as no surprise. What may be surprising, however, is the vulnerability of the SFPUC system to the type of catastrophic interruption caused by a major earthquake—an event that is considered likely in the next 30 years (Bay Area Economic Forum 2002).

As outlined above, the Hetch Hetchy water system supplies 85% of all SFPUC customers. This same system crosses at least five active earthquake faults. The numerous facilities located near these fault lines are highly vulnerable to failure in the event of a major earthquake. The primary components that make up the Hetch Hetchy system—the O'Shaughnessy Dam, its reservoir, pipeline, tunnel, and pump station—are all aging, with most between 60 and 80 years old. Despite this evident vulnerability, the existing facilities have not been upgraded to satisfy current seismic standards. In the event of a major earthquake or other catastrophy, the water supply system would most likely suffer significant failure, cutting off vast numbers of customers for between 10 and 30 days, and in some cases, as many as 60 days (Bay Area Economic Forum 2002).

Economic impacts of such an event would be enormous. Assuming a major earthquake on the San Andreas Fault (though several other likely scenarios exist), the losses have been estimated to be at least $28.7 billion. Both households and businesses would be left without water and unable to function. Moreover, limited supply would impede firefighting activities, thus increasing property losses. Combined, these types of losses would make it more difficult for the Bay Area economy to rebound and could spell irreversible long-term economic damage. An estimate of these losses has been developed, as shown in Table C.5. Clearly these broader social and economic impacts outweigh the financial burden of the $3.6 billion CIP (Bay Area Economic Forum 2002).

The economic and social implications of supply disruption are clearly enormous and potentially outweigh the cost of inaction—the CIP's total cost is $3.6 billion, and further, it includes projects that address more than just seismic vulnerability, as discussed below. Until recently, however, the value of reliable water supply during times of crisis had not been adequately assessed. Fortunately, this is changing, and the realization that sustained water supplies will be critical to the Bay Area's ability to cope with and recover from a major earthquake or other disaster is prompting concerted action and new investment.

Table C.5
Projected losses from water service interruption
(by type of seismic event, in billions of dollars)

	San Andreas Fault (7.9 event)	Hayward Fault (7.1 event)
Residential losses	$3.8	$1.5
Business losses		
Manufacturing	$4.35	$3.45
Wholesale/retail	$7.70	$5.60
Professional/scientific/technical	$1.60	$0.63
Lodging/food services	$0.54	$0.20
Total business losses	$14.2	$9.9
Fire damage (from inadequate water supplies)	$10.7	$5.8
Total estimated losses	$28.7	$17.2

Source: Based on Bay Area Economic Forum 2002.

Two examples of CIP projects that directly address the reliability of water delivery include:

- *Irvington Tunnel alternatives*: A new facility will be constructed to provide a backup for the Irvington Tunnel. The tunnel conveys Tuolumne River and Alameda watershed water to the Bay Division Pipelines and is key to water deliveries to customers across and around San Francisco Bay, the peninsula, and the city. The tunnel has several unreliable characteristics: it has not been inspected or maintained for over 30 years because it cannot be taken out of service, it has no operational alternative, its location between two faults is seismically unstable, and, because it conveys 85% of the SFPUC water supply, in the event of a tunnel failure that supply would be isolated from customers.
- *Summit Reservoir rehabilitation and seismic upgrade*: The Summit Reservoir, which stores 14 million gallons of water and supplies drinking water and fire protection to residents of the city's Twin Peaks area, will be rehabilitated and seismically upgraded to achieve current seismic standards and to meet all state health requirements.

The infrastructure upgrades in the CIP are clearly of great social value to the region, and in the event of a major seismic espisode, would provide value that is almost 10 times more than the costs to the SFPUC.

Reliability of Water Supply

Current and growing supply gap. With the Sierra Nevadas supplying abundant water into the Hetch Hetchy system and its series of reservoirs apparently ensuring supply, the SFPUC system was long considered a highly reliable source of water. Then, in 1986, a drought began that lasted six years and imposed significant strain on the system. Before it was over, reservoirs had been seriously depleted, a range of water services were impeded, and "extraordinary" water conservation and rationing measures had been implemented to reduce demand. Despite such efforts, the system had come close to running out of water, revealing its vulnerability to severe and prolonged drought (SFPUC and BAWUA 2000).

This experience served to reveal the higher value of water in times of drought, to which the SFPUC and its customers responded with lasting conservation measures. Indeed, the reductions achieved in per capita water use during the drought proved to be 30% below 1986 demand, and—as important—have proven to be lasting cuts. Users and suppliers alike recognized that the SFPUC's supply was inadequate to deal with a repeat of the six-year drought. Under conditions of climate change, where more frequent and intense droughts have been projected for the San Francisco region, the shortfall would only be more extreme and damaging. A clear need was seen for not only sustained and improved conservation measures but also significantly improved reliability of supply.

In the case of water supply, reliability is defined by the "magnitude and frequency of delivery deficiencies in dry years" (SFPUC and BAWUA 2000, p. 5). Currently, the SFPUC anticipates deficits of 10 to 15% roughly once every 10 years, during dry spells. According to the SFPUC's Water Supply Master Plan, "even with no increase in demand, unless the SFPUC system is improved, multiple consecutive years of deficiency can be expected" (SFPUC and BAWUA 2000). Because demand is likely to increase in coming years, the magnitude and frequency of deficiencies can be expected to follow suit.

The challenge facing the SFPUC is twofold: first to increase firm supply (defined as the level of water delivery that can be sustained over historical hydrologic conditions) to meet current demand and second to meet projected demand. Within the SFPUC system, firm supply is 239 mgd, averaged annually. However, the SFPUC current system supply is 260 mgd, suggesting three things: first, under conditions *worse* than the 1986-1992 drought, the system will not be able to maintain its current coverage of demands; at the same time, during dry (not necessarily drought) years, the system would not be able to meet its current 260 mgd level; and finally, growing demands will only exacerbate these shortfalls.

Estimates of future demand suggest that in 2030, dry year demand will rise to 303 mgd, and in 2050 to 310 mgd (SFPUC and BAWUA 2000). This suggests that the SFPUC will need to increase its firm supply by 71 mgd by 2050; of this amount, 21 mgd is needed now, while the remaining 50 mgd serves to fill the projected gap in future demand. It is this gap (71 mgd or 667,000 AF) that the SFPUC's April 2000 Water Supply Master Plan aimed to respond to by crafting and weighing a series of alternative activities, a number of which were ultimately included in the CIP, as discussed below.

CIP projects that directly address the reliability of water supply include (SFPUC 2002):

- *Calaveras Dam replacement*: To meet future demands, the SFPUC system needs to increase available water supply by 71 mgd through water transfers or additional storage. The Calaveras Dam currently collects and stores 97,000 AF of Alameda County watershed supplies, which are then treated and delivered to SFPUC regional water system customers. The dam replacement project will evaluate, plan, and then construct a new dam 200 feet higher than the existing dam to increase terminal storage and ensure seismic reliability.
- *New Northwest Reservoir*: The northeastern zone of San Francisco is susceptible to water supply problems because the area does not have a dependable storage reservoir. The new reservoir will provide water for fire protection and will ensure a pressurized supply of water to those neighborhoods that are currently served from other zones.

- *Recycled water project*: The SFPUC will implement recommendations of Recycled Water Master Plan, which would supply an additional 7 mgd, primarily for irrigation use.

High Quality Water

The main water quality concerns for the SFPUC system include threats of contamination and concerns related to the age of the system. The age of certain components in the SFPUC system presents potential threats to water quality, because infrastructure is vulnerable to failure and treatment plants are in need of upgrades. Additionally, the SFPUC system must continually address more stringent drinking water standards by improving its water treatment and storage facilities.

Routine activities to ensure high quality drinking water are part of the standard operation of the SFPUC system, and the costs of these activities are reflected in the system's water rates. Additionally, SFPUC implements water quality programs that are funded through the system's general operating costs, such as lead minimization outreach and an upcoming shift in water disinfectant from chlorine to chloramines (providing long-term health benefits).

The CIP also addresses water quality, including (SFPUC 2002):

- *Sunol Valley Water Treatment Plant improvements*: The Sunol Valley Water Treatment Plant is in need of significant improvements to ensure reliable service during periods of high demand. Both the operational and storage capacity of the plant will be increased through CIP projects, including construction of a new reservoir and the addition of new filters and basins.
- *Sunset Zone circulation improvement project*: The project will reduce the number of outlying zones dependent on Sunset Zone water by constructing new pipelines to provide alternative water supply to those outlying zones. This is necessary because the Sunset Reservoir water level required to supply water to all zones currently dependent on the source reduces circulation and causes elevated formation of trihalomethanes (THMs), a regulated disinfectant byproduct.

Affirmative Action Goals

SFPUC is committed to affirmative action, in particular supporting women and minority owned businesses in awarding contracts to implement the CIP.

Key Environmental Values in the SFPUC System

In addition to concerns related to consumptive uses, stakeholders in the region have also expressed values associated with nonconsumptive instream uses of the water in the SFPUC system, such as support of riparian species and habitats.

In the SFPUC system, projects such as acquisition and maintenance of large tracts of watershed lands and water recycling demonstrate the Commission's recognition of instream values. The CIP has one watershed and environmental project. This project is the enhancement of Alameda Creek Fishery, and is one element of a larger program to restore Alameda Creek, creating opportunities for steelhead trout and Chinook salmon passage and migration. To implement this program, a consortium of 11 local, state, and federal agencies, including the SFPUC, are funding and completing restoration projects (Bacher 2003). The SFPUC CIP Alameda Creek Fishery enhancement project will release water from Calaveras Dam and recapture the water in the upper Sunol Valley for establishing and enhancing fish populations. The project addresses both of the objectives of environmental values in the SFPUC system—restoring Chinook salmon and identifying and protecting wildlife corridors.

Restore Chinook Salmon

Over 100 species listed as federal endangered and threatened species are in the Bay Area, including the Chinook salmon, whose condition is closely related to the SFPUC water system (NRDC 2003). The salmon migrate up the Tuolumne River in the spring and fall. Shifts in land use, which eliminated areas of riparian habitat, and the 1970 construction of the New Don Pedro Dam, which altered streamflow, have contributed to a decline in the population of Chinook salmon. Following an evaluation of the decline in salmon population and available habitat and the other changes in the river channel's predevelopment state, the Federal Energy Regulatory Commission increased minimum streamflow requirements and the Tuolumne River Technical Advisory Committee researched river restoration planning and activities (McBain and Trush 1999).

Protect Environmental Value of Watersheds

Watershed management plans can be used to incorporate environmental concerns into watershed decision-making activities. Recently, the SFPUC published watershed management plans for the Alameda and Peninsula watersheds. The intention of the management plans is "to provide a policy framework for the SFPUC to make consistent decisions about the activities, practices, and procedures that are appropriate on SFPUC watershed lands." In addition to this broad approach, the plans also outline specific goals, activities, and actions for each watershed, including goals related to nonconsumptive uses. The environmental priorities for the watersheds include preserving and enhancing the ecological and cultural resources. One example to achieve these priorities is the identification and protection of wildlife corridors through management actions such as determining "primary wildlife movement corridors such as riparian corridors;" accommodating wildlife movement "when designing fencing, culverts, stream crossings, and underpasses;" identifying and restoring stream segments with excessive bank erosion; and maintaining "access for fish species of concern from reservoirs to upstream spawning grounds" (EDAW 2001).

Summary of Values in a TBL Framework

The values described above are summarized in Table C.6. The CIP clearly has implications beyond the financial value of water.

Table C.6
Selected CIP projects in a triple bottom line framework: The SFPUC "Report Card"

Value category	Value objective	Rating	Selected capital improvement program projects
Social	Reliability of water delivery	++	Irvington Tunnel Alternatives: A new facility will be constructed to provide a backup for the Irvington Tunnel.
		++	Summit Reservoir Rehabilitation and Seismic Upgrade: The reservoir will be rehabilitated and seismically upgraded to achieve current seismic standards and to meet all state health requirements.
	Reliability of water supply	++	Calaveras Dam Replacement: The dam replacement project will evaluate, plan, and then construct a new dam that will have increased terminal storage and will be seismically reliable.
		++	New Northwest Reservoir: A new reservoir will be constructed to alleviate the water supply problems of the northeastern zone of San Francisco.
		++	Recycled water project: The SFPUC will implement recommendations of the Recycled Water Master Plan.
	High quality water	++	Sunol Valley Water Treatment Plant Improvements: The operational and storage capacity of the plant will be increased through construction of a new reservoir and the addition of new filters and basins.
		++	Sunset Zone Circulation Improvement: The project will reduce the number of outlying zones dependent on Sunset Zone water, thus increasing circulation and reducing water contamination in the Sunset Reservoir.
Environmental	Restore Chinook salmon	+	Alameda Creek Fishery Enhancement: Project will release water from Calavaras Dam and recapture the water in the upper Sunol Valley to establish and enhance fish populations.
	Protection of environmental value of watersheds	+	
Economic	Efficient capital expenditure program	++	Efforts are being made to ensure that the projects of the CIP are being planned and implemented in an efficient manner. Efficiency achievements are currently unknown, since CIP is in its first year of implementation.
	Provide cost effective products and services	++	The CIP was influential in determining the need for an SFPUC-wide performance assessment program, which began in July 2003.

Key: + represents positive improvement.
++ represents very positive improvement.

LESSONS AND CONCLUSIONS

Several important lessons can be drawn from the review of the TBL approach more broadly, as well as the potential application of TBL to the San Francisco PUC as a case study on the value of water.

- Competing demands for scarce supplies, more stringent regulatory requirements, and growing customer quality and security expectations are creating increasing pressures on water utilities. To effectively address these multiple pressures and carry out their primary mission of supplying high quality water to sustain their community's health and economic well-being, water utilities could benefit from an explicit recognition of the multiple values of water and their contribution to these values.
- The TBL approach is a useful framework for illustrating the multidimensional role water plays in society. The TBL helps reveal and elevate the noneconomic, social, and environmental values associated with the provision of water services. Calculation of the full range of these social and environmental values is beyond the conventional value calculations of most water utilities. Use of a TBL approach can help ensure that relevant sustainable development considerations have been internalized in an organization's day-to-day functions and decision-making practices. The TBL can offer a framework for carrying out a performance assessment that helps identify and track key economic, social, and environmental objectives, not only for the utility, but for the public at large. This was demonstrated by the SFPUC using a referendum to have society weigh in on the social value of reliable water supply.
- As with many other entities, water utilities are implicitly moving toward a TBL approach. This movement in the SFPUC system is being driven by both the utility itself and by other stakeholders, including a concerned public. This is clearly demonstrated in the SFPUC CIP, particularly passage of the local bond, which was motivated primarily by public recognition of the social value of water supply. The SFPUC developed the multibillion dollar CIP to address seismic vulnerability and supply reliability problems and to ensure ongoing benefits of the social value of water, which has been estimated at almost an order of magnitude higher—in the tens of billions of dollars.
- The social and environmental value of water is often far greater than its immediate and most commonly recognized financial value based on water sales revenue. It is therefore imperative that utilities attempt to identify the social and environmental values of water in their system (quantitatively or qualitatively) so that their planning efforts and decision-making can reflect those values and ensure their continuation.

APPENDIX D
IPSWICH RIVER WATERSHED: A CASE STUDY FOR THE VALUE OF WATER

INTRODUCTION

The Ipswich River watershed is located in northeastern Massachusetts, comprising part or all of 22 towns and an area of 155 square miles. Historically, the river was the basis for economic development throughout the region, providing water, sustenance, power and transportation to early settlers and their descendants. Its abundant fisheries were economically important for centuries and the extensive clam beds in the estuary continue to be important today to the regional economy (IRWA 2003b, p. 11, par. 65). The watershed supplies drinking water to 330,000 people and thousands of businesses. The Ipswich has been highly stressed for years and was recently classified as America's third most endangered river by American Rivers in this conservation organization's annual report on rivers with acute threats. This designation is due primarily to seasonal extreme low-flow conditions caused by over-pumping groundwater for municipal water supply as well as wastewater transfers out of the basin. Though the region receives an average of more than 40 inches of rain annually, over the past decade portions of the river in the upper watershed have run completely dry for part of the summer each year, and low flows have also affected the lower watershed (Horsley & Witten 2002, p. 3-2).

The Ipswich River watershed provides an excellent example of a watershed experiencing land development and significant water use increases without regard either to sustainability or to the environmental consequences of the water management practices and decisions in the region.[1] The almost exclusive focus on meeting the growing consumptive demand for water, while ignoring the other functions and values of the river, has led to a crisis of no flow in large sections of the river during summer months, with serious consequences for the ecosystem, recreation, and shellfishing.

According to Kerry Mackin, Executive Director of the Ipswich River Watershed Association (IRWA), a key underlying factor is that permit holders pay nothing to withdraw water from the Ipswich basin. In response to these conditions, the IRWA and others filed a lawsuit in April of 2003 in Massachusetts Superior Court against the Massachusetts Department of Environmental Protection (DEP), alleging that the Department's issuance and administration of permits for water withdrawals violated the state's Water Management Act (G.L. c. 21G, § 1 et seq) and DEP's own regulations, causing "drastically reduced stream flows in the Ipswich River resulting in fish kills, habitat destruction, and severe ecosystem impairment" (IRWA 2003b, p. 1, par. 1). The relief sought includes: (1) a declaration that current withdrawals exceed the safe yield of the river, and (2) reductions in the permitted withdrawals of municipal suppliers to ensure the river and its resources are protected. This would obviously have profound impacts on local water suppliers and would force MA DEP, permit holders and others to recognize and plan for the multiple functions and values of water in the Ipswich. Some recognition of the broader value of water—beyond municipal supply—could have prevented this crisis. Approximately one month following the filing of the lawsuit, MA DEP completed long-awaited new water permits

---

1. Relative to other fast-growing regions of the country, for example, the Southwest, the Ipswich watershed, and much of the northeastern United States is characterized by mature development and less rapid growth.

for all municipalities that withdraw water from the basin except the towns of Ipswich and Reading (Ipswich and Reading are the only municipal users that have not sought to increase withdrawals over registered volumes). The updated permits "set mandatory water withdrawal restrictions on each community based on streamflow thresholds"—basin communities must implement mandatory water restriction measures when the river's flow level is 0.42 cubic feet per second per square mile ($cfsm^2$). These measures must include at minimum: "limited watering hours; no pool filling; no car washing; and use of hand-held hoses only" (MADEP 2003).

In late 2002 the Ipswich River Watershed Management Plan was completed. The Plan recognizes these broader functions and values and provides a series of options and recommendations to restore flow, improve environmental conditions, and use the river more sustainably (Horsley & Witten 2002). It is within this broader definition of value that we consider the Ipswich here.

This case study first presents the driving force behind the current crisis in the river. We then discuss the key decision makers in the watershed, followed by a description of the competing uses and associated values. The case study concludes with a section on how these values are shaping the future management of the Ipswich.

THE IMPACT OF GROWTH AND DEVELOPMENT ON THE IPSWICH RIVER

Beginning in the early 20th century, the communities of Massachusetts' North Shore (which includes the Ipswich River watershed) became increasingly connected to the urban Boston area through the development of commuter rail lines and highways. Over the years, these communities have become suburbs of metropolitan Boston. The increased access to the region initiated increased levels of development and land subdivision. Growth within the Ipswich watershed has occurred in the residential and the commercial sectors, and has generally been sprawl development—using large amounts of land rather than consolidated growth in dense developments. For example, in the town of Beverly, the population did not grow over a twenty-year period, yet significant open space was lost to development (750 acres, or 5% of the town acreage over the 20 years) (Harvard University Graduate School of Design 1999). Zoning regulations in the communities of the watershed have supported these development patterns, with limited or no regard for water resources protection.

These development patterns in the region have resulted in an increased demand on water resources, and have affected the overall hydrology. The patterns and characteristics of new developments result in greater water use than older developments. Many of the newer residential developments have larger lot sizes, and without requirements restricting water use, leading to increased summer water use for landscaping. In addition, some newer homes have swimming pools. In general, new developments "consume 50 percent more water than older ones; and [some suburban] communities use two to three times as much water in summer as during the rest of the year" (Glennon 2002, p. 109). Both residential and commercial growth have decreased the permeability of the watershed: developments and land conversions increase the amount of impermeable surfaces, which results in a decrease in the amount of water that is recharged naturally into the Ipswich system. Land conversion from rural to suburban and urban also results in loss of wildlife habitat.

Current Water Use

As noted above, over-pumping groundwater for municipal water supply and wastewater transfers out of the basin has led to severe seasonal (summer) low-flow periods in the Ipswich River. The average daily municipal withdrawals from the Ipswich during 1999 totaled 30.28 mgd. Authorized municipal withdrawals (approval is given from the state DEP via a permitting process) total 34.66 mgd.[2] Therefore, the water withdrawals are within the overall permitted levels. Although the permitting process is strict in terms of reviewing water supply and environmental concerns, the 1985 Water Management Act grandfathered existing diversions and pumping (now referred to as registered withdrawals), and often grants towns the right to more water than is actually available in the river. Water withdrawal permitting has created a situation within the Ipswich basin where some towns are "water-rich" while others are "water-poor." Some communities with rights to adequate water supplies have lower water rates and fewer water conservation programs. Other towns, with more stresses on their water supplies, have higher water rates. This situation has led to different consumption patterns in the various communities of the watershed (Glennon 2002, p. 107). Low-flow periods arise from the seasonal pattern of water withdrawals—the amount of water pumped in the summer is often two times higher than the annual average. These increased withdrawals, though within authorized levels, stress the groundwater supplies at a time when the natural system's capacity to provide water is the lowest (Horsley & Witten 2002). The situation of low and no flow is experienced most acutely in the upper watershed where groundwater pumping is the predominant source of water, but low flow is also present in the lower watershed sections of the river where towns use surface water sources.

Future Growth

Increased growth in Ipswich Basin communities will result in increased demands on the water supply. One way to calculate the potential additional supply that may be needed in the future is through a build-out analysis. A build-out analysis shows increased growth following complete development of all of a town's appropriately-zoned land. This assumption is a "worst case scenario" as it is extremely unlikely that all appropriately-zoned land in the region will be fully developed. A recent build-out analysis by the Metropolitan Area Planning Council (MAPC) projects that if the towns in the Ipswich watershed were to develop all remaining land zoned for development, an additional 6.1 mgd would be needed. Under the permits issued in May 2003, the DEP authorized water withdrawals totaling 34.66 mgd for the Ipswich watershed through 2009.[3]

Low water levels have implications for a variety of people and organizations, as well as the ecosystem within the Ipswich watershed. For example, low water levels have forced restrictions on municipal water withdrawals in several of the watershed's towns, forced decreases in commercial fishing and shellfishing, limited recreational activities, and altered the natural ecosystem dramatically. These types of consequences could affect the viability and clearly the sustainability of the region for future growth and economic development.

---

2. This includes Lynn, which is authorized to withdraw an average of 5.31 mgd over a 180 day period (December 1-May 31).

3. These permits are a decrease from the 37.92 mgd originally authorized through 2003. The revised permits are under appeal.

KEY DECISION MAKERS

Fourteen communities receive public water supply from sources within the Ipswich River watershed. Local governments in each of these towns play a role in water decisions, both in terms of those agencies directly responsible for water supply, as well as those with authority over land use and development. The following is a list of cities and towns that receive all or part of their water supplies from the watershed:

- Surface water: Beverly, Lynn (partial: 16%),[4] Salem
- Groundwater: Hamilton, Ipswich (partial: 20%), Lynnfield, North Reading, Reading, Topsfield, Wenham, Wilmington
- Surface water supplemented by groundwater: Danvers, Middleton, Peabody (partial: December-May, though the water stored in the reservoirs is used year-round).

Boxford and North Andover are also located in the Ipswich Basin. Boxford does not have a public water supply system and relies entirely on private wells. North Andover receives its water from outside the Basin, except for a small number of private wells.

Several state and federal government agencies also play a role in the decision-making process in the watershed. State agencies with a role include MA Department of Environmental Protection, MA Department of Conservation and Recreation (formerly the Department of Environmental Management), MA Water Resources Commission, and others, while federal agencies include the U.S. EPA and the USGS.

The MWRA, a regional wholesale water and wastewater utility, provides water to 47 communities and wastewater services to 43 communities in eastern and central Massachusetts (61 different communities in all). The Authority provides sewer service and/or back-up or partial water service to several communities within the Ipswich River Watershed. MWRA's role in the Ipswich is important not only because it facilitates the transfer of significant wastewater flows out of the watershed, but also because it is one of the major out-of-basin potential sources to augment local water supply.

Other environmental, economic, and recreational interests include the IRWA, the MA Audubon Society, local fishermen and shell fishermen, swimmers, canoeists and boaters, the Essex County Greenbelt Association, and others.

There are also stakeholders outside the Ipswich Basin that may be impacted by its water management decisions, such as those in the western part of the state in the Quabbin Reservoir watershed that export water to the MWRA system.

---

4. Percentage varies from year to year.

CONSIDERING COSTS AND VALUES OF WATER IN THE IPSWICH RIVER WATERSHED

There are many competing uses of water in the Ipswich basin, including municipal use for residents and businesses, agriculture, recreation such as swimming and boating, and maintenance of in-stream flows to provide ecological services such as habitat for fish and shellfish in the receiving estuary. Except for municipal withdrawals for consumptive use, many of these uses leave water in the river and allow for multiple in-stream uses. These uses can be evaluated and described quantitatively and/or qualitatively in terms of costs and values. In the following two sections, we describe the costs and values of water in the Ipswich basin, with the goal of presenting a full picture of both the costs and values of the uses of water in the watershed.

COSTS OF WATER IN THE IPSWICH RIVER WATERSHED

Costs to the Utilities

Except for private wells in some communities, water in the Ipswich basin is supplied by municipal water utilities. None of the utilities pay for the water they withdraw from either surface or groundwater sources in the watershed. The costs to the utilities are therefore only the standard financial costs associated with extraction, treatment, transport/distribution, infrastructure capital, operation and maintenance, as well as administrative costs. These costs are generally reflected in the water and wastewater rate system used throughout the watershed. Kerry Mackin of the IRWA suggests that charging utilities per volume of water withdrawn (i.e., imposing a charge on the extracted water resource itself) would be an appropriate way to reflect the value of the resource, to encourage more efficient water use, and to raise needed resources to implement conservation and other programs to reduce stress in the Ipswich watershed.

Municipal Charges to Users for Water

Each municipality sets the price it charges to consumers for water (and wastewater services) that it provides. Though the price does not necessarily represent the full direct cost of supplying the water (either because of municipal subsidies or because certain costs such as debt service may not be reflected), we use price here as a lower-bound proxy for supply cost.

The price charged by municipalities withdrawing from the Ipswich basin ranges from a \$19 per six months firm rate for up to 25,000 gallons in Wenham, to \$7.08 per 1,000 gallons for households in North Reading using more than 110 gallons per day.[5] An average household in Ipswich watershed communities paid \$377 for water in 2002, ranging from \$109 in Lynnfield Center to \$683 in North Reading. The per capita annual water cost ranges from \$21.95 in Ipswich (note that Ipswich obtains only 16% of its supplies from Ipswich Basin sources) to \$150.16 in North Reading. Table D.1 presents actual household water use and household costs by municipality. Figure D.1 presents water rates for each municipality in the basin.

---

5. Based on a summary spreadsheet provided by Kerry Mackin, IRWA, May 20, 2003.

Table D.1
Ipswich municipalities: Household water use and costs*

Municipality	Population served	Residential water use (2001, gallons per capita per day)†	Per capita annual water cost	Total annual water cost
Beverly	39,000	65	$52.48	$2,046,720
Danvers/Middleton‡	30,800	65	$88.26	$2,718,316
Hamilton	7,300	65	$62.26	$454,463
Ipswich	13,000	55	$21.95	$285,293
Lynn	80,000	65§	$5.68	$454,002
Lynnfield	3,000	65	$100.00	$300,000
North Reading	12,750	85	$150.16	$1,914,553
Peabody	46,500	65	$51.67	$2,402,797
Reading	22,600	55	$96.53	$2,181,487
Salem	37,926	85	$67.12	$2,545,573
Topsfield	1,650	65	$106.76	$176,158
Wenham	3,405	85	$38.00	$129,390
Wilmington	21,326	65	$81.85	$1,745,560

*Water use adjusted to represent only Ipswich Basin supplies.
†Mackin and Wagner 2002, p. 11-13.
‡The Town of Danvers provides water to the Town of Middleton. Reports filed with the DEP indicate that residential water use in Middleton is significantly higher than residential water use in Danvers.
§Only permitted for Ipswich supplies in the summer months.

In 1999 average daily municipal withdrawals totaled 30.28 mgd. Assuming a price to consumers of $3.35 per 1,000 gallons (a weighted average of the water rates as published in 2002), the minimum direct cost of water to residents and businesses is about $101,438 per day, or more than $37 million per year. This figure represents the direct payments to water suppliers of water sales in the watershed based on the municipal rate structures described above.

Municipal Charges to Users for Wastewater

The amount of water transferred out of the Ipswich basin each day averaged 23.54 million gallons in 1999. Water transfer out of the basin includes "wastewater that is transported via sewers to treatment plants, and then discharged into Salem Sound, Boston Harbor, or the Atlantic Ocean. The end result is that, of the water pumped from the Ipswich River watershed sources by municipalities, almost 80% is discharged outside the Ipswich River watershed" (Horsley & Witten 2002, p. 4-10). Moreover, some sewers receive stormwater flow and can also transfer that water out of the basin, resulting in less groundwater recharge. The only municipality in the basin that operates an in-basin wastewater treatment system is Ipswich.

The communities that have municipal sewer systems charge for the provision of wastewater services. The price charged by municipalities in the basin for wastewater services ranges from $3.04 per 1,000 gallons in Wilmington to $7.14 per 1,000 gallons in Reading. Figure D.2 presents sewer rates for each municipality in the basin that is sewered (Hamilton, part of Ipswich, Lynnfield, Middleton, North Reading, Topsfield, and Wenham all have on-side septic systems).

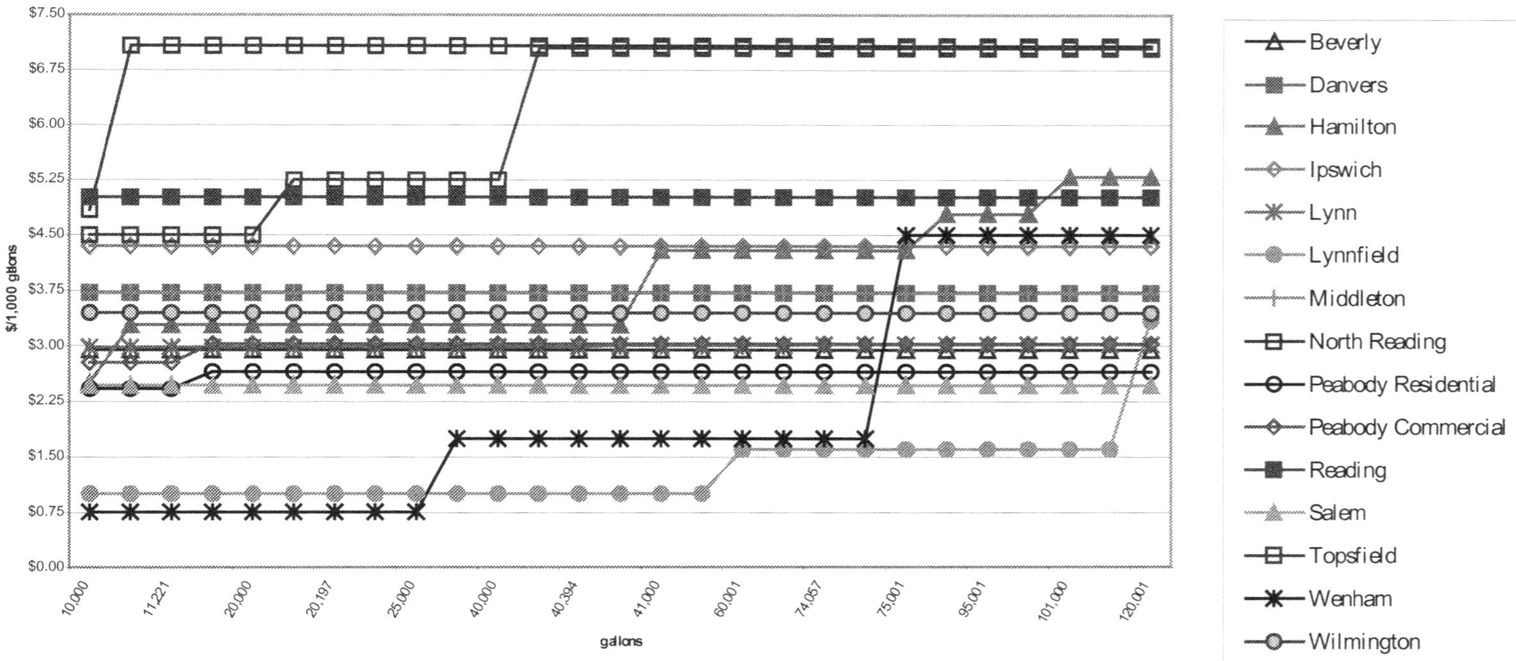

Source: Based on Tighe & Bond (2002) Water Rate Survey.
Massachusetts communities. Note: Hamilton's water rates increase further for water consumption above the 120,000-gallon maximum. The rates are $5.79 per 1,000 gallons (401,000 to 800,000 gal); $6.29 (801,000 to 1,000,000 gal); and $6.79 (1,000,000 gal and over). North Reading's rates are based on quarterly billing

Figure D.1 Ipswich River watershed communities: Water rates

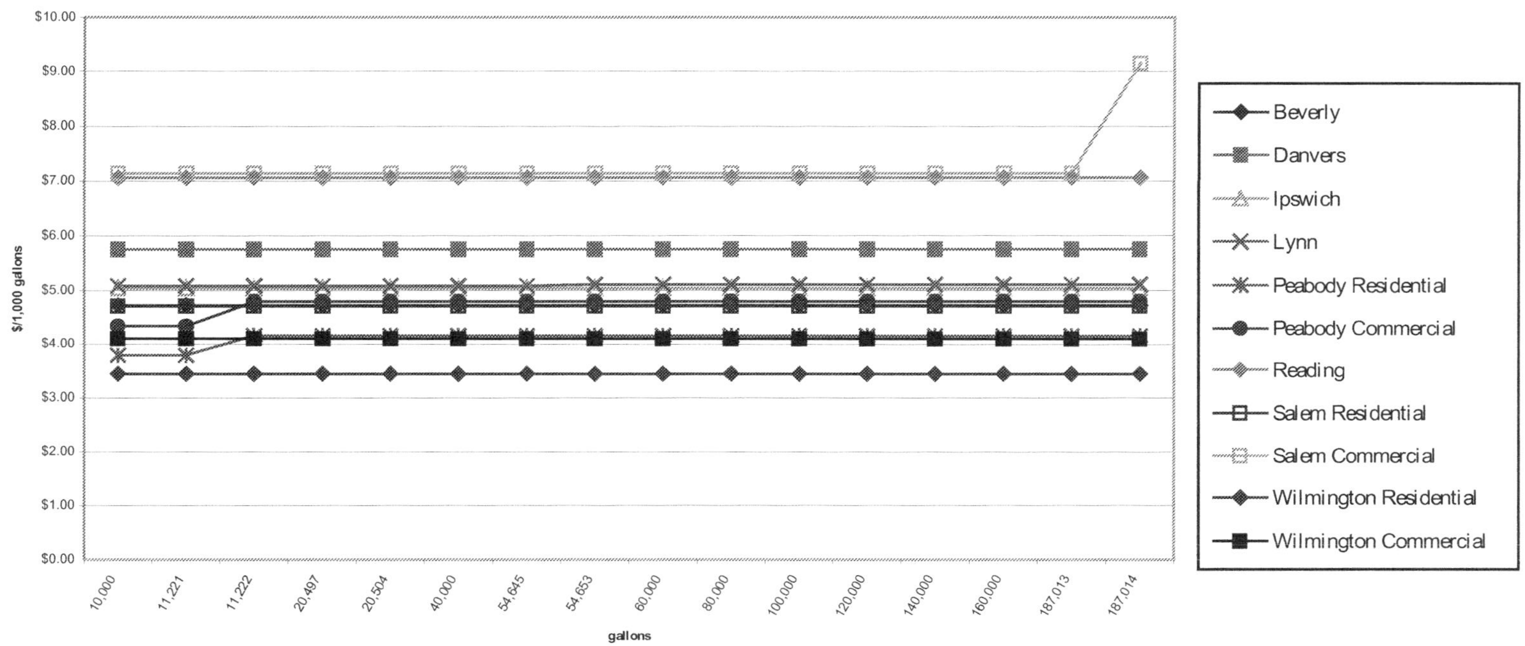

Source: Based on Tighe & Bond (2002) Sewer Rate Survey. Massachusetts communities.
Note: Lynn's sewer rates increase slightly to $5.12 for wastewater consumption above 2,103,866 gallons.

Figure D.2 Ipswich River watershed communities: Sewer rates

In 1999 average daily municipal withdrawals for sewered towns totaled 27.89 mgd. Assuming a price to consumers of $4.85 per 1,000 gallons (a weighted average of the published 2002 wastewater rates), the minimum direct cost of wastewater to residents and businesses is about $135,267 per day, or more than $49.4 million per year.

Other Costs

The current management of water in the Ipswich, which results in extreme no-flow and low-flow conditions during summer months, clearly has implications for the value of water for instream uses such as shellfishing, fishing, swimming, and other recreation and for the loss of value for properties adjacent to the river. The loss of these values from the way water is managed is in effect an unrecognized cost to society in the form of "externalities." The following section presents estimates for these externalities, and discusses these values in more detail.

VALUES OF WATER IN THE IPSWICH RIVER WATERSHED

Direct Value of Municipal Consumptive Uses

As discussed above, the direct value to users as measured by municipal revenue for water and wastewater in the Ipswich basin totals approximately $86.4 million per year ($236,705 per day). This amount should be considered the lower bound of the full value of water to users. Not included in this amount is the customers' willingness to pay for their water and wastewater, which is higher than the charge rates and which demonstrates a more complete valuation of water and wastewater services by residential and commercial users in the watershed.

Similarly, the undervaluation of wastewater is evident by the present condition in which almost 80% of the basin's water is discharged outside the Ipswich River watershed. Treated wastewater is viewed as something to dispose of, rather than a valued resource critically important to the maintenance of in-stream flows and aquifer recharge. Although wastewater does not appear to be valued in terms of its relationship to sustaining flow in the watershed, the provision of wastewater services are valued, as shown through the wastewater charge rates described above.

Shellfish

The shellfish industry in the watershed has been threatened by point and nonpoint source pollution in the watershed, exacerbated by low flows, which has caused closures of wells, shellfish beds, and beaches. About one-third of the shellfish beds are closed at all times and the remaining beds are closed following large rainstorms. These closures severely affect the area's shellfishing industry. Commercial shellfish harvest value between 1996 and 2001 averaged almost $1 million per year (high: $1,222,000 in 1999, low: $522,000 in 2000, most recent [2001]: $666,000), while the value of recreational shellfish harvests during this period averaged about $110,000 per year (with $134,000 in 2001) (Kent 2003).

Fish Habitat

Low flow conditions in conjunction with the 30 dams on the river's main stem and tributaries have significantly altered the aquatic ecosystem and the river corridor area. The river's ecosystem is currently similar to that of a pond because of the low flow, warmer water temperatures, and dams. This shift has reduced the biodiversity of the river, and many of the original species (e.g., trout) are no longer present in the river. Values associated with a healthy aquatic ecosystem are related to robust habitats, fishing opportunities, and improved water quality. While the value of maintaining flow for healthy fish populations is estimated below, other ecosystem values have not been quantitatively assessed for the Ipswich. Note that the values discussed here do not explicitly quantify the value of fish kills experienced in recent years due to the extreme low flows in certain areas.

In 2002 the IRWA conducted a resource economics study of the Ipswich River, including a "willingness to pay" contingent valuation survey. The goal of the survey was to gain information about how residents of the watershed communities use and value the Ipswich River. The written survey included background information about the Ipswich River and its role as a public drinking water source, its role in supporting ecological values and other uses, and the threats it faces. More than 2,800 surveys were mailed to a random list of households in 14 communities, and about 500 were sent to IRWA members in 22 communities. Of these, 218 (less than 8%) of the random surveys were completed and returned and 154 (about 30%) of the members completed the surveys.

Respondents were asked the amount of additional funds their household would be willing to pay each year if that money were dedicated to restoring flows needed to sustain a healthy fish population in the Ipswich River. Response options were $0 to $100 ($10 increments) or other. Of those surveyed, 86% of random respondents and 91% of member respondents answered the question. About 77% of random respondents and 96% of member respondents were willing to pay a non-zero amount. While the amount households were willing to pay ranged from zero to $100 or more ($500 in the case of one IRWA member), on average random respondents were willing to pay $31 per household per year to "restore flows needed to sustain healthy fish populations" while IRWA members were willing to pay $55 (excluding the $500 outlier). Because survey respondents were self-selected, the values given may be inflated and not representative of the opinions of the general population in the watershed. In addition, contingent valuation surveys can generate overestimates of WTP because of the hypothetical nature of the payment. On the other hand, these potential biases may be counteracted by the phrasing of the willingness to pay question: respondents were only asked about their willingness to pay for restoring flows needed to sustain a healthy fish population, not about their willingness to pay for other types of identified values. For example, another identified value is restoring flows needed to support recreational opportunities such as canoeing and swimming. The detailed results of the survey's willingness to pay question are presented in Figure D.3.

If we assume these results are representative of all households in the watershed, applying the average random respondent's willingness to pay of $31 per year to the roughly 100,000 households results in a total value of about $3.1 million per year to restore flows needed to sustain healthy fish populations.

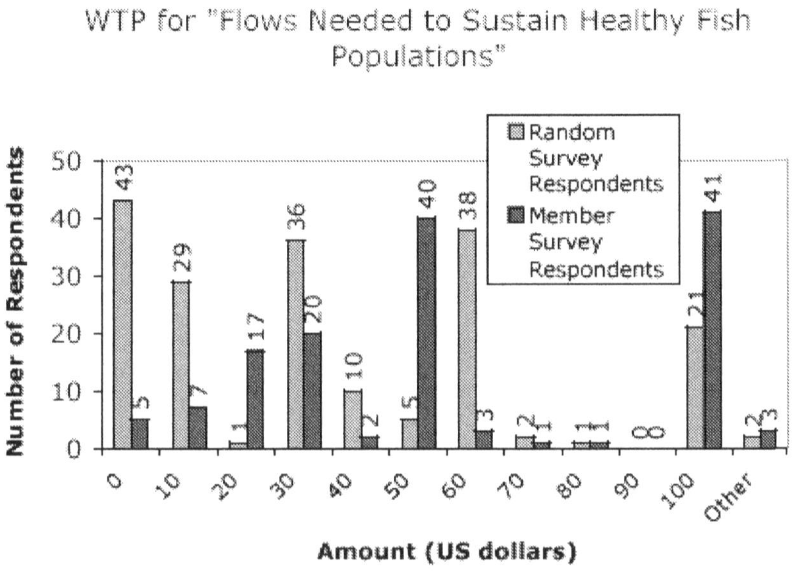

Source: IRWA 2003b, p. 17.

Figure D.3 Willingness to pay for "flows needed to sustain healthy fish populations" (dollars per household per year)

Recreation and Instream Uses

The Ipswich River has been a major recreational resource in eastern Massachusetts for many years. As described in the IWRA lawsuit: "The river is essential to the region's growing eco-tourism industry, attracting canoeists, kayakers, bird watchers, naturalists, anglers, and hunters, as well as hikers, bicyclists, cross-country skiers and others who use the river and its environs. Beaches at the mouth of the river attract residents of Eastern Massachusetts and tourists from around the world. Portions of the watershed are part of the Essex National Heritage Area" (IRWA et al. 2003). The river was historically one of New England's leading trout streams; however the native trout population is so diminished that in recent years the Massachusetts Department of Fisheries, Wildlife and Environmental Law Enforcement has stocked the river with trout to provide a supply of fish. In the tidal and estuary portions of the watershed, shellfishing, fishing, and swimming are popular activities. However, low flow in portions of the river and pollution in tidal areas has caused a decline in the potential for these recreational activities, resulting in economic losses for the region.

The WTP survey described above included a section asking respondents to indicate the importance they place on different uses of water, questions on the respondents' outdoor recreational use of the watershed, and a set of questions on their willingness to pay for healthy fish habitat. The rank given to each of the nine uses of the Ipswich River by both the random and Watershed Association member respondents is provided in Table D.2. The ranking system comprised a range from 1 (very important) to 5 (not important).

Table D.2
Rank of watershed uses by respondents

Ordinal rank	Random respondents	Member respondents
1	Supporting a healthy environment (average rank 1.4)	Tie: Supporting a healthy environment (1.2) and providing wildlife habitat (1.2)
2	Providing drinking water (1.5)	
3	Providing wildlife habitat (1.6)	Tie: Providing drinking water (1.4) and providing scenic views, rural character, open space (1.4)
4	Providing scenic views, rural character, open space (1.8)	
5	Providing recreation (2.1)	Providing recreation (1.7)
6	Supporting fish/shellfish industry (2.2)	Supporting fish/shellfish industry (2.1)
7	Providing water for agriculture (2.3)	Providing water for agriculture (2.2)
8	Providing water for business/industry (2.9)	Providing water for business/industry (3.1)
9	Providing water for lawns (3.6)	Providing water for lawns (4.2)

The results show strong similarities between the rankings by random survey respondents and those by Watershed Association members, particularly in ranking the essential and historical human uses of the watershed (drinking water, fish/shellfish industry, agriculture), while members were more likely to rank the intrinsic uses of water (rural character, habitat, recreation) higher than random respondents. A noteworthy observation is that supporting a healthy environment received the highest ranking from both random respondents and IRWA members, even slightly higher than drinking water, yet these environmental benefits are rarely quantified or considered in water management decision-making. Other high ranking responses from both groups were providing wildlife habitat, and providing scenic views and rural character, followed by providing recreation, supporting the fish/shellfish industry, and providing water for agriculture. The lowest ranking uses according to both sets of respondents were providing water for business/industry, and providing water for lawns.[6]

Popular recreational activities that survey respondents participate in within the watershed include: walking/hiking (62% random respondents, 85% member respondents), canoeing/ kayaking (38% random, 71% member), and bird watching/nature observations (30% random, 58% member).

Property Values

In 2001, the average single-family home property value for communities in the watershed was about $265,000 (low: Lynn, $139,200; high: Wenham, $385,500) (Boston Globe 2002). Property values of land close to rivers and other water bodies, including those in the Ipswich watershed, are generally higher than surrounding land, because such parcels are "attractive for residential and commercial development" (Horsley & Witten 2002, p. 2-1). The decline in river flows and healthy river habitats has prompted unease among some residents of the watershed: "Long-time residents of the upper Ipswich watershed express concern that stagnant conditions or no-flows of the river affect the desirability of riverfront properties, and may affect adversely

6. Although business/industry and lawn watering ranked low among Ipswich watershed respondents, these have been major drivers of efforts in other Massachusetts regions, such as Brockton and Stoughton, to augment supplies.

property values. Fish kills, which have occurred every two years recently, are not only a huge environmental loss, but create visual and odor problems" (Horsley & Witten 2002, p. 3-9). The low flow conditions in the watershed present similar potential impacts to certain commercial property values (those in close proximity to the river) in the region.

Intrinsic Values

A final component of the full value of water in the Ipswich basin is intrinsic values—the significant but difficult-to-quantify values associated with aesthetics, regional character, and biodiversity. These values were expressed through the results of the willingness to pay survey, described above, and are also shown in the mission statement of the IRWA, which is "dedicated to protecting water quality and quantity, fish and wildlife habitat, recreational opportunities and scenic values" of the watershed (IRWA 2003a).

THE VALUE OF WATER AS A FACTOR IN DECISION-MAKING IN THE WATERSHED

Management Options

Most stakeholders and decision-makers within the Ipswich watershed now broadly acknowledge the problem of low flows and its implications. There is a growing recognition that in order to provide high quality water in the future, immediate actions must be taken. The low flow/no flow periods represent an unsustainable situation where more water is demanded than can be supplied. According to the Ipswich River Watershed Management Plan, which cites USGS data, an additional 14.4 mgd are required in the entire watershed during the low-flow season to restore a more natural flow to the river, estimated at 39.5 mgd, one that would support a healthy aquatic ecosystem and provide high quality water for municipal uses. The streamflow deficits are not uniform throughout the watershed. In the upper watershed, specifically the headwater communities of North Reading and Wilmington, the low/no-flow conditions are more acute as a result of groundwater pumping for water supplies. Recognizing this condition, the Watershed Management Plan presented the streamflow deficits by sub-watershed (Table D.3).

There are a range of options for improving management of the system and restoring flow to the Ipswich. Each of these options has certain costs and benefits associated with it, though these have not been calculated for most options. Part of the challenge is that non-consumptive

Table D.3
Streamflow deficits by sub-watershed

Streamflow	Headwaters	Upper watershed (cumulative)	Entire watershed (cumulative)
Healthy summer streamflows	5.69 mgd	13.9 mgd	39.5 mgd
Average summer monthly medians (July-September)	n/a	4.9 mgd	25.1 mgd
Estimated deficit	5.0 mgd	9.0 mgd	14.4 mgd

Measured at the following streamflow gages: Reach 8 (headwaters), South Middleton (upper watershed), Ipswich (entire watershed).

use values have not been rigorously quantified. Acknowledging the different uses and estimating the associated values for each use not only broadens understanding and appreciation of the water resources of the Ipswich, it also informs decision-makers on how best to manage the resource. Below, we present the six management options identified in the Ipswich River Watershed Management Plan, focusing primarily on the costs and benefits of the first three—water conservation, purchases from outside the basin, and reducing exports of wastewater.

Option 1: Water Conservation

Improve water conservation measures and implement limits on withdrawals of water, especially during the critical flow periods. Currently, the demand on groundwater supplies is highest when the groundwater is at its lowest level, in the summer months. The demand in the summer months increases because of increased residential lawn watering and commercial irrigation such as golf course watering. Year-round and seasonal water conservation measures would increase the supply of groundwater. One method of improving water conservation is to implement rate structures that reward the low-volume customers, such as increasing block rates and summer surcharges for increased use above winter levels. For seasonal rates to be effective, they must be coupled with more frequent (i.e., monthly) billing based on actual consumption. Currently, half of the communities in the basin charge a flat rate per gallon, regardless of the volume used, and none have implemented seasonal surcharges with monthly billing (note that in high-income areas, water demand may be less price elastic, and other measures, such as restrictive ordinances, may be necessary).

Costs. The costs associated with this option include costs of developing and implementing a water conservation program (including education, retrofit or technology improvement projects, and technical assistance). The town of Reading, located in the headwaters of the Ipswich Watershed, has initiated a four-year $1 million water conservation program to achieve an estimated annual water use reduction of 0.19 mgd. The town estimates that the water conservation program will cost $10 per household. Based on estimates of the amount of water conserved via the program, the cost for conserving 1,000 gallons in Reading is $3.60 (2003 USD).

Applying the cost per gallon conserved for Reading's water conservation program to the water conservation objectives for the entire watershed (from the Watershed Management Plan), the annual cost for water conservation would be $985,500 in the headwaters, $315,360 in the other upper watershed communities, and $5,794,740 in the lower watershed communities, resulting in a total annual cost to the watershed of $7,095,600 (2003 USD).

Benefits. In Reading, the water conservation activities are estimated to conserve 0.19 mgd[7] and include: residential water audits, a rebate program for water-saving devices, a water conservation retrofit program, retrofits of municipal buildings, general public education and outreach, public education and outreach to schools and to large users, a system-wide annual water audit, and leak detection (Camp Dresser & McKee 2003, p. 3-1). The Watershed Management Plan, in its summary of conservation objectives from proposed management strategies, indicates the following demand reductions for water conservation programs (Table D.4):

---

7. Wilmington, the other Ipswich headwaters town, has proposed a water conservation program that would save between 0.58 and 0.84 mgd on an annual average basis (S E A Consultants 2003, p. 421).

Table D.4
Benefits of water conservation objectives (gallons reduced)

	Headwaters	Upper watershed (cumulative)	Entire watershed (cumulative)
Water conservation objectives	0.75 mgd	0.99 mgd	5.4 mgd

Source: Based on Horsley & Witten 2002, p. 6-19.

The water conservation objectives are based on a conservative estimate of a 15% reduction in water demand. This estimate does not account for the significant increases in consumption in Ipswich communities during the summer months (communities use 25%-160% more water in the peak months than the winter period) (Horsley & Witten 2002, p. 6–4). It is during this period when conservation efforts would likely be even more effective and necessary.

Option 2: Purchase Water From Sources Outside of the Basin

Some communities already draw water from outside the basin (e.g., from the MWRA). In the upper reaches of the river, where the low flow situation is especially severe, the use of streamside wells is a major cause of the decline in flow. As stated in the Watershed Management Plan, "Water withdrawals from the Ipswich River, especially "streamside" groundwater wells, are a major factor causing the low-flow problems of the river. Replacing these sources of water with less damaging supplies is essential to restore the river to health. The first priority is to reduce or eliminate the use of the most damaging wells during low-flow periods" (Horsley & Witten 2002, p. 6–10). There are many issues associated with interbasin water transfers, including the need for regulatory approval,[8] the cost of the transfer (including ecological impacts in the exporting basin), and the implications of increasing demand on the source systems.

Costs. Reading has applied for admission to the MWRA system to supplement its groundwater supplies in the summer months. During these months, Reading's groundwater pumping, though within authorized levels, has resulted in conditions of low and no flow in the Ipswich River. In 1999, the Town initiated a study "to develop a long-term water supply strategy that was less vulnerable to contamination than the existing supply… and [that addressed] the Town's contribution to a reduction of flow in the Ipswich River" (Camp Dresser & Mckee 2003). The study examined several options for alternative water supply. The cost of purchasing water from the MWRA was calculated to be approximately $2.18 per 1,000 gallons.[9] This cost is in addition to the existing municipal water rate [note that this additional cost for purchase of MWRA water would be distributed over all water sales in the Reading system (local and MWRA sources), thereby reducing the increase to municipal rates]. This is an explicit recognition of the

8. Under the Massachusetts Interbasin Transfer Act, communities requesting the transfer of water from an external basin must gain approval from the Massachusetts Water Resources Commission. The transfers that existed prior to the Act were grandfathered in. In order for new transfers to gain approval, the community requesting the transfer must "demonstrate not only that it has a real need but also that it has explored all other alternatives, including conservation. Withdrawals must also ensure reasonable in-stream flow in the river from which the water is diverted" (Glennon 2002, p. 105).

9. Reading's MWRA purchase cost was calculated using the average annual cost of the option (derived from the 20-year NPV value of the purchase option - $9,538,240) and the amount of water purchased (0.6 mgd).

value of maintaining in-stream flows, which if implemented implies expenditures by the town approaching $10 million over twenty years.

Wilmington, the other headwater town in the Ipswich basin, recently completed a draft Comprehensive Water Resource Management Plan and Environmental Impact Report to inventory its current water supply and stormwater and wastewater management systems, to evaluate alternatives for those systems, and as part of its application for admission to the MWRA system. Wilmington currently experiences a deficit in water supply as a result of contaminated wells, and as noted earlier, as a headwater community its groundwater withdrawals are responsible for some of the low flows in the Ipswich basin. As in Reading, the Wilmington report examined the use of MWRA water supplies. The MWRA option was the only option that alone would provide the Town with sufficient water to meet its deficit (currently 0.88 mgd, projected to be as much as 1.5 mgd in 2025) while providing benefits of reduced withdrawals in the Ipswich basin (S E A Consultants 2003, p. 419). In Wilmington, the cost of purchasing water from the MWRA was calculated to be $3.48 per 1,000 gallons (based on an annual purchase of 1.5 mgd). This cost is in addition to the existing municipal water rate. As in Reading, Wilmington's proposed purchase of MWRA water is a clear acknowledgement of the value of maintaining in-stream flows.

The total cost of MWRA water purchases in the communities of Reading and Wilmington is calculated by applying the cost for MWRA water to each of the communities' deficits during the low-flow season identified in the Management Plan (2.4 mgd in Reading and 4 mgd in Wilmington). The total annual cost for water purchase would be $1.97 million in the headwaters (2003 USD) (the annual cost calculation uses the water deficit during the summer months, the only period during the year when water would be purchased, and the water rate).

Benefits. According to Reading's application for admission to the MWRA waterworks system, MWRA water will supplement the town's existing groundwater supplies and allow it to reduce withdrawals from the Ipswich during summer months from 2.2 mgd, to 1.0 mgd, thereby reducing demands on the river during this critical low-flow period.

The Watershed Management Plan is more ambitious and assumes Reading withdrawals from the Ipswich during low-flow periods are eliminated and fully replaced with water purchased from MWRA. The Plan calculates the benefits of proposed management strategies, identifies only the headwater communities of Wilmington and Reading, which are entirely supplied by the Ipswich River via groundwater wells, as those where the benefits of alternative water sources, such as purchasing water from outside the basin, should be achieved. The total proposed benefit for the summer months is 6.4 mgd, which includes 2.4 mgd from Reading and 4 mgd from Wilmington.

Option 3: Reduce Exports of Water Via Sewers

Currently, almost 80% of the water that is withdrawn from the Ipswich river is not returned to the basin because municipalities have sewer systems that transfer the water to treatment facilities out of the basin. If more of the wastewater were treated and discharged within the basin, it could recharge groundwater and possibly surface water resources. However, the process of wastewater treatment and management is difficult and can be costly, and raises important concerns about pollution and threats to water quality. Nonetheless, the development of effective, safe, in-basin wastewater treatment is an important management option to consider for recharging groundwater and ultimately contributing to the restoration of flows in the Ipswich.

Costs. As mentioned earlier, the town of Reading recently completed an application to MWRA for admission to the waterworks system to supplement the Town's existing groundwater supplies during the summer months. As part of this application, the Town was required to complete an Environmental Impact Report (EIR) addressing several relevant issues relating to the proposed purchase, including the issue of wastewater export out of the basin. Reading currently sends its wastewater to the MWRA Deer Island Plant outside the Ipswich watershed. The capital cost of constructing a new wastewater treatment plant in Reading that would enable the use of wastewater for groundwater recharge was estimated at between \$15 and \$30 million, and as asserted in the EIR, would also require the development of a large tract of land (Camp Dresser & McKee 2003, p. 6-7). The Town did not explore this option in great detail, and without further information regarding additional costs such as operations and maintenance, it is difficult to compare this option to other options where costs are expressed in per gallon cost terms.

Benefits. The wastewater treatment facility examined during the development of Reading's application to MWRA would have reclaimed 50 to 100% of the Town's wastewater flow, which was 2.05 mgd in 1999. Similarly, the proposed strategies for reducing wastewater exports outlined in the Watershed Management Plan are estimated to provide 2.0 mgd over 10 years in the headwaters and 4.7 mgd in the entire basin.

Option 4: Improve Stormwater Management

Development in the basin has changed land cover, creating more impervious surface. This prevents infiltration of precipitation and stormwater, and as a result limits recharge to groundwater supplies. Potential improvements to stormwater management include the design and installation of infiltration systems to collect and infiltrate stormwater from impervious surfaces. These systems are now required by Massachusetts state policy for new development, however the policy does not apply to existing development or to single-family homes and small housing developments, the dominant form of development in this region in recent decades. A shift in the policy to include these properties would increase the amount of recharge to groundwater, and this in turn would enhance instream flows.

Option 5: Increase Water Storage

Expanding the current water storage system, (including increased infiltration, wetlands restoration, and reservoir construction or expansion), particularly in the upper basin, would provide more water for use during low flow and drier periods. Diverting more surface water, however, affects the river flow (and consequently the aquatic habitat and composition) and presents economic and environmental challenges related to the size and location of the storage facility. Small-scale storage methods, including rainwater harvesting can also be used to augment water supplies.

Option 6: Improve Land Use Planning

The projected growth in water demand due to sprawl development and population growth may significantly increase the stress on the water system. In order to achieve the goals of increased flow, improved river health, and sufficient supply, it will be necessary to address water resource protection and supply through improved land use planning and zoning, including increased protection of critical watershed lands. Encouraging more-dense development in urban areas and town centers, rather than new subdivisions on undeveloped land, and prioritizing investment of public resources for infrastructure in these more dense areas will protect critical watershed lands, preserve more open space and pervious surfaces, and help constrain water demand growth.

Option 7: Desalination

While not addressed in detail in the Ipswich River Watershed Management Plan, private investors undertook preliminary feasibility investigations to identify the potential for a desalination plant on the North Shore. Presentations by Aquagenics, Inc. (Woburn Mass.) and Hanson Murphy Associates (Brockton, Mass.) indicated that the cost of desalinated water would range from $2 to $6 per thousand gallons.

Alternative Economic Analysis of Management Options

Estimating the value of water for its various uses in the Ipswich River watershed is challenging and complex. One attempt to address this complexity is the Ipswich River Watershed Economic Analysis included as appendix B of the Ipswich River Watershed Management Plan prepared by Horsley & Witten (2002). This document describes the results of a modeling exercise to assess the trade-offs between the economic value of water withdrawals (for residential, commercial, industrial, and agricultural use), versus the economic losses that result from low-flow events. The model for this analysis was developed by Denise Jarvinen, formerly of the Marine Policy Center of Woods Hole Oceanographic Institute, and was applied to various well withdrawal and wastewater return scenarios developed by USGS. The physical model reflects water quantities withdrawn for residential, commercial, industrial, and agricultural consumption, wastewater return flows, and water remaining in the river. Based on factors derived from economic studies of the value of water in other basins, dollar values were calculated for these quantities in the Ipswich.

In the economic modeling, several assumptions were made to simplify the valuation analysis. Among the key assumptions was that reduced withdrawals from the Ipswich are fully replaced by water from out-of-basin sources. If out-of-basin sources were available at the same cost as existing withdrawals, there would be no change to the value of water to consumers and changes in overall value would reflect only the change in value of water left in the river. Of course this assumption is not realistic, as we would expect alternative sources of water to be priced higher than the existing in-basin sources. By comparing changes in the value of water associated with different quantities of water withdrawals with the value of water left flowing in the Ipswich, we can calculate the relative economic gains or losses of alternative water management scenarios.

Using the very conservative estimates of the value of instream flows ($3.61-$12.11 per household per year based on studies from other basins, rather than the WTP survey estimate of $31 per household per year in the Ipswich, just to restore flows for fish habitat), the modeling analysis shows mixed results in terms of the impact of reducing withdrawals on net economic losses. The scenario with no seasonal groundwater withdrawals by Reading and Wilmington is the most cost effective for watershed consumers. Using the upper bound value for instream flows, net economic benefits (the change in the value of instream flow minus the change in consumer value of residential water use) were calculated to be between $1.73 million per year and $1.36 million per year depending on the assumed increase in costs for replacement water (0–10%). Other scenarios show a range of results with net benefits ranging from a negative $2 million (i.e., a net loss) to a positive $1.7 million, depending on assumptions concerning the value of instream flows, the incremental cost of replacement water, and the relative cost of alternative in-basin wastewater treatment. If this modeling analysis were repeated with the higher in-stream values identified in the Ipswich willingness-to-pay survey, probably most (if not all) scenarios studied would have positive net benefit economic results.

Though technically an appendix to the Management Plan, this economic analysis was completed and presented to Ipswich River watershed communities after the main Plan. As a result, to date it has not been directly relied on in specific water management decisions, though according to the Plan's primary author it is serving as an important information resource in evaluating alternative habitat restoration options.

CONCLUSIONS

Land development and growth in water use in the Ipswich River watershed over the last few decades has resulted in extreme low-flow conditions during peak-demand summer months. Contributing to this situation has been the limited value placed on the Ipswich basin's water, and an ineffective permitting system governing withdrawals. Municipalities and other permitees continue to withdraw more than 30 million gallons of water per day for free (i.e., at no charge for the water itself), and projections indicate significant future growth in the demand for water. While the withdrawals have economic value to the water-extracting communities, others suffer economic consequences due to the resulting low-flow conditions. As noted earlier in this report, Kerry Mackin of the IRWA suggests that water suppliers obtaining water from the basin be assessed a per gallon fee. This would place an explicit price on the resource, help encourage conservation, and help fund restoration measures. Requiring communities to pay for water resources as an asset would also help correct the damage caused by the extreme low-flow conditions (i.e., a surcharge on water withdrawals or other means could be a method to hold communities financially accountable for this damage).

Based on hydrological modeling performed by USGS, in order to restore adequate flows withdrawals must be reduced by an estimated 14.4 mgd (during summer months). A series of management options have been identified in the Ipswich River Watershed Management Plan (October 2002) to restore adequate flows in the river that maintain habitat and other ecological and recreational uses. Options include enhanced water conservation, purchases from out-of-basin sources, improved stormwater management, increasing in-basin wastewater treatment, increased water storage, and improved land use planning. To date, no systematic economic analysis has been conducted to assess the net economic impact of the various management options. One economic analysis that applied very conservative (i.e., low) instream use values derived from other basins to the Ipswich found various management alternatives result in a wide range of economic impacts (from $2.1 million in net losses per year to $1.7 million in net benefits), with the results highly dependent on assumptions about in-stream values, the incremental cost of alternative out-of-basin sources, and the relative cost of additional in-basin wastewater treatment.

While conventional views of the value of water often only recognize the direct value for consumptive use as shown through the municipal charges to users for water and wastewater, it is important to recognize the full value of these uses as well as the values of instream uses that provide ecological and habitat services, a range of recreational services, as well as aesthetic values. Thus, in addition to the value of withdrawals, our simplified analysis has also identified values for shellfish and for maintaining adequate flows for fish habitat in the Ipswich River watershed. It is important to note that different elements of watershed value (i.e., water withdrawals, fishing, aesthetics) are not fully independent. Thus, valuation of separate uses and functions may not be strictly additive.

The values summarized in Table D.5 do not include quantification of the values for other habitat and recreational services or aesthetic values. These remain an important challenge, but should certainly be recognized at least qualitatively in assessing management options for the Ipswich watershed. What is clear, however, is that these values are substantial and have not been adequately factored into decisions in the past.

Table D.5
Summary of costs and values for water in the Ipswich Watershed

Use	Cost per year (million USD)	Value per year (million USD)	Source
Residential, commercial, industrial, agricultural water supply	27.5	37+	Municipal charges to users for water (lower-bound figure)
Wastewater services	49.4	49.4+	Municipal charges to users for wastewater (lower-bound figure)
Fish habitat	—	3.1	Willingness to pay survey of residents
Shellfish	1.0	1.1	Shellfish harvest value includes recreational harvest
Other habitat/recreation	—	++	Highly valued on resident survey
Property values	—	++	Closely linked with character of region; IRWA lawsuit
Other (aesthetic)	—	++	Highly valued on resident survey

++: Positive value, perhaps a significant (high) value, but not quantifiable in current analysis.

APPENDIX E
CASE STUDY—THE VALUE OF WATER SUPPLY INITIATIVES IN THE TAMPA BAY REGION

INTRODUCTION

The Tampa Bay region on Florida's west coast has switched from its conventional system-level reliance on mostly groundwater wells to using expensive new water supply approaches, including a desalination plant built by the regional wholesale supplier, Tampa Bay Water. The reasons for this change in approach are predominantly twofold:

- To avoid environmental and related consequences of over-pumping groundwater
- To provide more reliable sources of water in face of increasing populations and periodic droughts.

The decision to move ahead with expensive new supply approaches provides meaningful evidence of the value of water, across several contexts, which are explored below. Even with costs for new sources being many times higher than those from prior conventional supplies, the new sources proved to be economically acceptable by the communities impacted because of the values they generated or preserved.

NEED FOR NEW WATER SUPPLIES

The Tampa region has historically relied on groundwater from the Floridan Aquifer as the primary source for potable water, with the lone exception of the City of Tampa, which has made use of the Hillsborough River. The permitted capacity of the region's wellfields was reduced from 192 mgd to 158 mgd in 1998, and was further reduced at the end of 2002 to 121 mgd. At the end of 2007, only 90 mgd can be withdrawn from the wellfield, a reduction of over 53%. Clearly, with a growing population and a decreasing water supply from its usual main source, the region is faced with the need to find substitute supplies.

The Environmental Impact of Overpumping

Overextraction of groundwater has had serious environmental consequences for the region, including lowered lake levels, the creation of sinkholes, and stressed wetland areas. The U.S. Fish and Wildlife Service lists 13 species of plants and animals as either threatened or endangered in Pinellas (see Table E.1), Pasco, and Hillsborough counties. Many of these species require a water rich habitat that can be potentially damaged by continued dewatering of the area caused by overpumping of the aquifer.

Hydrologic studies, conducted by Northern Tampa Bay Water Resources Assessment Project (NTB-WRAP) in the Northern Tampa Bay Region, indicate a hydrologic connection between the upper Floridan Aquifer and the surficial aquifer (Hazen and Sawyer 1998). Although factors such as rainfall, runoff, and evapotranspiration contribute to the decline of the

Table E.1
Threatened and endangered species in Pinellas County, Florida

Category	Species common name	Status
Mammals	West Indian manatee	E
Birds	Bald eagle	T
	Piping plover	T
	Florida scrub-jay	T
	Wood stork	E
	Red-cockaded woodpecker	E
Fish	Gulf sturgeon	T
Reptiles	Eastern indigo snake	T
	Green sea turtle	E
	Kemp's Ridley sea turtle	E
	Leatherback sea turtle	E
	Loggerhead sea turtle	T
Plants	Florida golden aster	E

Source: USFWS (1998), North Florida Office.
Key code E = endangered, T = threatened.

water level of the surficial aquifer, groundwater pumping is believed to be a significant factor. The decrease in water levels in the surficial aquifer can in turn negatively impact water levels in surface water features such as streams, lakes, estuaries, and wetlands.

The Impact of Lowered Water Levels on Lakes

The most significant decline in lake levels in the past 30 years has occurred in the regions of the public supply wellfields. Although public access to these areas is often restricted, they provide critical habitat for many important plant and animal species. The littoral zone of these lakes provides support for many amphibians, migratory birds, and other plant and animal species that are a critical part of the food chain (Hazen and Sawyer 1998). "Studies have shown that there can be as much as 500% more species diversity in the area near the water's edge compared to adjoining uplands" (Shoreland Connection 2003).

Shallow depths, warm temperatures, and wide littoral zones often characterize Florida lakes. The littoral zone is the area in and along the shoreline that is shallow enough to allow light to reach the bottom sediment and the organisms that grow there. Because of these characteristics, small changes in water temperature and depth make these lakes particularly susceptible to increases in the rate of eutrophication. Although eutrophication occurs as part of the natural aging process of a lake, hypereutrophication (high rate of eutrophication) may result in dieback of desirable plant and animal species, proliferation of less desirable species, and fluctuations in oxygen levels and water quality. Elevation of water temperatures accelerates the process of biodegradation while at the same time decreasing the levels of dissolved oxygen in the water necessary to maintain the overall health of the system (City of Tallahassee 2003).

In addition, lake drawdown can have a significant impact on species diversity as less mobile species become stranded in plant material and depressions and mobile species become more concentrated as water levels drop. The extent to which the drawdown of lake levels impacts the wildlife is dependent on the morphology of the lake but may result in an accelerated accumulation of unconsolidated bottom sediments, decline in dissolved oxygen, nutrient enrichment, vegetational changes, and reduction in fish and wildlife populations (Hazen and Sawyer 1998).

The Impact of Lowered Water Levels on Wetlands

Concern is growing over the decline in the potentiometric surface of the Upper Floridan Aquifer due to groundwater pumping and its impact on wetlands and those species dependent on them for survival. By 1991, of the 595 plant and animal species listed as threatened or endangered by the U.S. Fish and Wildlife Service, 43% were considered wetland dependent (Flynn 1996). In addition to providing habitat for species dependent on wetlands for food, shelter, and reproduction, wetlands also improve water quality, provide flood mitigation, recharge groundwater, and create a transitional zone between uplands and open water. Other effects of groundwater pumping on wetlands range from a decrease in the diversity and abundance of wetland-dependent wildlife to soil subsidence and the collapse of mature cypress domes.

Many of the species that inhabit wetlands have very specific habitat requirements. Large, interconnected wetlands with minimal fluctuation in water levels provide the necessary habitat requirements for many fish species, provide adequate territories for larger animals, and prevent the inbreeding or isolation from suitable partners that may occur when wetlands are reduced in size. Other species benefit from the intermittent hydroperiods that small, isolated wetlands often experience. Semi-terrestrial species such as frogs, toads, and salamanders use isolated wetlands for reproduction since the ephemeral nature of these areas reduces the number of fish and other species that prey on the eggs and larvae (Schaefer and Liebertz 2003). Water depth and intermittent hydroperiods are critical for other species as well; the specialized feeding technique of the endangered wood stork is most effective in water that is 6 to 10 inches deep and where fish are concentrated by dry-season water levels. Water levels affect not only the many species that inhabit the wetlands year round but also those species that use them seasonally. Many birds use wetland areas to rest and feed during long migrations; they also provide breeding and nesting habitat for a large number of permanent residents.

A study performed by J.P. Heaney on environmental valuation for the St. Johns River Water Management District in Florida looked at methods for establishing values for wetlands. From an ecological perspective, property values may not reflect the true value of wetlands to the public (Farber and Costanza 1987). Wetland uses are diverse and include flood and erosion control, water quality maintenance, habitat protection and enhancement, and areas for aesthetic and recreational activities. While it is possible to establish an estimate of wetland values, these values are specific to the individual wetland and the functions they perform. The value of wetlands, as for any other land use, can be expected to vary widely depending on the demand for these services (Heaney 2003). For the case study performed for the St. John's River Water Management District, the benefit provided by the wetland for flood control was $5 per acre, whereas it provided a recreational benefit of $1,620 per acre.

The Impact of Lowered Water Levels on Streams and Estuaries

Tampa Bay Estuary is Florida's largest open water estuary and home to a host of species living in this very unique habitat. Nearly 10% of Florida's endangered manatee population live and breed in the estuary. It provides nesting areas for 40,000 pairs of 25 species of birds annually. The shallow seagrass beds and mangroves are used by a variety of ecologically and commercially important fish species such as snook, redfish, and the spotted sea trout for spawning and nurseries. Many of these species require the lower salinity levels provided by the estuaries. It is estimated that estuaries provide habitat for 95% of the fish harvested commercially from the Gulf of Mexico (American Oceans 2003), with a 1994 market value of $806 million (Hoppe 2003). Decline in water quality, overfishing, and habitat loss resulted in a 38% decrease in the fish harvested from the Tampa Bay area between 1964 and 1990 (American Oceans 2003).

The watershed area that feeds the estuary is approximately 2,200 square miles. It covers most of Pinellas, Hillsborough, and Manatee counties as well as parts of Polk, Pasco, and Sarasota counties. Upstream activities and changes in inflow to Tampa Bay from several major area streams and rivers have the potential to negatively influence water quality and estuarine health in Tampa Bay. Many estuarine dependent species are very sensitive to fluctuations in salinity levels, which can result in the rapid proliferation of some species and significant dieoff in others. Adequate streamflows not only help maintain the balance in salinity levels in the estuary but also transport vital upland nutrients and deposit sediments along the shoreline, thereby reducing erosion. While some of the fluctuations in streamflows are the result of precipitation, drought, evaporation, and other climatic conditions, anthropogenic influences on streamflows often magnify the effect.

Water quality, particularly along coastal areas, may be compromised as groundwater pumping lowers the potentiometric pressure of the aquifer, resulting in saltwater intrusion that can render the water unpotable. Contamination of a freshwater aquifer will not resolve immediately with the cessation of pumping and the restoration of potentiometric levels; subsidence of the saltwater interface occurs at the rate of groundwater velocity and will often take decades to return to normal.

The Impact of Groundwater Pumping on Uplands

Many species that use wetland areas seasonally for nesting and spawning rely on uplands the remainder of the year. Groundwater pumping can dramatically reduce the availability of surficial water to the root zone of upland trees and plants, making them more susceptible to loss from fire, diseases, and pests. Uplands perform several critical functions. As streams and rivers move through uplands, they carry a supply of minerals and nutrients essential to the health of wetlands and estuaries into which they flow. At the same time they serve as a filter for contaminants and reduce soil erosion (Hazen and Sawyer 1998).

Population increases frequently result in the increased use of uplands for development and agriculture. As upland vegetation is cleared or replaced with large areas of impermeable surfaces, wildlife habitat is destroyed and the amount and rate of runoff increase. Increased runoff increases the rate of erosion and deposition of sediment in adjacent wetlands. Contaminants that would be retained or filtered by the upland soil are carried to the wetlands in the runoff creating an additional burden on already stressed wetlands. Agricultural byproducts such as fertilizer contribute to eutrophication of adjacent wetlands, and as water withdrawals for irrigation and human consumption increase, low-lying wetlands adjacent to these areas are more likely to experience shortened hydroperiods, fluctuations in water levels, and damage to wetland habitat.

The Impact of Groundwater Pumping on Land Subsidence

The unique geology throughout Florida makes many areas susceptible to land subsidence. Land subsidence occurs when subsurface support provided by groundwater is removed and land elevation drops. Over millions of years, groundwater movement has created fissures and caverns in the limestone or other carbonate rock that underlies much of Florida. Known as karst, these rock layers are frequently covered by relatively thin layers of sand, clay, or silt. In northern Pinellas County and parts of Pasco County, the limestone is often no more than 15 feet below the surface. Absence of a stiff clay confining layer, combined with a variety of natural phenomena and human activities, makes this area particularly prone to land subsidence.

Subsidence may be very rapid or can take place over a number of years; areas of subsidence range from a few square feet to areas covering several miles. Subsidence that is very sudden often follows extended periods of heavy rainfall. The increased weight from the saturated soil causes the thin roof of the dewatered underlying limestone cavern to collapse. Subsidence that takes place over a long period of time is characterized by a gradual depression of the land surface as soil is slowly washed into the underlying cavities. Drought and groundwater pumping contribute to subsidence by reducing the water levels that provide buoyant vertical support to the limestone structures (Hazen and Sawyer 1998).

Structural damage to homes, businesses, property, and public structures due to subsidence results in millions of dollars in insurance claims each year. Between 1990 and 1991 there were 323 claims in the northern Tampa Bay area alone. These claims totaled approximately $12 million in 1991.

Losses to the Recreational Industry from Groundwater Pumping

Maintaining water levels in streams, lakes, rivers, estuaries, and surficial aquifers is beneficial not only to the environment but also to the economy. In 2001, more than 3.2 million wildlife watchers spent as much as $1.6 billion on food, lodging, and equipment in Florida to observe, feed, and photograph wildlife, according to an economic analysis released by the Florida Fish and Wildlife Conservation Commission (Hoppe 2003).

Fishing and boating are also popular recreational pastimes and contribute substantially to the local economy. Lowered lake levels affect boat access, water quality, and fish populations and may have substantial economic impact on lakefront property values. Local businesses who provide recreational goods and services to the many residents and vacationers who come to the area each year to fish, sail, swim, dive, and waterski may be adversely affected as well. In 1991, recreational fishing in Florida generated approximately $3.5 billion in economic output and employed over 58,000 people (American Oceans 2003).

The Effect of Groundwater Pumping on Property Values

Water levels and water quality have the potential to affect property owners and the real estate market throughout many regions in Florida. While it may be more difficult to attribute lowered lake levels outside the public supply wellfields directly to groundwater pumping, the environmental and economic impacts of lowered lake levels are significant nonetheless. As lake size decreases, its ability to provide the recreational and aesthetic benefits to lakefront property owners also decreases. A study performed on property values in South Carolina found that sales values of lakefront properties decreased by approximately 3.1% with each 1 foot drop in lake levels (Hazen and Sawyer 1998).

A study on environmental valuation for the St. Johns River Water Management District in Florida found that lake values can be estimated based on its ability to generate income and enhance nearby property values (Heaney 2003). Valuations of lakefront property were based on estimates of property values per front foot, obtained from the county tax assessor's office and combined with hydrologic and water quality information to determine their economic impact. Heaney (2003) found surface area has an even greater impact on lake values than does water depth. The criteria given by the National Ecology Research Center in 1990 found that typical water-based recreational activities such as swimming, canoeing, waterskiing, and power boating require a maximum water depth of only 9 feet. On the other hand, the sales price of lakefront property where the lake surface area was still 96% of its original area was $904 per front foot, whereas property along a nearby lake whose surface area was reduced to 23% of its original area was valued at only $204 per front foot.

Water quality of Florida lakes is measured using the Trophic State Index (TSI) and is based on the levels of chlorophyll, nitrogen, phosphorus, and other nutrients essential for plant life. TSI levels range from 0 to 100. Readings from 0 to 59 are considered to be good, 60 to 69 are considered fair, and readings from 70 to 100 are considered poor. Heaney's study found that the annual income for a highly degraded lake (TSI 89) was $24 per lake surface acre, whereas income for a nearby lake with good water quality (TSI 40) was nearly $1,000 per lake surface acre, an increase of over 40 times (Heaney 2003).

The Impact of Groundwater Pumping on Agriculture

Similar impacts have been felt by the agricultural industry. As groundwater withdrawals increase, many of the surface water bodies that are used to supply livestock water and irrigate crops and pastures have dried up or are substantially reduced in size. When individual wells are used to supply water, there have been problems ranging from complaints of sediment or other contaminants in the water to reduction or complete loss of water supply. Profits decrease as local water supplies become limited and the cost of alternative water supplies, food sources, and foregone production increases.

NEW SOURCES TO OFFSET PUMPING LIMITATIONS

While it is not possible to put an exact figure on the value of maintaining groundwater levels, clearly the potential damage from continued overpumping is immense. Because of this, strict limits have been put in place on the use of groundwater that significantly scale its use back from usual levels. Tampa Bay Water (TBW) is a governmental agency funded through wholesale water sales to its member governments—Hillsborough, Pasco, and Pinellas counties, and the cities of New Port Richey, St. Petersburg, and Tampa. TBW is within the jurisdiction of the Southwest Florida Water Management District (SWFWMD). TBW serves approximately 85% of the more than 2.4 million people who live in the region. TBW was formed to comply with the groundwater pumping restrictions. The TBW Board approved a series of Master Water Plan projects, including:

System Configuration I
91 mgd to meet supply needs through 2007
Estimated capital cost—$609 million

- 25 mgd seawater desalination plant on Tampa Bay (expandable to 35 mgd)
- 60 mgd enhanced surface water system (Tampa Bay Regional Reservoir and treatment facility)
- 6 mgd groundwater project

System Configuration II and III
49 mgd to meet supply needs through 2014
Estimated capital cost—$700 million

- 25 mgd seawater desalination on the Gulf of Mexico (Gulf Coast Desalination)
- 5 mgd brackish water desalination (Mid-Pinellas Brackish Water Desalination)
- 10.5 mgd groundwater (Cone Ranch Wellfield—10 mgd, Eagles Well—0.5 mgd)
- 3.5 mgd re-use of industrial process water (Crystals International Water Supply)
- 1.0 mgd reclaimed water exchange (Cargill Reclaimed Exchange)
- Hillsborough River augmentation during drought conditions (Morris Bridge Sink)

Of the combined 140 mgd supply to be brought on line over the next 10 years, at least 55 mgd (39%) will come from desalination, currently one of the most expensive water treatment options available. It is also noteworthy that the unit cost of configurations II and III is over twice that of configuration I. These are capital costs and do not include annual O&M or interest expenses.

In the Tampa area, the damage caused by overpumping of the Floridan Aquifer is significant enough to effectively eliminate this as a source for major new supplies. These damages include damage to wildlife and the environment and destruction of land and property caused by subsidence. To avoid these consequences, the SWFWMD has committed $183 million to the System Configuration I program, with an additional $57 million to come from the federal government for the Tampa Bay Regional Reservoir project. The remaining $369 million will be funded through utility bond issuances and paid for by higher retail water rates, as detailed below.

These water supply projects are viewed as "essential" components necessary to improve environmental conditions and sustain the economic vitality of the Tampa Bay region (Tampa Bay Water 2001). Clearly the water supplies and community as a whole in the region made a determination that the benefits of reduced reliance on groundwater must be at least as great as the added cost of the new supply options, otherwise they would not have imposed or accepted the higher costs.

Benefits

The value of Tampa's new water supply initiative includes economic, social, and environmental benefits. The economic value of the desalination project alone is substantial for the entire region. An economic impact report on the desalination plant prepared by the Tampa Bay Regional Planning Council (1999) concluded, "In the most rudimentary analysis, the project (itself) creates 119 permanent jobs and generates $482 million in Gross Regional Product over

30 years." In addition, the economic analysis estimated that the economic damages caused by a water deficit of 25 mgd would amount to over 10,000 jobs and $24 billion in gross regional product. This clearly, is the upper end of the value of water to the Tampa area since is ascribes all of the damages to the economy to the water deficit as if water were the only limiting factor in the region's growth. But it does, nonetheless, show an interesting data point in the consideration of the value of water to a region, and when one compares this value with the capital costs for the water supply configurations described above, which range from $609 million to $700 million, it shows the relative magnitude of the economic value verses the cost of new supplies to the region.

The positive social impacts of Tampa's new water supply initiative include maintaining the lifestyles to which residents have become accustomed, continued quality of life in the region, capacity for economic development and growth, and a stabilized regional economy and, by extension, social structure. The environmental benefits of Tampa's water supply initiative include protection of groundwater levels and associated discharges to key surface bodies. The initiative also reduces the likelihood of sinkhole formation. Associated improvements to surface water include improvements to lakes, rivers, and estuaries. These bodies of water support important ecosystems and are critical habitat for birds, fish, and wildlife. They are also used by the public for fishing, boating, wildlife viewing, etc. A reduced reliance on groundwater pumping decreases saltwater intrusion into fresh water sources and may reduce the formation of dangerous sinkholes.

DESCRIPTION OF DESALINATION PROJECT

The 25 mgd Tampa Bay Seawater Desalination (TBSD) plant began producing water (4.9 mgd) in March 2003. This $110 million facility is a key component of the System Configuration I program and will be the largest reverse osmosis seawater desalination facility in North America. The facility is located in Apollo Beach on Tampa Bay property owned by the Big Bend Power Station site operated by the Tampa Electric Company (TECO).

History and Ownership of Project

The TBSD plant is owned by Tampa Bay Water. It was built by Covanta Tampa Construction and will be operated by Covanta Tampa Bay (or a replacement entity) in a public/private partnership. The plant was originally designed and owned by Poseidon Resources, but in May 2002 Tampa Bay Water exercised its option to purchase the plant, while retaining the services of Covanta and Poseidon.

Physical Characteristics of Plant

The TBSD plant uses a reverse osmosis process with a semipermeable membrane process. The feedwater for the plant is obtained from water used to cool Tampa Electric's Big Bend power plant. This water is first filtered to remove debris, suspended solids, and bacteria before the reverse osmosis process.

Using a two-stage process, the filtered cooling water is forced, under pressure, through a semipermeable membrane system where the freshwater is separated from saltwater. This leaves a supply of fresh water as well as concentrated saltwater. The finished fresh water is then moved to a regional facilities site where it is blended with water from other sources before being introduced into the Tampa Bay Water distribution system. The concentrated saltwater is discharged back into Tampa Bay along with the remainder of the power plant cooling water. In this process it is diluted by a ratio of approximately 70:1. The dilution with the cooling water helps mitigate the impact of the concentrate on the salinity of Tampa Bay.

The plant will have a capacity of 25 mgd when completed, and will contribute approximately 10% of the total Tampa Bay supply. It had an initial capacity of 4.9 mgd when it began producing potable water in March 2003. The plant has been designed so that the capacity can be increased to 35 mgd.

Costs for Desal Water

Table E.2 shows an analysis of the costs for the construction and operation of the first 25 mgd of desalination capacity for the Tampa Master Water Plan. This table includes both capital costs to construct the plant and annual operations and maintenance costs. The plant cost $110 million to construct. Of this, $26 million is for pipelines, $67 million for the plant equipment and construction, and $18 million for design and other professional services. The pipeline costs for the plant reflect sharing of intake and brine disposal with the power plant, which reduces these costs substantially. The major operational expense for the plant is electricity, which is estimated to total $69 million over the 15 year planning period. Electrical costs are also reduced by the use of warm water from the power plant, which uses less energy for reverse osmosis. The next largest expense is for general maintenance, chemicals and membrane, and wages. The total operation costs for the plant over the first 15 years of its life are estimated to be $143 million in current dollars.

When all expenses are brought back to a present worth, at a 5% discount rate, the total costs for the desalination plant come to just under $197 million. This is equivalent to a total annual cost of $19 million. This represents the average annual cost to build, maintain, and operate the plant over a 15 year period. The plant is rated at 25 mgd, which amounts to 9,125,000 kgal/year. If we allow for down time of 25% (perhaps conservative, given anticipated production at 80% of capacity), this reduces its actual expected yield to 7,300,000 kgal/year, which results in an average cost per thousand gallons of production of $2.60.

A major point of controversy surrounding the actual costs for the desalination water is the amount of energy required to produce the finished water and whether charges being paid for that energy reflect actual costs or have been subsidized. Seawater desalination plants can consume up to 30,000 kWh/AF of product water. This is equivalent to 92 kWh/kgal. At $0.07/kWh, this comes to $6.44/kgal just for energy. The cost estimate in Table E.2 shows an annual energy cost of over $4 million. If the plant produces 7.3 million kgal per year, this equals only $0.63/kgal, which is an order of magnitude lower than $6.44/kgal. Despite these controversies, the official estimates of costs remain close to those shown in Table E.2.

Table E.2
Cost analysis for 25 mgd desalination project in Tampa Bay (\$millions)

Year	15 year total	Base year	Annually in years 2-15
Construction			
Pipeline	$25.70	$25.70	
Plant equipment	$60.00	$60.00	
Plant cons.	$6.70	$6.70	
Profession services	$17.60	$17.60	
Subtotal	$110.00	$110.00	
Operation			
Chems and membranes	$22.64	$0.43	$0.87
Electric	$69.42	$4.16	$4.22
Operator wages	$19.06	$0.68	$0.70
Maintenance/other	$32.00	$1.75	$1.80
Subtotal	$143.12	$7.03	$7.59
Total	$246.60	$117.03	$7.59
Present worth	$197.12		
Discount rate	5%		
Annual cost of plant	$18.99		
Rated capacity (kgal/year)	9,125,000		
Capacity factor	80%		
Actual capacity	7,300,000		
Actual cost/kgal	$2.60		

Source: Based on data from Tampa Bay Regional Planning Council (1999, Table 2).
Note: Construction costs prorated to \$110 million from \$84.5 million.

Price for Desal Water

According to Tampa Bay Water, the price they will pay for desalinated water produced by Tampa Bay Desal will be between \$2.02 and \$2.49/kgal for the next 30 years. This will be reduced to \$1.58 and \$1.88/kgal by a subsidy paid by SWFWMD on certain capital costs (these prices are based on overviews of the Desal project provided at www.tampabaywater.org). Pinellas County Utilities will purchase blended water from Tampa Bay Water at the wholesale price of \$1.91/kgal in 2003. It is noteworthy that these prices charged by the wholesale utility appear to be below the price of producing the water, and could be substantially below these costs depending on the outcome of several operating variables, as discussed above.

VALUE OF WATER

The value that is placed on water is the highest price that a customer will pay for water intended for a specific end use. Tampa Bay Water sponsored a study of the value of water within its service area that provides some very good insights into the problem of estimating value and some specific data applicable to the Tampa Bay area (Oscher Consulting 1999).

Elasticities of Demand and Value for Water

The price elasticity of demand is the amount by which the current quantity of water demand will change—at the margin—in response to a unit change in the current price. For example, if the elasticity of demand is -0.2, then a 1% increase in the price of water will result in a decrease of 0.2% in its demand. The elasticity of value, on the other hand, is the inverse of the elasticity of demand, and is the amount that a unit change in the supply of water will affect the customer's estimate of the value of the next unit of consumption, at the margin. Using the same example, if the elasticity of demand is -0.2 then the elasticity of value is -5, and a 1% reduction in supply may cause customers to increase their estimate of the value of water by 5%. Thus the lower the elasticity of demand the higher the elasticity of value, and the greater resistance there is in changes to demand in response to changes in the price of water because there are few low valued uses of water relative to its price. Oscher Consulting (1999) used -5 for its estimate of the elasticity of value of water for residential uses in Tampa Bay area.

The study references 32 papers and reports that provide estimates of elasticities of demand and value. Table E.3 shows the elasticities used in the Oscher Consulting study. These range from a low of -1 for exterior uses to -10 for indoor institutional uses.

The economic model created for Tampa uses these elasticities along with the planning information on future water demands and water supplies to estimate the value added to the economy from eliminating shortages, as shown in Figure E.1. These shortages equal the differences between the current permitted water supply for the Tampa area and the projected water demand. Over the period from 2002 to 2010 these are projected to total 136,192,000 kgal of water. Estimates of the value of this water (based on a value added analysis) range from $500 million to $1.7 billion. This equates to a unit value of new water supplies to the Tampa area ranging from $3.67 to $12.75/kgal. Compared to these values, the cost of desalinated water at $2.68/kgal appears quite reasonable.

This is one example of how a regional study was done to place a value on water in use within its service area. The key to estimating the value of lost water is having reliable estimates of the elasticities of value for water users of different categories. Tampa Bay Water developed elasticities for the five categories shown in Table E.3. Demand models were used to estimate future unconstrained demand for water for each category. Resulting shortages were evaluated using the elasticity functions. This exercise shows that, if the assumptions used for the analysis are correct, then the value of water in use by the customers of the Tampa Bay Water may be from 136% to 475% of the cost of production water using desalination.

Table E.3
Elasticities for water use for Tampa Bay

Category	Demand elasticity	Value elasticity*
Single family interior	-0.2	-5
Multi family interior	-0.15	-6.7
Commercial interior	-0.5	-2
Institutional interior	-0.1	-10
Exterior	-1.0	-1

Source: Developed from Tampa Bay Regional Planning Council (2000).
*Defined by Oscher Consulting (1999) as the inverse of price elasticity of demand.

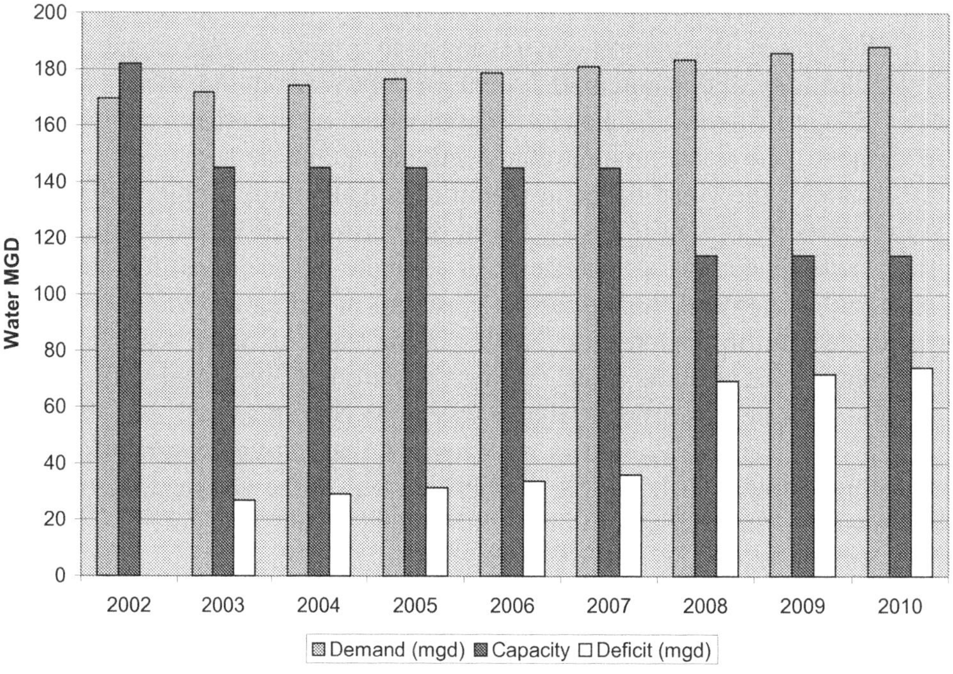

Source: Based on Tampa Bay Water (2000).

Figure E.1 Demand and supply for TBW

CONCLUSIONS

The Tampa Bay region is faced with a major realignment of its water supply system. This is being driven by the continued growth in the area population and the unavoidable need to move away from reliance on groundwater. In the face of this, the regional water provider has developed new water supply options that include desalination and new surface supplies. This involves switching from supplies that have essentially no costs (other than the cost of pumping) to ones that will cost at least $2.60/kgal to produce.

These new costs, while significant, are relatively small in comparison to the overall value of reliable water supply to the regional economy, which has been estimated to range from $3.67 to $12.75/kgal. The Tampa Bay Water study has also provided a good example of how econometric methods, coupled with good end use and elasticity data, can be used to estimate the value of water in a regional economy.

APPENDIX F: INSTREAM AND EXTRACTIVE USE VALUES GENERATE CONSIDERABLE TOTAL VALUE FOR LOCAL RIVER WATER USERS (VERDE RIVER, ARIZ.)

INTRODUCTION

Background on the Verde River

The Verde River is located in north-central Arizona. It is a tributary to the Salt River and is part of the Colorado River System. Water from the Verde provides the Phoenix area with water for municipal uses and irrigation. The main population centers along the upper reaches are in the Verde Valley near Cottonwood, Camp Verde, Clarkdale, and Sedona. Gila and Maricopa are located along the middle reaches. The Verde River joins the Salt River just before reaching the Phoenix area (see Figure F.1).

The Verde River is one of Arizona's major perennial rivers and a treasured resource in the central part of the state. The river is over 120 miles in length and comprises roughly 5% of the state's perennial streams. The Verde is Arizona's only national wild and scenic river.

During recent years, this region has experienced a rapid growth in population and a concurrent increase in water-resources development. The population of the major cities and towns within the Verde Watershed has more than doubled in the last 20 years and is projected to more than double again within the next 50 years. Municipal water usage has increased by more than 39% over the last eight years and at the present rate of growth will increase by more than 400% over the next 50 years (AZ Department of Water Resources 2000). The increase in population is closely associated with the growing interest in this area as a retirement location. The Verde River, numerous lakes, Indian ruins, and spectacular scenery make this a popular tourist and recreation area.

The Concept of Complementary and Competing Water Uses

Values derived from water in rivers may be classified into 2 groups: (1) values obtained from instream flows (e.g., nurturing the ecology and providing opportunities for recreation); and (2) values obtained by extractive uses (e.g., providing municipalities with water to supply residential and other customers). Under the current patterns of use in the Verde River, the majority of these values are complementary, rather than competitive. This is due to the water generating instream values upstream, and providing municipal water from stretches downstream. However, some competing uses do exist such as upstream demands for municipal/domestic, irrigation, industrial, and livestock.

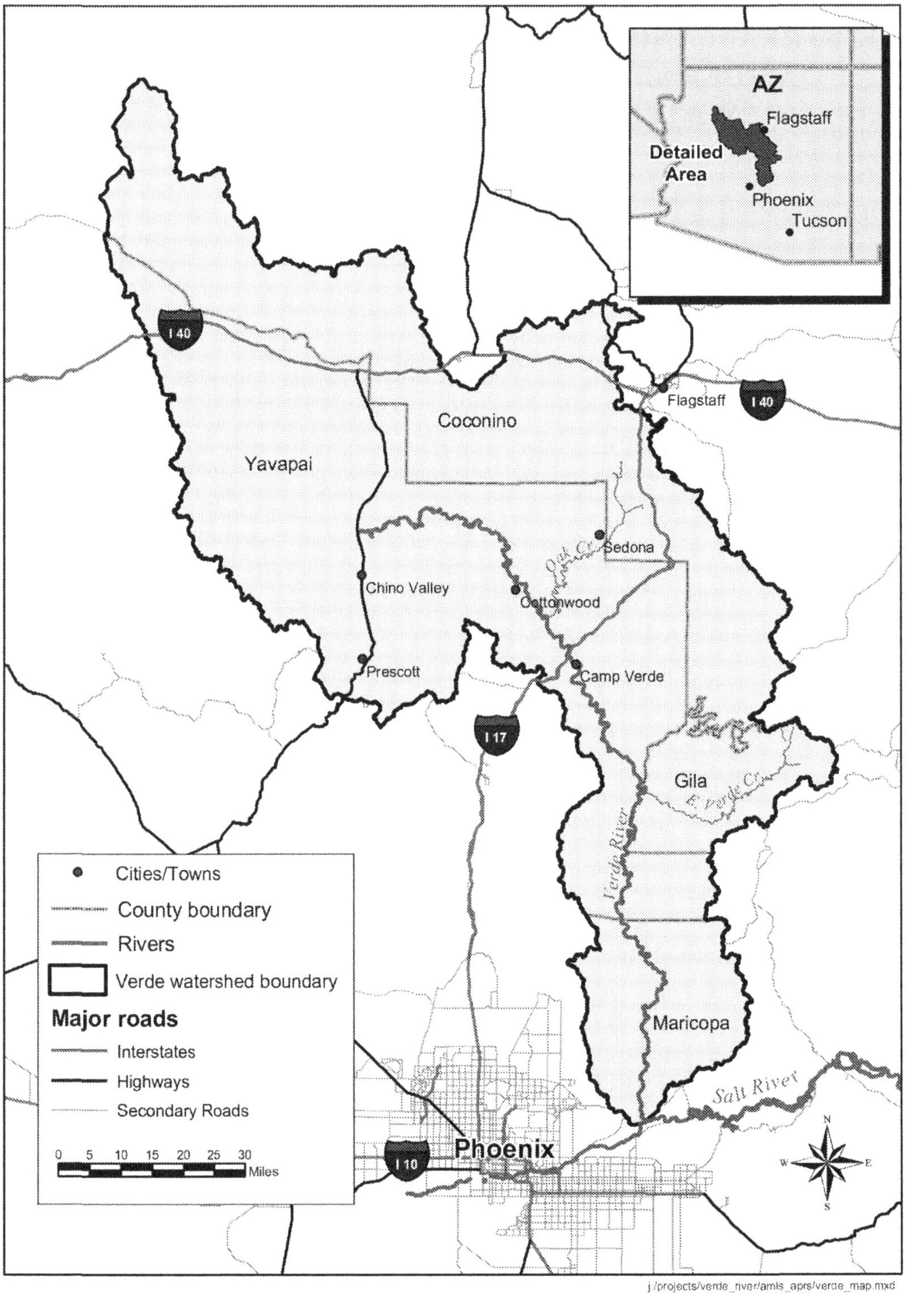

Figure F.1 Map of the Verde River

A key issue for the Verde watershed—and the values it generates—is the future use and management of its headwaters, especially the Little Chino Valley groundwaters that are the source of springs that constitute the Verde's headwaters. Future extraction of these groundwaters may cause any significant impacts on baseflow levels of the Verde River itself. Increasing water demands, given potential rates of population growth and development, could be detrimental to both the instream and downstream extractive use values. Hence an issue for long-term water resource planning is the need to balance demands for headwater-feeding aquifers against the values currently generated by the spring-based flows downstream.

The Little Chino sub-basin is in the Prescott Active Management Area (AMA), meaning that under Arizona state law, Prescott can tap these groundwaters to meet its future needs, even though moving the water to Prescott would export it from the basin (no other entity has a right to export Little Chino waters out of the basin). The basin has experienced significant groundwater declines in some areas and these declines have reduced flow in Del Rio Springs, the perennial headwaters of the Verde River (AZ Department of Water Resources 2000). Similar effects on other springs could be seen in the future with unplanned continued development. Prescott is a rapidly growing city with no long-term surface water supply. Without another source, the increasing demand might force Prescott to look to the Verde River itself for supply, in addition to continuing to deplete the groundwater which provides the Verde River headwaters (U.S. Water News Online 2003).

Little is known about how much groundwater is actually in storage in many areas of the Verde Watershed or about how water use in the Upper Verde may affect the continued availability of water for the Verde Valley, which depends on Verde River flows. These issues have caused a great deal of concern, expressed by water users within the Upper and Middle Verde areas as well as by downstream users of Verde River water, about the future availability and reliability of surface water and groundwater within the Verde Watershed (AZ Department of Water Resources 2000).

This report examines both the upstream and downstream values of the Verde River water. This document discusses instream use and nonuse values of the Verde River upstream and the values of the water downstream as it provides water supply to municipalities, and presents conclusions.

Overview of Upstream and Downstream Values for the Verde River

The Upper Verde River begins in north central Arizona, and flows south to Bartlett Lake. Free flowing for most of its length, it provides lush riparian habitat and abundant wildlife, including several threatened and endangered species. This unique ecological system provides several nonuse values, such as the preservation of endangered species. The abundant wildlife combines with spectacular scenery to provide diverse recreational opportunities. Boating, fishing, hunting, swimming, picnicking, camping, wildlife viewing (including birdwatching), photography, and sightseeing are all popular activities. Some of the Upper Verde water is used to supply agricultural, domestic, and industrial uses near the river.

Downstream, the water from the Verde River is an important source of municipal and agricultural water supply for metropolitan Phoenix (including Scottsdale) and surrounding rural areas. The water also is used for power production. The Salt River Project maintains two dams on the Verde River. The uppermost of these, Horseshoe Dam, backs up an impoundment known as Horseshoe Reservoir, which can hold up to 131,427 acre-feet of water (an acre-foot of water is the amount which will cover an acre of land to the depth of one foot). Bartlett Dam, which forms Bartlett Reservoir, is below Horseshoe. Bartlett Reservoir can hold up to 178,186 acre-feet of water. Both of these dams were built for power production, and neither has any flood control space.

ESTIMATES OF INSTREAM VALUES IN THE UPPER VERDE RIVER

Overview of Upstream Use and Nonuse Values of the Verde River

Values of environmental resources may be classified into 2 groups: (1) attributes that contribute to economic welfare from consumptive or nonconsumptive use of the resource (e.g., contributions to the well-being of humans who derive use value from market or nonmarket goods such as a rafting trip or recreational fisheries); and (2) nonuse values (e.g., bequest and existence values, including stewardship motives to preserve threatened and endangered species). As previously stated, the upper reaches of the Verde River include both use (e.g., recreation) and nonuse (e.g., endangered species) values and these are discussed in further detail below.

Use Values

Some Verde River water from upper reaches is diverted for agricultural, domestic and industrial uses. Water supplies for these uses come from diversion of surface waters from the mainstem and tributaries of the Verde River and from groundwater pumping. The total amounts are not quantified in this study, as it is assumed for the purposes of this research that the recreational use comprises the largest use values for the Verde River.

The river provides a large volume of recreational use. The uppermost reach of the river is primarily canyon bound and provides a wilderness experience, while recreational use increases as one heads downstream with more public access points, day use sites (including primitive and those with facilities such as picnic tables and toilets) and camping locations along the river banks.

The Verde offers some of the best rafting, canoeing, and kayaking in the state. Some of the waters flow at or near Class Three whitewater (a desired class for rafting and kayaking). Visitors enjoy classic Colorado Plateau scenery.

The Verde River is also known for good fishing. Smallmouth bass, largemouth bass, flathead catfish, channel catfish, yellow bullhead, roundtail chub, and carp. Rainbow trout are stocked seasonally from about November to March in the Verde River within the Verde Valley. The river is open all year for fishing. There are unlimited bag limits on bass and catfish, and there is a daily bag limit for trout of 6 and 1 for chub (with a minimum size of 13 inches).

Other recreational values of the Verde River stem from swimming, picnicking, camping, wildlife viewing (including birdwatching), and sightseeing. Resident and migratory birds abound throughout the river and river otter and beavers are occasionally seen.

Other users of the Verde River may come on a 4-hour scenic excursion offered by Verde Canyon Railroad. The train goes through the North Verde River Canyon and Sycamore Wilderness Area and passes Sinagua Indian ruins and desert flora and fauna. This historic route takes tourists along the floor of the otherwise inaccessible Upper Verde River Canyon—which is referred to as Arizona's other "grand canyon." Likely sitings of bald eagle, blue herons and deer are also advertised as features of the excursions. Instream flows of the Verde River contribute to the habitat and aesthetics of this area in many ways, making it more attractive for the visitors coming by train.

Nonuse Values

Nonuse or passive use impacts arise when individuals value an environmental resource apart from any past, present, or anticipated future use of the resource in question. Such passive use values have been categorized in several ways in the economic literature, typically embracing the concepts of existence (stewardship) and bequest (intergenerational equity) motives. Existence value is the satisfaction gained from knowing that a particular species or entire ecosystem continues to exist and function. Bequest value is the satisfaction gained from knowing that protection today will provide future generations with a particular species or ecosystem. Passive use values also may include the concept that some ecological services are valuable apart from any human uses or motives. Examples of these ecological services may include improved reproductive success for aquatic and terrestrial wildlife, increased diversity of aquatic and terrestrial species, and improved conditions for recovery of T&E species.

The preservation of T&E species is likely to be the basis for the majority of nonuse values on the Verde River, as several T&E species inhabit the aquatic and riparian habitats of the Verde River:

- The spikedace (a native minnow) is listed as threatened.
- The razorback sucker (a large-river species that is found only in the Colorado River basin) is listed as endangered.
- The southwestern willow flycatcher (a small bird whose breeding range includes Arizona and adjacent states) is listed as endangered.
- The Colorado pikeminnow (the largest minnow in North America and one of the largest in the world) is listed as endangered, though within the Verde River, it is considered an "experimental non-essential" population under Section 10J of the Endangered Species Act, where they are treated as a threatened species, except for the purposes of Section 7 of the Endangered Species Act, where they are treated as a proposed species.
- The bald eagle is listed as threatened.
- The roundtail chub (a fish similar to trout in appearance and found only in the Colorado River basin) is listed as a Forest Service Sensitive species and as a Species of Concern for the state of Arizona.

Other Values (Historical, Cultural, and Special Designation Sites)

Information gained from limited surveys along the Verde River shows the area contains abundant historic and cultural values. The Verde River Valley has long been known for its wealth of prehistoric and historic sites dating back to 12,000 B.C. Hundreds of archaeological sites have been found or are suspected to be present in the River corridor. Most of these sites, depending on their condition, are significant and eligible for the National Register of Historic Places, either individually or as part of a group.

These exceptional historic and cultural resources helped earn the Verde River its designation as a Wild and Scenic River. The Wild and Scenic Rivers Act of 1968 requires a determination that a river and its immediate environments possess one or more specific "outstandingly remarkable values" (ORVs) before that river corridor can be considered for designation as a National Wild and Scenic River (NPS 2004). The Environmental Impact Statement for the Verde River, approved in 1981, found that this portion of the River corridor contained outstandingly remarkable scenic, fish and wildlife, and historic and cultural values. The Act also requires that, once established, the Wild and Scenic River must be administered in such a manner as to protect and enhance the ORVs while minimizing impacts to private landowners and existing land uses.

Congress designated a portion of the Verde River as a National Wild & Scenic River through the Arizona Wilderness Act of 1984. The Verde Wild & Scenic River (VWSR) is located within the administrative boundaries of the Coconino, Prescott, and Tonto National Forests in Gila and Yavapai Counties of central Arizona. The wild and scenic river area is generally ½ mile wide, ¼ mile each side of, and parallel to the river, and contains approximately 5,692 acres (NPS 2004).

Factors affecting wildlife habitat along the VWSR include upstream uses and diversion of water; increased settlement, human structures and other watershed conditions which may increase peak flows; off-highway vehicle use; camping and related recreation impacts; and livestock grazing (NPS 2004).

The greatest threats to maintaining instream flows in the Wild and Scenic River Area come from accelerated ground water pumping to support the rapid population growth upstream from the Wild and Scenic River corridor and from water exchanges that would allow upstream diversions. In an effort to protect the water dependent resources of the Verde River, the Tonto, Coconino, and Prescott National Forests filed an instream flow water right application in 1985. The Forest Service subsequently received a certificate for the beneficial use of wildlife, including fish, that requires minimum annual volumes at select points along the river. Protected monthly flows at these gauges range from 10 to 65 cfs at the upper gauge to 70 to 135 cfs at the lower gauge. The Forest Service has also applied for instream flow water rights for other sections of the river: on the East Verde River, Fossil Creek, and Red Creek.

Techniques for Estimating Values (Benefits Transfer)

Values for environmental resources can be estimated in primary research through the use of revealed preference or stated preference techniques. However, primary research is rarely feasible within budgeting, scheduling, and the other constraints.

A secondary (or expedited) method for valuing water and related environmental resource services is known as benefits transfer (BT), and BT often is the preferred (or only feasible) approach to valuing environmental resources when time and budget constraints do not allow for primary research. Most upstream values of the Verde River may be estimated using the BT approach.

The BT approach involves taking the results of existing valuation studies and transferring them to another context, e.g., a different geographic area or policy context. Estimates for use and nonuse values may be derived, for example, using BT by applying an annual WTP estimate per household (e.g., Mitchell and Carson 1986, 1989) to all the households with motives for the resource in question. The challenges in this approach include defining the appropriate "market" for the impacted site (e.g., what are the boundaries for defining how many households apply), as well as matching the primary research scenario to the site in question. The approach is widely accepted as a suitable method, when implemented correctly, for estimating the use and nonuse benefits of environmental resources.

A literature has emerged to help guide practitioners in the implementation of the BT approach (e.g., Bingham et al. 1992, Desvouges, Naughton, and Parsons 1992, U.S. EPA 2000a). Criteria for studies to be used for BT include:

- Studies should be based on natural resources that are similar to those being evaluated
- Studies should be based on areas that have user populations similar to the study area
- Studies should be based on sound research techniques (e.g., more recent studies are preferred to older ones, studies from peer-reviewed journals are favored).

Only in ideal circumstances do the environmental resource and policy variables of the original study very closely match those of the analytic situation to which a policy or regulatory analyst may wish to extrapolate study results. Despite discrepancies, this approach often provides useful insights for valuing environmental resources.

As another potential approach, nonuse or total values may be deduced (under limited circumstances) using restoration-based costs as a proxy for the value of the change in stocks. For example, for T&E species, the costs of restoration programs and various resource use restrictions may indicate the revealed preference value of preserving the species.

Estimation of Instream Use Values of the Verde River, Using Benefits Transfer

Description of Applicable Use Studies for Benefits Transfer

One database of BT values has several applicable values for use in this case study. This database, developed by Loomis (1998a), is a comprehensive collection of 701 recreational nonmarket valuation studies and values obtained in these studies from the period 1970 through 1998, but focusing on the late 1980s and 1990s. This database contains "user day" values for a range of recreational activities. The activities that are relevant for application to the Verde River are fishing, swimming, float boating, hunting, wildlife watching, camping, picnicking, and sightseeing.

Calculation of Use Values Using Benefits Transfer

While both attributable to Verde River instream flows, some recreation activities may be considered "instream activities" and some "near-stream activities." Tables F.1 and F.2 show participation rates and value estimates of participation in "instream activities" (fishing, swimming, and boating) and "near-stream activities" (hunting, wildlife watching, camping, picnicking, and sightseeing), respectively, for the entire state of Arizona, for the counties included in the Verde River Watershed (Yavapai, Coconino, Gila, and Maricopa counties), and for the Verde River area (a subset of the watershed counties). The total value of instream flows for "instream activities" for the Verde River are estimated to be greater than $10.2 million (2002 USD) annually, and "near-stream activities" are estimated to be greater than $24.8 million (2002 USD) per year. Thus, it is estimated that instream flows in the Verde River generate over $35.0 million in value per year to resident recreationalists in the counties containing the Verde River.

Table F.1
Valuing instream flows of the Verde River: Participation in and value of "instream activities"

	Days of participation in Arizona* (#)	Days of participation in Verde counties† (#)	Mean user-day value‡ (2002 USD)	Arizona value (2002 USD)	Verde counties value (2002 USD)	Verde River value§ (2002 USD)
Fishing	4,689,000	3,127,254	$36.90	$173.0 mil	$115.4 mil	$10.2 mil
Swimming	Unknown	Unknown	$28.00	Unknown	Unknown	Unknown
Float boating	Unknown	Unknown	$43.05	Unknown	Unknown	Unknown
Total	>4,689,000	>3,127,254		>$173.0 mil	>$115.4 mil	>$10.2 mil

* *Source*: USFWS 1998; residents and nonresidents.
†Yavapai, Coconino, Gila, and Maricopa counties contain the Verde River and their combined population is 67% of the state population (U.S. Census Bureau 2003) and thus 67% is applied to the state participation estimates to obtain the combined counties' estimate.
‡*Source*: Intermountain region, Loomis 1998a, 1998b.
§Verde River values are an approximated percent of the Verde Counties values. The percent used is the percent of the Verde County land area that is in the Verde River watershed. This is believed to be an underestimate, since the Verde River provides exceptional recreational qualities and draw more recreational visitors than other locations in these counties.

Table F.2
Valuing instream flows of the Verde River: Participation in and value of "near-stream activities"

	Days of participation in Arizona* (#)	Days of participation in Verde counties† (#)	Mean user-day value‡ (2002 USD)	Arizona value (2002 USD)	Verde counties value§ (2002 USD)	Verde River value (2002 USD)
Hunting**	1,630,000	1,087,103	$34.26	$55.8 mil	$37.2 mil	$3.3 mil
Wildlife watching††	9,447,000	6,300,526	$38.70	$365.6 mil	$243.8 mil	$21.5 mil
Camping	Unknown	Unknown	$28.50	Unknown	Unknown	Unknown
Picnicking	Unknown	Unknown	$26.10	Unknown	Unknown	Unknown
Sightseeing	Unknown	Unknown	$27.97	Unknown	Unknown	Unknown
Total	>11.1 mil	>609,235		>$421.4 mil	>$281.1 mil	>$24.8 mil

**Source*: USFWS 1998; residents and nonresidents.
†Yavapai, Coconino, Gila, and Maricopa counties contain the Verde River and their combined population is 67% of the state population (U.S. Census Bureau 2003) and thus 67% is applied to the state participation estimates to obtain the combined counties' estimate.
‡*Source*: Intermountain region, Loomis 1998a, 1998b.
§Verde River values are an approximated percent of the Verde counties values. The percent used is the percent of the Verde County land area that is in the Verde River watershed. This is believed to be an underestimate, since the Verde River provides exceptional recreational qualities and draw more recreational visitors than other locations in these counties.
**Mean hunting value applied here is an average of Loomis' values for small game, big game and waterfowl hunting.
††Includes observing, photographing, and feeding wildlife.

Instream Nonuse Values of the Verde River (Benefits Transfer)

Description of Applicable Nonuse Studies for Benefits Transfer

Most of the nonuse values of the Verde River are comprised of the values people have expressed for preserving the several species dependent on the river which are T&E. Most of the T&E species do not have consumptive uses and do not have non-consumptive uses (e.g., as a subject for photography), therefore nonuse motives are primarily the principal source of benefits estimates for these T&E species. Adverse impacts to these unique species would be locally or globally irreversible, leading to extinction being a relevant concern. Irreversible adverse impacts on unique resources are not a necessary condition for the presence of significant nonuse values, but these attributes (e.g., uniqueness, irreversibility, and regional, national, or international significance) would generally be expected to generate relatively high nonuse values (Carson et al. 1996).

For valuing T&E species, the number of candidate studies is frequently small and the variation of characteristics among the studies substantial, and thus requiring a more careful consideration of analytic situations in the original and case studies. When only one or a few studies are available, an analyst evaluates their transferability based on technical criteria articulated by Desvouges, Naughton, and Parsons (1992). The analyst first identifies the T&E species in question and the type of environmental scenario, and then selects from a pool of available studies the appropriate WTP values for protecting those species.

Some of the relevant studies that have conducted surveys on what people are willing to pay to prevent a species from going extinct are summarized below.

- Boyle and Bishop (1987) elicited an average state-wide bid of $6.88 (2002 USD) per household to preserve the striped shiner, a state-listed endangered minnow with no direct use value in Wisconsin (the striped shiner is state listed as an endangered species, but not listed federally). This study makes clear that it is possible that people are concerned about the survival of all species, even obscure ones (Bishop and Welsh 1992).
- Cummings, Ganderton, and McGuckin (1994), found that taxpayers would be willing to pay an average of $10.24 (2002 USD) annually to preserve the federally-listed endangered Colorado squawfish in New Mexico.
- Berrens, Ganderton, and Silva (1996) found that preservation of the endangered silvery minnow in New Mexico would be worth an average of $9.26 (2002 USD) per household, state-wide, per year, after converting the value to an equivalent basis as Boyle and Bishop (1987) and Cummings, Ganderton, and McGuckin (1994). This value was obtained by applying a 5% discount rate on a 5 year program using a 30-year payment period. If a 3% discount rate was used, the annual value is $7.67 and if a 7% discount rate is applied, the annual value is $10.85.
- In addition, there is empirical evidence that willingness to pay is greater when multiple T&E species are applicable, but that values for multiple T&E species are less than a strict multiple of the value for a single species (e.g., Loomis and Ekstrand 1997).

It is also necessary to note that the economic benefits of preserving T&E species estimated to date reflect a human-centered view (e.g., most often valuing present concerns more than future), and that these estimates may not be appropriate when T&E species are involved because extinction is irreversible.

Calculation of Nonuse Values Using Benefits Transfer

Once individuals' WTP for protecting T&E species or increasing their population is developed, the next step is the estimation of the value of preserving special status species. The analyst should apply the estimated WTP value to the relevant population groups to estimate the total value of improving protection of T&E species. The relevant population may include area residents, regional population, or, in exceptional cases (e.g., bald eagle), the U.S. population.

From the previous section, a range of values may be created: a low of $6.88 (2002 USD) per household (for the striped shiner) to a high of $10.24 (2002 USD) per taxpayer (for the Colorado squawfish). This range of value may be applied to the number of 2001 households of Arizona of 1.9 million (U.S. Census Bureau 2003), to obtain annual contributions ranging from $13.1 to $19.5 million to preserve one of these species. The high end of this range is likely an underestimate of the high end of household value, since estimate used is per taxpayer, and a household contains more than one taxpayer. Applying this to the 6 species that are listed above as T&E species along the Verde River, Arizona households presumably would be willing to contribute some amount higher than $13.1 million to $19.5 million annually to preserve the multiple species.

For comparisons purposes, at the highest end of the spectrum, citizens of Wisconsin were found to be willing to pay an average of $41.57 (2002 USD) to preserve the bald eagle (Boyle and Bishop 1987). Applying this to the number of 2001 households of Arizona of 1.9 million, it can be deduced that Arizona households might be willing to contribute $79.0 million per year to preserve the bald eagle alone. However, not all of that amount could be associated with Verde River flows, since bald eagle habitat in Arizona extends well beyond the site.

Opportunity Costs of Protecting Endangered Species

Another measure of economic values of protecting special status species is opportunity cost—the foregone economic opportunities due to restrictions on the use of property due to listings, designation of critical habitat, and recovery plans, often imposed by authority of the ESA. Opportunity costs include the reduced economic profit from restricted or altered development projects including agriculture production, timber harvesting, minerals extraction, and recreation activities; wages lost by displaced workers who remain unemployed or who are re-employed at lower pay; lower consumer surplus due to higher prices; and lower county property and severance tax revenue. However, it must be noted that "costs" and "economic impacts" do not necessarily reveal "value" and, hence, readers need to consider the important distinction between these terms and concepts.

A discussion of various protective programs are discussed in this section for illustrative purposes only, and although they entail costs and economic impacts, these measures should not be taken to necessarily represent the value of protection. They are merely presented here to display that notion that society does value protection, and that in protecting resources some costs are incurred. Whether the value (benefits) of the resource protecting government actions outweigh the costs is not always clear. This section discusses measures of economic activity, which is a methodologically different category than the measures of welfare used in the rest of this chapter. Thus, the two methods are not directly comparable, but offer another way of examining the issues.

The majority of the management actions along the Verde River have revolved around endangered species management. The designated critical habitat (CH) in the Verde River for razorback sucker extends from Perkinsville to Horseshoe Reservoir. The Verde River is also designated CH for the spikedace (*Meda fulgida*) and loach minnow (*Rhininchthys cobitis*) from Sullivan Dam to Fossil Creek (NPS 2004). The U.S. Forest Service has been involved in consultation with the Fish and Wildlife Service since the listings and designations of critical habitat. Consultation has resulted in management changes, mainly in livestock grazing activities. Drift and gap fences have been built on the Brown Springs Allotment to restrict/hinder livestock use along the Verde River. Several water access points have been closed. Also, several water sources were improved or developed in the uplands to draw livestock away from the river, and livestock grazing has been excluded along several sections (NPS 2004).

One may use the U.S. Fish and Wildlife Service's framework of "without critical habitat" (baseline) and "with critical habitat" for assessing economic benefits and costs as well as for assessing potential social and community impacts. Economic costs and benefits are measured by using changes in producer and consumer surpluses which are the economically valid measures of changes in social welfare. Regional economic benefits and costs include local (generally city and county) or state level changes in employment, income, or tax revenues. Analysis can be conducted over a time frame consistent with the likely expected period until the species of concern are delisted (typically 20 to 40 years), and quantitative results can be presented using both net present value estimates and annualized values. Social and community impacts may be significant for local communities; in some cases they may not be quantifiable. Impacts on small businesses may also be considered under social and community impacts.

Brookshire, McKee, and Watts (1994) conducted an economic analysis of CH designation for the Colorado River Basin for 4 endangered Colorado River fishes. One of the species included in the study which resides in the Verde River is the razorback sucker (the other 3 species are the bonytail, the humpback chub, and the Colorado squawfish). The study used economic input-output (IO) models to determine CH-imposed changes in economic activity in the agricultural, hydroelectric, mining/gas/oil, recreation, municipal, and industrial sectors. The study only examined economic impacts of the 4 species collectively. These impacts were estimated to have a negative annualized impact on economic activity in Arizona of $1.4 million (in 2002 USD). As stated at the beginning of this section, it must be noted that this impact is a measure of economic activity and not a measure of welfare, and thus cannot be compared to other values estimated in this chapter.

Other protective actions taken by the Arizona Game and Fish Department include reintroduction programs of razorback suckers and Colorado pikeminnow in the Verde River. These species were extirpated from the Verde River Basin and are part of a reintroduction program to recover the species in the state. Arizona is expending efforts to protect antelope habitat in the Upper Verde area. The U.S. Forest Service is in the process of acquiring land "to include the maximum antelope habitat advocated by the Arizona Game and Fish Department," through the Yavapai Ranch Land Exchange Act of 2002. All of these protection programs have costs and associated economic impacts, but estimates of their magnitude are unknown.

Table F.3
Annual value of instream water of the Verde River

	Low	High
Use values	>$35.0 mil	>$35.0 mil
Nonuse values	$13.1 mil	$19.5 mil
Other values (recreational values from swimming, boating, camping, picnicking, and sightseeing, the designation of the Verde River as a Wild and Scenic River, historical and cultural values, and other nonuse values)	+	+
Total	+$48.1 mil	+$54.5 mil

Summary of Instream Values of the Verde River

Table F.3 shows the annual value estimates of use and nonuse values of instream flows for the upper reaches of the Verde River area. Use and nonuse values total from more than $48.1 million to $54.5 million annually. This estimate does not include other recreation values (for swimming, boating, camping, picnicking, sightseeing), historical and cultural values, other nonuse values (e.g., existence and bequest motives) aside from preserving T&E species, and additional values from the Wild and Scenic River designation.

ESTIMATES OF DOWNSTREAM WATER SUPPLY VALUES OF THE VERDE RIVER

Water Sources for Phoenix Metro Area

Municipal water users in the Phoenix area are the primary downstream water users of the Verde River. There are three main water supply options for these users—Salt River Project (SRP) water (which includes use of Salt and Verde River waters), imported Central Arizona Project (CAP) water, and groundwater. SRP waters are fully appropriated, and are generally the cheapest source of high quality water in the Phoenix area. Any loss of Verde River waters would therefore need to be replaced by groundwater, CAP waters, or some innovative measures (e.g., water reuse).

Extractive Use Values for Municipalities and Other Users of Lower Verde River Water

The cost to replace a water source is commonly used to value water rights. Replacement cost for the next best option is taken in lieu of the potential willingness to pay for new supplies. This section examines the cost of Phoenix's next best option for acquiring water for municipal and agricultural uses.

One major source of water in the Phoenix area is groundwater. However, it is unlikely that Phoenix and surrounding areas will be able to look at groundwater sources for further supplies. Groundwater use in Arizona is governed by the Arizona Groundwater Management Act (AGMA). The AGMA requires cities to limit groundwater use to their renewable supply of groundwater by 2025, and to prove that they have an assured water supply for 100 years that is consistent with management goals and conservation standards for the Active Management Area in which the water user is located.

CAP water is the other major source available in the Phoenix area, and is generally the marginal source. Arizona water utilities can purchase a subcontract for CAP water from another entity in Arizona that has a contract and cannot use the water. There are currently declined/terminated subcontracts in Arizona that are available for purchase.

The City of Scottsdale has been active in acquiring additional CAP supplies. Subcontracts to water users in the Gila River Indian Community have been identified as the next source for possible acquisition by Scottsdale. The cost is estimated between $1,500 and $2,000 per acre foot to acquire the rights (Beth Miller, Scottsdale Water Department, personal communication, November 11, 2002). Scottsdale also pays operations and maintenance charges to the CAP. These charges are projected to grow from $66 per acre-foot in 2003 to $77 per acre-foot in 2007, although costs projected for future years are considered advisory and such projections made in the past have generally been adjusted downwards. Assuming that these charges reach $75 per acre-foot in the year projected (2006) and then stay constant at $75 over the next 100 years, the net present value of O&M charges would be approximately $1,000/AF, assuming an 8% discount rate. Adding the net present value O&M cost to the purchase cost, the total estimate for the city of Scottsdale to acquire additional CAP water would range from $2,500 to $3,000 per acre foot.

There are many assumptions made in this calculation. It assumes that treatment and distribution costs for CAP water would be the same as for SRP water, and ignores any transaction costs. This cost also has not been adjusted for reliability or constraints in the ability to deliver CAP water to all Phoenix area users currently able to use SRP water. In addition, drought risk to CAP deliveries is expected to grow over time. CAP water shortages are not expected to be a significant possibility for Arizona users of Colorado River water until at least 2025, when all lower basin users will be consistently using their full allotment. However, the probability of drought shortages for municipal users is expected to be about 5% after 2025.

The upper end of the estimate, $3,000 per acre-foot, generally matches an estimate derived for the replacement value of Salt River Valley Water Users in a water rights appraisal submitted to the Chino Valley Irrigation District (Sommers 1997). That estimate was based on an extensive study of the cost of replacement supplies by Phoenix area cities and SRP. The value of $3,000 per acre-foot was considered a conservative estimate based on the marginal net present value of new water supplies ranging from $2,500 to $4,000 per acre-foot in Arizona, and $2,000 to $10,000 per acre-foot elsewhere in the West (Sommers 1997). As another upper boundary comparison, the Tusayan community in Arizona can have water costs as high as $16,000 an acre-foot because they must supplement their supply with water trucked-in (Verde River Watershed Association 2000).

The base flows at the headwaters of the Verde River are 25 cubic feet per second, which is the equivalent of roughly 18,000 acre-feet per year. To obtain an annual value of these headwater flows, the $3,000 one time fee per acre-foot figure may be converted to an annual value. $3,000 per acre-foot at a 7% discount rate corresponds to an annualized value of $196 per acre-foot per year. Applying this to the 18,000 acre-feet per year of flows at the source, it may be deduced that the headwaters are worth over $3.5 million per year to the municipal users of SRP waters in the Phoenix area.

CONCLUSIONS

Value of Verde River Water

Combining the recreational use values of the Verde River, nonuse values of instream flows for the preservation of T&E species, and downstream municipal values of the flows originating at the headwaters, sums to a range from $51.6 million to $58.0 million per year (see Table F.4). This value pertains only to the flows in the Verde that originate at its headwaters (i.e., the springs fed by the groundwater in the Little Chino Basin), and do not include values for inflows that enter the main stem below the upper river. This value range is an underestimate as it does not include additional values experienced from:

- Recreational values from swimming, boating, camping, picnicking, and sightseeing
- The designation of the Verde River as a Wild and Scenic River
- Historical and cultural values
- Other nonuse values (e.g., existence and bequest motives) aside from the values stemming from the preservation of T&E species.

This valuation method also does not attempt to measure the value of the water actually used as it makes it way down the Verde River—for irrigation, drinking water, power generation, and other industrial/commercial uses such as dairy, golf, sand and gravel, water bottling, and fish hatcheries (AZ Department of Water Resources 2000).

Also, it should be mentioned that the natural water uses within the Verde Watershed are an important factor related to water loss along both tributaries and the main channel of the Verde River (AZ Department of Water Resources 2000). These water losses include evaporation of running surface water, evapotranspiration (ET) from naturally occurring vegetation in riparian areas (which is the single largest use of water in the Verde Watershed area), and diversions into wetland areas such as Tavasci Marsh and Pecks Lake (AZ Department of Water Resources 2000).

Table F.4
Annual value of water of the Verde River

	Low	High
Upstream instream values	>$48.1 mil	>$54.5 mil
Downstream municipal values	$3.5 mil	$3.5 mil
Other values (recreational values from swimming, boating, camping, picnicking, and sightseeing, the designation of the Verde River as a Wild and Scenic River, historical and cultural values, and other nonuse values)	+	+
Total	>$51.6 mil	>$58.0 mil

Implications

This study shows that instream flows are highly valuable, and not necessarily incompatible with downstream extraction for agricultural, residential, commercial and industrial uses. Verde River water provides large positive economic impacts from these downstream uses (not estimated in this study), and the instream flows of the upper reaches of the Verde River are estimated to be worth over $50 million a year.

These values also provide a benchmark for considering whether extraction of water from the source groundwater basin and upper river, by Prescott, may impose costs (foregone values) that outweigh the value realized by new users in Prescott. While this illustration has not attempted to value the water if used in Prescott, nor assess the extent to which Prescott extractions might reduce flows and adversely impact instream and extractive uses downstream, it is evident that any impact on these downstream values could amount to a sizable loss of economic value.

APPENDIX G
RECLAIMED WATER PROVIDES VALUE AS A RELIABLE SUPPLY FOR LOCAL BUSINESSES (SAN DIEGO, CALIF.)

The San Diego area has a relatively strong, high-value biotechnology employment base. Water often serves as a critical ingredient to the industry for process, cooling, sanitary, and other purposes. Water shortages can curtail production and even deter firms from locating in the area.

A study of water shortages (*Cost of Industrial Water Shortages*, CUWA 1991) shows that in response to a hypothetical 30% water supply reduction, about half of high-tech firms would consider locating plant expansions and new plant production facilities outside of California. This study also documented a direct link between water shortages and job and production losses.

In response to this situation, the city of San Diego developed its "Guaranteed Water for Industry Program." The program exempts research and development or industrial manufacturing firms from mandatory water restrictions in times of drought. To qualify for the program, the business must be located in an area where reclaimed water is available; use reclaimed water on its premises to the fullest extent possible; and, participate in all applicable city water conservation programs.

One area where companies are taking advantage of this program and insuring reliable water supplies is in the Torrey Pines area near the North City Water Reclamation Plant (NCWRP) in San Diego. This area has been dubbed "Biotech Hill" and has over 2.7 million square feet of office and laboratory space. The NCWRP is a state-of-the-art facility that can treat up to 30 million gallons of wastewater per day generated by northern San Diego communities. It has over 45 miles of distribution pipelines for reclaimed water.

The first participant in the Guaranteed Water for Industry Program was R.W. Johnson Pharmaceutical Research Institute, a member of the Johnson & Johnson family of companies. The facility was completed in 1999 and uses an average of 22 acre-feet per year of reclaimed water for landscape irrigation and its cooling system.

Another participant is PharMingen, a biotechnology company built on a diversified technology base covering immunology, cell biology, neurosciences, molecular biology and protein expression systems. PharMingen is the fourth largest biotechnology employer in city of San Diego. Reclaimed water is used for irrigation and cooling at an 80,000 square-foot facility on a six-acre campus. It is also used for irrigation and indoor plumbing (toilet flushing, urinal flushing, and priming of floor drain traps) at another 58,000 square-foot facility containing office space and custom designed, state-of-the-art laboratories. This PharMingen site is the first in San Diego to be dual plumbed to use reclaimed water.[1]

VALUE OF RECYCLED WATER FROM THE COMPANY PERSPECTIVE

Recycled water provides substantial drought insurance to companies participating in the Guaranteed Water for Industry Program such as R.W. Johnson Pharmaceutical Research Institute and PharMingen. This security can motivate companies to locate and expand in San Diego.

1. Based on interviews with Bob Schulz (858) 812-8850, Director of Facility Operations for PharMingen and Shirley Harry (619) 533-5356, city of San Diego.

Although water is not a large factor in the production chain, water cutbacks due to drought can be disruptive to business operations.

Recycled water is less expensive to the customer. The city of San Diego has offered a program to pay for all qualified capital costs related to the retrofit of existing facilities to receive recycled water. At the PharMingen site, for example, the city's reimbursements totaled $26,208.

Recycled water also costs less on a volumetric basis. In 2003, the city charged recycled water customers $350 per acre-foot. This charge is less than half the $712 cost being collected from similar customers for potable water.

Hence, these participating companies receive water this is both significantly cheaper and of higher reliability. Reliability is defined here as the ability to receive water during drought periods. Another form of reliability is the engineering reliability of the system. PharMingen has experienced short-term disturbances related to distribution system interruptions.

VALUE OF RECYCLED WATER FROM THE WATER RETAILER PERSPECTIVE

The city receives substantial benefits from water recycling. One benefit is that recycling serves an important role in its strategy to diversify the mix of water supplies available to its customers. A majority of the water delivered to the region is imported from far away sources such as the Colorado River and Northern California. These sources can dramatically fluctuate with weather and the long distribution lines can be subject to interruption (e.g., earthquake). Recycling provides more local control.

Another important benefit is associated with reduced disposal of wastewater flows. The U.S. EPA found San Diego to be in violation of wastewater discharges to the Pacific Ocean. Recycling projects are part of the solution identified by both the city and U.S. EPA.

Water recycling can also be the most cost effective solution in quenching increased water demands, especially when water reliability is factored into the equation. The city uses a cost-of-service philosophy whereby recycling costs are recovered from recycling customers. The $350 per acre-foot charge is based on such a cost-accounting philosophy.

We note, however, that this relatively low cost is achieved by two circumstances. First, the San Diego County Water Authority (Authority) and the MWD jointly provide financial incentives to stimulate the use of recycled water in their service areas. In 2003, the incentives total $400 per acre-foot delivered: $150 from the Authority and $250 from the MWD. The other circumstance is that wastewater customers pick up all treatment costs as wastewater flows already need to be treated to a high level before outfall into the ocean. Recycling customers only pay for the distribution costs from the plant to the end customers.

If recycled water was not available, the city would need to tap relatively high-cost alternative water supplies to procure the high reliability water key customers require. Recently the Authority evaluated desalination projects as a new source of supply. The delivered costs of this water, that has a high degree of reliability like recycled water, could be upward of $1,000 per acre-foot.

VALUE OF RECYCLED WATER FROM THE REGIONAL PERSPECTIVE

The costs and yields of water recycling projects need to be factored into the total mix of water supply and demand projects available to a region, a task undertaken by the Authority in this case. The Authority was established in 1944 and supplies from 75 to 95% of the region's needs, depending primarily upon annual surface water runoff into local reservoirs. The Authority's boundaries extend from the border with Mexico in the south, to Orange and Riverside counties in the north, and from the Pacific Ocean to the foothills that terminate the coastal plain in the east. With a total of 908,959 acres (1,420.3 square miles), the Authority's service area encompasses the western third of San Diego County.

The Authority is faced with ever increasing demands on the water supply system. With the increasing uncertainty regarding the availability of imported supplies, particularly from the State Water Project through the Bay Delta, a strategic part of the Authority's water resources planning is identification of alternative and supplemental supply sources. These supplemental supply sources will extend the limited potable water supplies available to meet the growing demands within San Diego County. As part of the overall plan that was adopted by the Authority as part of their Water Resources Plan, there is a need to identify local supply sources and increase water use efficiency.

Water recycling is an important component of the area's local water resources. A number of agencies in the San Diego region implement and are expanding their water recycling projects. Currently, about 13,700 AF of recycled water is beneficially reused within the Authority's service area annually. This number is projected to increase to over 53,000 acre-feet per year by 2020.

Information about specific future regional recycling projects is shown in Table G.1 (*San Diego County Water Authority Regional Recycled Water System Study*, RBF Consulting 2002).

Table G.1
Summary of strategies—most probable regional demands

Strategy	Yield (afy)	Capital cost ($)	Annual cost ($)	Wholesale cost ($/af)	No. of agencies served
1. Pendleton Groundwater Recharge	4,000	$9,594,000	$1,436,000	$359	1
2. Pendleton Golf Course	800	$5,444,000	$682,000	$853	2
3. F/O Pipeline Conversion	4,700	$14,181,000	$1,910,000	$406	3
4. Shadowridge Pipeline Conversion	1,500	$6,623,000	$908,000	$605	2
5. Vallecitos Extension	2,000	$11,501,000	$1,444,000	$722	2
6. Carlsbad/Olivenhain/ San Dieguito	6,910	$32,787,000	$4,377,000	$633	8
7. Escondido/Rincon/ San Pasqual	1,700	$14,041,000	$1,641,000	$965	4
8. Escondido/Padre Dam/ Helix/San Diego/Sweetwater	14,200	$74,368,000	$8,423,000	$593	7
9. Tijuana Valley	1,300	$7,636,000	$952,000	$732	2

The Regional Recycled Water System Study provides an evaluation of the technical feasibility of regional recycled water system strategies. The study provides an evaluation of the opportunities, considerations and costs for the regional strategies. A number of important implementation issues were identified during the course of the study. Advancement of the regional system will require resolution of the following implementation issues:

1. Identification of avoided water and wastewater costs that may result from implementation of the regional strategies. These costs may result from avoided disposal costs and reduced potable water costs.
2. Identification of the benefit that increased recycled water supply will have on the region's water supply mix and long term reliability.
3. Allocation of costs and benefits among project participants and beneficiaries, including establishment of a project baseline, as appropriate. Further analysis will be required to determine the cost sharing requirements for the regional system improvements.
4. Identification of institutional arrangements necessary for operation, management and administration of regional facilities.
5. Identification of external funding sources. Several potential funding sources have been identified; however, other sources may become available as new bonds and funding opportunities are developed.

Beyond these direct regional benefits, large-scale recycling can also provide significant improvements to aquatic habitats impacted by waters imported by San Diego. For example, enhanced instream flows in the Sacramento-San Joaquin Delta region can improve and increase aquatic and terrestrial habitats and improve ecological functions to support sustainable populations of diverse and valuable plant and animal species. These enhancements generate nonmarket goods and services, and nonuse (passive use) and indirect use values. The photo in Figure G.1 provides an example of how San Diego communicated its reclaimed water program values.

Figure G.1 San Diego communication on reclaimed water value

APPENDIX H
DEVELOPING AN APPLES-TO-APPLES COMPARISON OF WATER SUPPLY OPTIONS FOR A WATER-SHORT REGION (EL PASO, TEXAS)

El Paso Water Utility currently utilizes both groundwater and surface water resources. Groundwater is currently pumped from the Hueco and Mesilla Bolsons (aquifers). Pumping in the Hueco Bolson peaked in 1989, and has been slowed after concerns were raised over the rate of depletion of the resource. Non-sustainable groundwater mining of the Hueco and Mesilla Bolsons was, in effect, taking water away from other current users (spatially, in Mexico and New Mexico) as well as future users (temporally, leaving higher costs and diminishing water yields to future El Paso residents).

SURFACE WATER RIGHTS

EPWU has acquired surface water rights to the Rio Grande Project from the El Paso County Water Improvement District #1 (EPCWID #1, or the District). The project is run by the U.S. Bureau of Reclamation, and almost all water has historically been used for irrigation purposes. Surface water is supplied from the EPCWID irrigation canals at the Robertson/Umbenhauer and Jonathan Rogers water treatment plants. Under normal conditions, these plants are operated for 8 months out of the year. Water from the Rio Grande is not available to El Paso in the other months, or is not of sufficient quality.

Several contracts govern existing surface water sources available to EPWU. A 1941 contract between the city of El Paso, USBR, and the District allows EPWU to acquire a maximum of 2,000 acres of water rights lands within the EPCWID #1 boundaries, whose water could be used for municipal purposes. Water appurtenant to acres owned by EPWU in excess of the 2,000 acres is one of the sources of water, up to 1,000 acres maximum, under the 2001 Third Party Implementing Contract. New water supplied under this contract is the water saved from the American Canal Extension Project completed in April 2000. A 1962 contract allowed EPWU to lease an unlimited number of surface water rights from parties with tracts 2 acres or less within the City limits. A 1989 contract allows EPWU to sell potable water on a wholesale basis to the Lower Valley Water District (LVWD) in exchange for water rights LVWD leases under a November 1988 contract. A total of 66,831 AF are available to EPWU under these contracts at full allotment. EPWU has reached contract limits on additional surface water under the 1941 contract. Water leases under the 1962 and 1989 contracts are not capped, but the quantity of lands that become available per year is limited.

One option for obtaining surface water is an adjudication settlement of a 1949 permit. EPWU is in the process of reaching a MOU with the District for a settlement of 11,000 acre-feet of water (Option A for purposes of this analysis). There is no charge for the water itself under this agreement. Included in this settlement is a $15 per AF charge from the District and a $5 per AF charge from the Bureau of Reclamation. Total charges under this MOU would be $20 per AF.

As another surface water option for securing a large volume of additional surface water supplies, EPWU is considering large-scale acquisition of surface water rights from the Rio Grande. A March 1998 contract between USBR and the District allows the District to enter into long-term agreements with third parties such as EPWU for the delivery of Rio Grande Project water for uses other than irrigation. EPWU has signed such an agreement with the District that expires in 2010-2011. The price of water under this agreement is \$208 per AF with scheduled price escalations each year. This water would also be subject to a \$15 per AF charge from the District, and a \$5 per AF charge from the Bureau of Reclamation. The total cost for this option is \$228 per AF (\$208 + 20 per AF).

For either surface water option, however, this is only a portion of the total cost to provide this water to EPWU's customers. Treatment of this water is expected to be accomplished at the Johnathan Rogers and Canal Street treatment plants. Use of the plants depends in part on the location of the diversion, but a majority of the treatment is expected at the Johnathan Rogers plant. Johnathan Rogers has recently been expanded, with annualized capital costs at \$140 per AF (annualized at 5.0% over 20 years), and operation and maintenance expenses totaling \$111 per AF. Thus total treatment costs are approximately \$251 per AF. Transmission facilities required to connect water leaving the expanded facility with the existing distribution system total \$10.5 million, adding an additional \$58 per AF. The combined cost of acquisition and treatment of water under the MOU at Johnathan Rogers is \$329 per AF (\$20 + \$251 + \$58). The combined cost of treating water under subcontract to the District is \$537 per AF (\$208 + \$20 + \$251 + \$58) (Option B). Figure H.1 shows the incremental costs of providing surface water to EPWU customers under Option B.

DESALINATION

One source of potable water to come on-line in the near future is desalination of brackish (highly mineralized) groundwaters from the Hueco Bolson. A joint desalination plant is being built with Fort Bliss (a local military base), and is scheduled to be running by the spring of 2006. The reverse osmosis (RO) plant is designed with a capacity of 27.5 million gallons per day (mgd) to deliver 27.0 mgd or approximately 30,250 AF per year. Existing wells are to be rehabilitated and 16 new brackish water wells are to be installed. A 7.5 mgd pump station is included along with a finished water transmission line. Brine disposal will be handled by deep well injection. Based on annualizing the capital costs over a 20 year period at 5%, and including O&M costs, the cost of obtaining 30,250 AF per year of desalinized water is \$518 per AF.

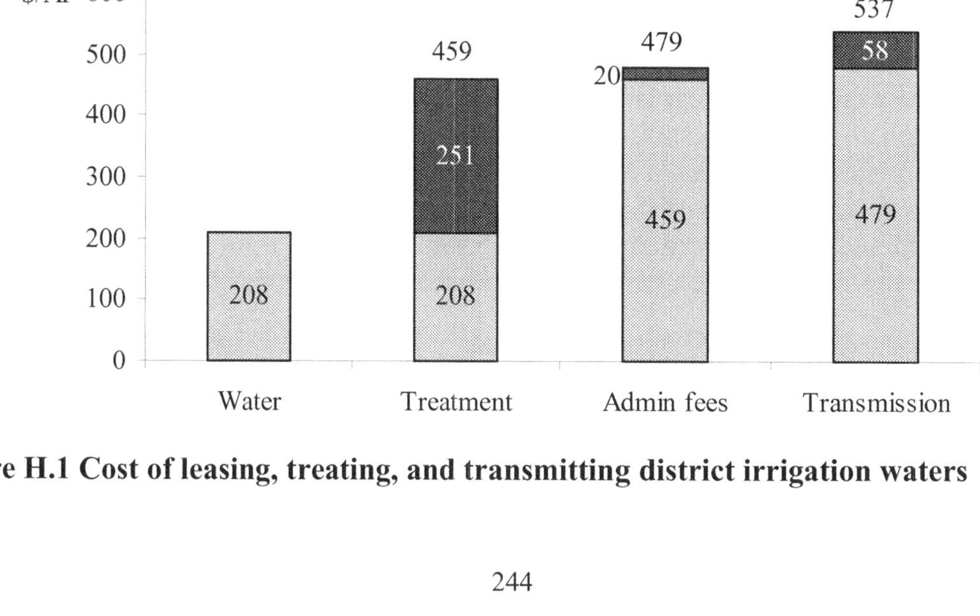

Figure H.1 Cost of leasing, treating, and transmitting district irrigation waters

There are several uncertainties associated with this estimate. For instance, it is not clear whether disposal of brine wastes via injection well will be accepted by regulatory authorities. Other options for disposal of brine waste such as a pond liner system may be less expensive, while other options may be more expensive. We include an uncertainty factor of plus and minus 20% to the cost estimate. The projected costs of desalination are then $414 to $622 per AF.

GROUNDWATER IMPORTATION

EPWU continues to evaluate the option of importing groundwater from other parts of west Texas. EPWU has acquired several west Texas ranches: the 21,000 acre Wild Horse Ranch 110 miles from El Paso, the 25,000 acre Antelope Valley Ranch 150 miles from El Paso, and the 8,800 acre Diablo Farms property approximately 80 miles from El Paso. Water from Dell City, 70 miles from El Paso where a desalination plant is to be constructed, is also being evaluated.

The scenario analyzed is the acquisition, treatment and conveyance costs associated with providing water from the Diablo Farms and Dell City. Diablo Farms was purchased for $13.3 million from an irrigator to acquire an estimated 16,700 AF per year of groundwater. For the first 15 years EPWU will lease the property back to the irrigator, allowing the irrigator to take up to 11,000 AF/year. EPWU will not use groundwater from this source for the first 15 years, and assuming a 75-year supply of groundwater, the present value cost of this supply is $60 per AF after subtracting out the revenue from the lease-back. The acquisition cost for the Dell City property is approximately $18.8 million, or $186 per AF, again assuming a 75-year supply that will not be used until approximately 2020, yielding approximately 11,100 acre feet per year. The combined acquisition cost for 38,800 AF per year from these sources is $246 per AF.

The estimated cost of treating and conveying the approximately 38,800 AF of water through the desalination plant at Dell City would be $2,054 per AF ($1,314 per AF capital, $740 O&M), based on annualizing capital costs over a 20 year period at 5%. The conveyance part of this cost is based on laying a 87" pipe for 111 miles and a 60" pipe for the remainder of the distance to Diablo Farms. The water also would be desalinated in Dell City, turning approximately 2300 TDS feed water into 500 TDS finished water.

The combined acquisition, treatment and conveyance cost for this supply is $2,300 per AF. There are several uncertainties with this estimate. Desalination costs may be understated to the extent that brine removal becomes more difficult and costly to manage than anticipated. There also may be considerably higher (or lower) costs associated with acquiring rights of way and environmental impact clearance than is currently assumed in the estimate for conveyance costs. Therefore, an uncertainty factor of plus and minus 20% is applied to the cost estimate, to derive a range of $1,840 to $2,760 per AF.

RECLAMATION

Reclamation (water recycling) provides another method to augment EPWU's water supply. To the extent that use of reclaimed water is a substitute for potable waters, then each AF of reclamation water produced and used represents an AF of new surface water that will not need to be acquired, conveyed, and treated.

The cost per AF of reclaimed water was estimated using accounting data on costs and production levels for the Northwest reclamation plant (it does not include costs or production associated with other EPWU reclamation facilities). There are several issues associated with this analysis that deserve careful attention and interpretation:

1. The Northwest system is the least cost-effective of the EPWU reclamation facilities. As such, the results below reflect above-average costs per AF of reclaimed water produced by EPWU.
2. A significant portion of the costs associated with reclamation are linked to the need to develop dedicated distribution lines and associated pump and reservoir systems. Reclamation source water is wastewater that would otherwise still be treated prior to discharge; hence, the incremental costs of additional treatment to render the wastewater suitable for reclamation are relatively modest. In contrast, the distribution costs for reclaimed water are in large measure responsible for the relatively high cost per AF results derived below.
3. While distribution-related expenses for reclaimed water are a necessary expense of water reuse as currently practiced, it is important to net out distribution-related costs savings for the potable supply system. Specifically, by using dedicated reclaimed water distribution systems to serve peak demand customers, EPWU is able to forego expansions to the potable supply distribution network. The analysis should therefore net out the avoided costs of any potable system expansions that otherwise would be needed to meet demand absent the reclamation program.
4. If the other water supply options evaluated in this report are costed without accounting for any major conveyance and distribution expenses, then the comparison between those source options and reclamation will be skewed so as to make reclamation appear less economic than it really is. For example, new surface water supplies to feed the water treatment plant will require conveyance over a considerable distance, and the costs of such conveyance is not included in the analysis so far.

Given the above factors, the accounting data provided by EPWU for the Northwest facility were converted to total annual production costs in the following manner:

1. Capital outlays were converted into annual debt service payments. The total capital outlay for the facility was $25.8 million. Debt service on this amount, assuming a 5% average borrowing cost for EPWU and a typical repayment timeframe (and equipment lifetime) of 20 years, would be $2.07 million per year.
2. Netting out the cost of potable supply distribution upgrades that otherwise would have been required in the Coronado area, the appropriate portion of the annualized capital outlays is approximately $1.5 million per year. The cost savings were approximately $5 million (as per memorandum from David Brosman to Nick Costanzo, April 3, 2000).
3. O&M costs of $0.3238 per ccf; or $140 per AF are based on the level of production and are an estimate of cost to add future production. The portion of these costs that is attributable to distribution (rather than incremental treatment beyond wastewater discharge) should not be counted in these calculations, because absent the reclaimed water source, the alternative supply also would require distribution (i.e., pumping) to the customers. However, estimates of the share of O&M costs associated with distribution are not currently available. Hence, the O&M costs shown here include

distribution costs and, as such, overstate the cost of the reclamation as a substitute for other source waters.

4. At current production levels (firm demand for 2003) of 670 AF (216 million gallons per year), the total annualized share of the cost of production is $1.6 million.

Costs per AF are then derived using alternate production levels. The costs per AF decrease with increased production, because the fixed cost (capital expense) is spread over a greater volume of water. The results are as follows:

- For near-term firm demand (670 AF/year), the cost is $2,337 per AF. Since this reflects high per unit costs associated with facility (and market) start up, this estimate is not a useful benchmark for this analysis.
- For planned demand (1,750 AF per year), the cost is $981 per AF.
- For demand at capacity (10,400 AF per year), the cost is $282 per AF.

Clearly, the cost per AF is highly dependent on the level of production, which in turn is driven by the level of demand for reclaimed water. The costs become very competitive to other supply options as production increases. In addition, given that we have not been able to properly segregate or otherwise account for some distribution and conveyance-related expenses, the reclamation costs provided above are overstated compared to some other supply augmentation options presented elsewhere in this report.

Aside from what reclaimed water will cost, it also is important to recognize the amount of water that reclamation can supply. The discussion above has focused exclusively on the Northwest facility, and omits the other reclamation plants that exist in the EPWU system. To gain a broader perspective, one needs to recognize that EPWU uses approximately 120,000 AF per year, of which half is required for return flows. This in effect places a cap on reclamation potential of approximately 60,000 AF per year. Of this 60,000 AF potential, there are physical constraints on reclamation water production due to infrastructure (i.e., treatment and distribution facilities). According to EPWU staff, this effectively caps the volume of reclamation water that can be delivered to approximately 35,000 to 40,000 AF per year (with a best guess estimate of 38,000 AF per year). As a point of comparison, EPWU's recent reclaimed water use was about 14,000 AF, and the utility's medium-term goal is for reclamation use to grow to approximately 21,000 AF per year. This reveals that anticipated near-term demand is what will limit reclamation use (as demand is well below EPWU's existing ability to supply reclaimed water).

CONSERVATION PROGRAMS

Water conservation measures that have a long-term effect on water demand and consumption can be considered an alternative supply source. EPWU has an aggressive water conservation program with a portfolio of water conservation measure that produce long-term water savings. Below are listed some of these options, with a calculation of their cost per acre-foot to supply. Table H.1 shows the details on each calculation.

Table H.1
Water conservation program unit costs and water savings

	Residential showerhead replacement	Scenario 1	Scenario 2	Scenario 3	Scenario 4	Notes
A	Unit cost to EPWU	$2.13	$2.13	$2.13	$2.13	$1.72 per SH + $0.41 admin
B	Persons per dwelling unit	3.2	3.2	3.2	3.2	2000 census
C	Showers per dwelling unit delivered	2.0	2.0	2.0	2.0	1.7 shower/unit
D	Persons per shower	1.33	1.33	1.33	1.33	B/C * 83% >age 9
E	Minutes showering per day per person	7.8	7.8	10.0	10	Scenario 1, 2 = AwwaRF
F	Gallons savings per minute	0.6	0.6	1.0	2.5	2.3 gpm low-flow showerhead
G	Water savings per showerhead (GPD)	6.2	6.2	13.3	33.2	D*E*F
H	Remaining expected life of replaced SH (years)	***5.0***	***7.0***	***7.0***	***7.0***	
I	Water savings over lifetime (AF)	0.0301	0.0403	0.0861	0.2152	Present value over life @ 5%
J	$/AF	$71	$53	$25	$10	A/I
K	Dwelling units - single family + duplex 2003	144,980	144,980	144,980	144,980	From billing records
L	Total showerheads 1999	289,960	289,960	289,960	289,960	C*K
M	Low-flow showerheads already installed (10%)	28,996	28,996	28,996	28,996	Unknown
N	High-flow showerheads installed (90%)	260,964	260,964	260,964	260,964	L-M
O	Extra savings over lifetime if all low-flow (AF)	7,866	10,513	22,463	56,157	
P	Water savings if all low flow (AF per year)	1,817	1,817	3,882	9,705	Note limited time horizon
	Turf replacement incentive (SF)					
A	Unit cost to EPWU per Ft2	***$1.00***	***$1.00***	***$1.00***	***$1.00***	Per Ft2 of turfgrass removed
B	Front lawn size replaced (Ft2)	1,000	1,000	1,000	1,000	Assumed average
C	Cost to EPWU per site	$1,000	$1,000	$1,000	$1,000	A*B
D	Lawn watering per FT2 per year (feet)	6.2	6.6	6.9	7.2	ETo x Kc (0.6) / DU (0.46)
E	Replacement watering per FT2 per year (feet)	1.2	1.1	1.1	1.0	ETo * Xeriscape Coefficient (0.2)
F	Water savings per FT2 per year (feet)	5.0	5.5	5.8	6.2	D-E

(continued)

Table H.1 (Continued)

	Residential showerhead replacement	Scenario 1	Scenario 2	Scenario 3	Scenario 4	Notes
G	Water savings (GPD) per year	102.5	112.7	118.3	127.1	B*(F*7.48 gallons/Ft3)/365
H	Remaining expected life of replaced turf (years)	20	20	20	20	
I	Water savings over lifetime (AF)	1.4304	1.5734	1.6521	1.7737	Present value over life @ 5%
J	$/AF	$699	$636	$605	$564	C/I
K	Dwelling units - single family + duplex 2003	144,980	144,980	144,980	144,980	From billing records
L	Total front lawns (100%)	144,980	144,980	144,980	144,980	
M	% of front lawns potentially replaced	50%	50%	50%	50%	
N	Potential front lawns to be replaced	72,490	72,490	72,490	72,490	L-M
O	Extra savings over lifetime if all xeriscape (AF)	103,688	114,056	119,759	128,573	
P	Water savings if all participate (AF per year)	8,320	9,152	9,610	10,317	20 year horizon
H-axis washing machine rebate (SF)						
A	Unit cost to EPWU	$200	$200	$200	$200	$200 per washer + $0 Admin
B	Persons per dwelling unit	3.2	3.2	3.2	3.2	2000 Census
C	Washers per dwelling unit	1.0	1.0	1.0	1.0	
D	Persons per washer	3.2	3.2	3.2	3.2	B/C
E	Loads per day	1.0	1.0	1.3	1.0	Thelma
F	Water savings per load (GPD)	21.5	15.8	21.5	21.5	Scenario 2 and 4 = AwwaRF
G	Water savings per washer (GPD)	21.5	15.8	28.0	21.5	E*F
H	Remaining expected life of replaced SH (years)	***10***	***20***	***20***	***20***	
I	Water savings over lifetime (AF)	0.1860	0.2199	0.3902	0.3001	Present value over life @ 5%
J	$/AF	$1,075	$910	$513	$666	A/I
K	Dwelling units - single family + duplex 2003	144,980	144,980	144,980	144,980	From billing records
L	Total washers 2003	144,980	144,980	144,980	144,980	C*K
M	H-axis washers already installed	0	0	0	0	Unknown: small #

(continued)

Table H.1 (Continued)

	Residential showerhead replacement	Scenario 1	Scenario 2	Scenario 3	Scenario 4	Notes
N	Non-H-axis washers installed	144,980	144,980	144,980	144,980	L-M
O	Extra savings over lifetime if all H-axis (AF)	26,961	31,876	56,567	43,513	
P	Water savings if all H-axis (AF per year)	3,492	2,558	4,539	3,492	Note limited time horizon
ULFT single family replacement program						
A	Unit cost to EPWU	$103	$103	$103	$103	$75 rebate + admin
B	Persons per dwelling unit	3.2	3.2	3.2	3.2	2000 Census
C	Toilets per dwelling unit	2.0	2.0	2.0	2.0	EBMUD
D	Persons per toilet	1.60	1.60	1.60	1.60	B/C
E	Flushes per person per day	5.0	4.0	5.0	5.0	Scenario 2 = Brown & Caldwell
F	Gallon savings per flush	2.1	2.5	2.1	2.5	Scenario 1, 3 = AwwaRF
G	Water savings per ULFT (GPD)	16.8	16.0	16.8	20.0	D*E*F
H	Remaining expected life of replaced toilet (years)	***5***	***10***	***10***	***15***	
I	Water savings over lifetime (AF)	0.0815	0.1384	0.1453	0.2325	Present value over life @ 5%
J	$/AF	$1,266	$745	$710	$444	A/I
K	Total dwelling units 1992 (includes duplexes)	126,553	126,553	126,553	126,553	From billing records
L	Total 1992 toilets	253,106	253,106	253,106	253,106	C*K
M	ULFTs installed in pre-1993 housing	75,299	75,299	75,299	75,299	4.25% natural replacement rate
N	Non-ULFTs in pre-1993 housing	177,807	177,807	177,807	177,807	L-M
O	Extra savings over lifetime if all ULFTs (AF)	14,487	24,607	25,837	41,346	Savings diminish over time
P	Water savings if all ULFT (AF per year)	3,346	3,187	3,346	3,983	Note limited time horizon
Refrigerated air rebate						
	Unit cost to EPWU	$200	$200	$200	$200	
	Water savings (GPD)	81.5	81.5	81.5	81.5	
	Number of use days	213	213	213	213	

(continued)

Table H.1 (Continued)

Residential showerhead replacement	Scenario 1	Scenario 2	Scenario 3	Scenario 4	Notes
Water savings gallons	17,360	17,360	15,000	17,360	
Remaining expected life of replaced cooler	10	12	15	15	
Water savings over lifetime (AF)	$0.4114	$0.4722	$0.4778	$0.5530	Present value over life @ 5%
$/AF	$486	$424	$419	$362	
Dwelling units - single family + duplex 2003	144,980	144,980	144,980	144,980	
Saturation fraction of evaporative coolers	0.58	0.58	0.79	0.79	Cooler only, cooler and AC Tucson
Coolers per dwelling	1	1	1	1	Unknown
Total coolers	84,088	84,088	114,534	114,534	
% coolers potential replaced	30%	50%	30%	50%	Assumed
Potential coolers replaced	25,227	42,044	34,360	57,267	
Potential savings (AF per year)	1,344	2,240	1,582	3,051	

Low-Flow Showerheads

This potential program would consist of wide distribution of showerheads rated 2.3 gallons per minute. EPWU customers would be provided with a coupon that they could redeem at several sites for up to two free showerheads. The relatively low cost of this program is based on EPWU making a bulk purchase of the showerheads at $1.72 per showerhead from a manufacturer, and its relatively low distribution cost of $0.41/showerhead. Our best estimate of the return of this program is $53 per AF, based on a seven-year showerhead useful life, a 0.6 gallon per minute water savings, and a 100% installation rate.

Variations in our assumptions show that the program's return may range between $10 and $71 per AF. If we assume that the gallons saved per minute increased to 2.5 (Scenario 4), the program return is $10 per AF. If the assumed average useful life of the new shower heads is shortened from seven years to five (Scenario 1), the program return is $71 per AF. The potential water savings from all single family customers installing these showerheads is estimated to be at least 1,817 AFY.

Refrigerated Air Rebate

EPWU teams with EL Paso Electric to provide an incentive to customers to replace evaporative water cooling systems with refrigerated air systems in their existing or new homes. Converting an evaporative cooling system to refrigerated air has been estimated to save 17,000 gallons per year. EPWU provides $200 of a $300 total rebate to customers. Assuming a 15 year remaining life on the evaporative cooler replaced, the program return is $362 per AF. However, the total estimated life of a cooler using water with 500 parts per million of total dissolved solids is 18.3 years (based on appliance lifetime functions for remaining lifetime at 500 ppm TDS). If the average remaining life of coolers replaced is 10 years, then the program costs rise to $486 per AF.

Turf Replacement

This conservation program pays established residential customers and established commercial and industrial customers to replace turfgrass with low water using landscapes. This analysis concentrates only on the residential customers. The rebate offered is $1.00 per square foot of turfgrass replaced, of which $0.13 is allocated to administrative costs. It is assumed that irrigation would decrease by 5.8 feet per year over the turf replacement area. This assumes that customers would match their water application with xeric water needs—which would most likely require some education and behavioral changes. Our best estimate of the return of the turf replacement program is $605 per AF.

The scenarios show the net feet of water savings per year varying from 5 to 6.2. At 5 feet of net water savings, the program return is $699 per AF. At 6.2 feet of net water savings, the program return is $564 per AF. If half of the homes served by EPWU participate in the program and replace an average of 1,000 square feet of turf, then EPWU water use would decrease by about 9,600 AFY.

Horizontal-Axis Washer

Horizontal-axis (front loading) clothes washers use significantly less water per load, along with saving electricity or gas, and have received good performance reviews. EPWU offers a $200 rebate to customers purchasing these relatively expensive machines (typically $600 to $1,200). Assuming all customers purchasing horizontal-axis washers and receiving a rebate would not have purchased machines without the rebate, the program cost is $666 per AF.

Changing assumptions in loads per day, water saving per load, and levels of free ridership produces a range of estimated program returns, as shown in Table H.1. If loads per day are assumed to be 1.3 instead of 1.0, then the program cost is $513 per AF. If half of the customers would have purchased the horizontal-axis washers without the rebate (i.e., program free rider causing effective life to drop to 10 years as shown in Scenario 1), then the EPWU cost increases to $1,075. We have no estimate of the level of free-ridership to expect with this program.

Single-Family Ultra-Low-Flush Toilets

EPWU currently offers a rebate to customers for replacing older toilets designed to flush 3.5 gallons or more with an ultra-low-flush toilet (ULFT) designed to flush 1.6 gallons or less. Customers receive a rebate of 75% of the cost of the toilet up to a ceiling of $100. We estimate EPWU's return from the ULFT rebate to be $1,266.

Alternative scenarios from Table H.1 show the impact of altering the effective remaining life of the non-ULFTs that are replaced under the program. If the remaining life were extended out to 10 years, the cost would drop to $710 per AF (Scenario 3). The cost drops to $444 per AF with a 15-year effective life (and higher gallons per flush estimate).

SUMMARY

Figure H.2 shows the range of water supply options examined in this study. Some of the costs of acquiring Rio Grande through the District are shown first, followed by desalination of local groundwaters, and groundwater importation (and desalination) and, finally, by the water conservation water reuse options. The figure shows that in general the costs of future options will be high especially in obtaining larger amounts of water through importation and desalination of groundwater.

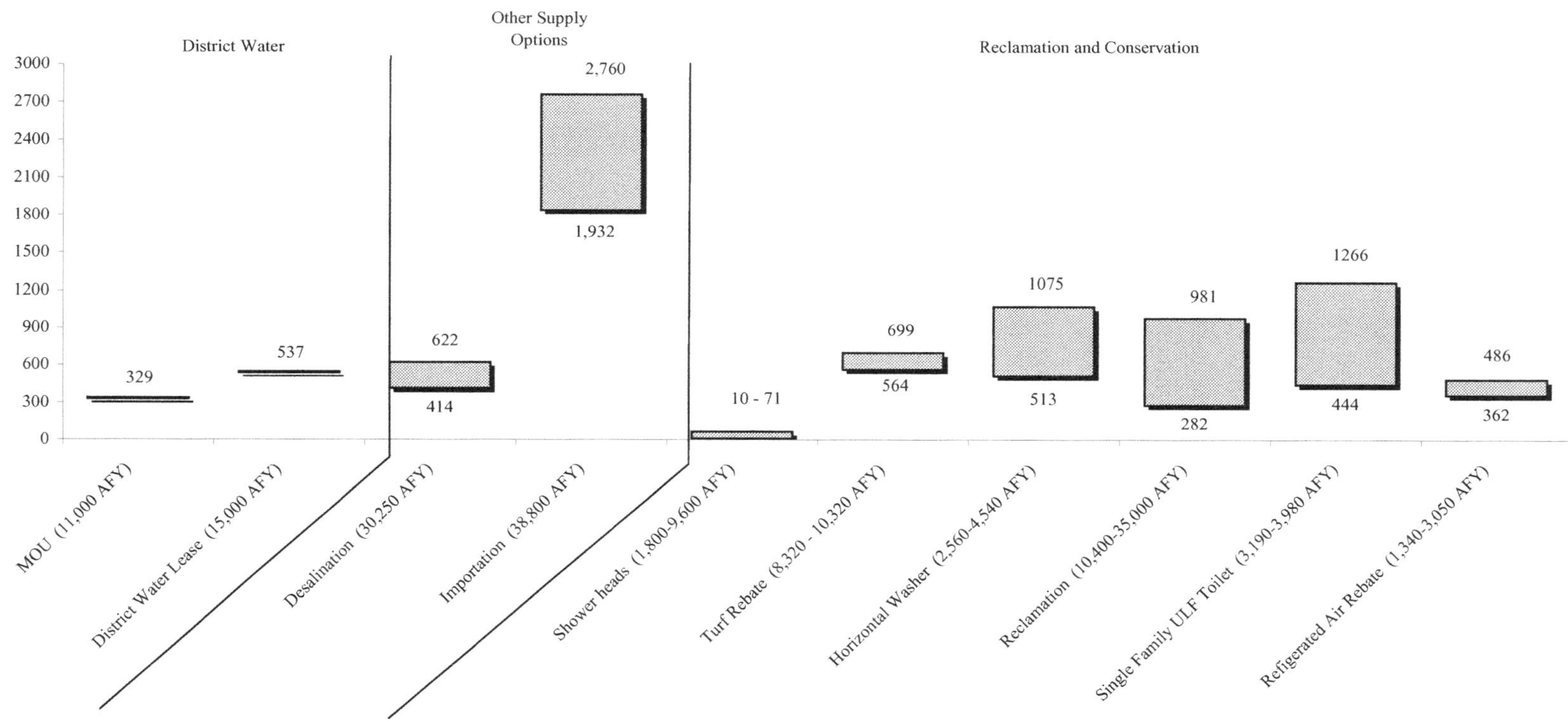

Figure H.2 Water supply options

APPENDIX I
INSTREAM FLOWS GENERATE SIZABLE ECONOMIC VALUES FOR REGIONAL ECONOMICS BASED ON OUTDOOR RECREATION AND TOURISM (VAIL, COLO.)

WATER MANAGEMENT ISSUES RELATED TO RECREATIONAL INSTREAM USE VALUES

Instream recreational uses of water, such as for recreational purposes, can be highly valuable. These instream values accrue not only to those engaged in the recreational activities (e.g., whitewater boaters), but can also provide value for spectators and for the local and regional economy. When water resource managers are considering instream recreational uses, they may need to make decisions about whether certain investments are economically justified (e.g., is it worth the expense to develop access points and whitewater course features to attract and enable recreators to use the facility?). Broader water resource allocation issues also may arise, such as whether there are tradeoffs with other water uses when waters committed to instream uses may preclude or limit upstream extractions. In the latter case, managers need to ascertain the comparative value of the instream uses against potentially impacted upstream extractive uses. This case study provides an illustration of how much value can be generated by instream whitewater boating, based on experience at Gore Creek, in Vail, Colo.

BACKGROUND

In the autumn of 2000, the town of Vail, Colo. constructed the Vail Whitewater Park, a kayaking and canoe course. The park is located near downtown Vail on Gore Creek, which is a tributary to the Eagle River. The park consists of three control structures along approximately 300 feet of the existing channel and provides recreational opportunities for beginner to advanced boaters. This case study provides a cursory summary of the order of magnitude of the economic benefits associated with the water rights used to supply the kayak course. The findings reported here are based on a 2002 review of available information on kayaking and other activities associated with the park, economic valuation literature on kayaking, and user data.

The results of this case study demonstrate that water diverted for recreational use at the site generates considerable economic benefit to the boaters using the course, other interested parties, and the community as a whole. As summarized later in this case study, in Table I.5, the total annual beneficial value generated by the waters diverted within the Vail Whitewater Park is estimated to be greater than \$1.8 million (in 2001 dollars), not including several benefits that could not be quantified or valued within the present study's constraints. Over a 20 year period, the present value of all current and anticipated future benefits provided by the waters diverted within the kayak course would be over \$20 million (in 2001 dollars).

ESTIMATION OF TOTAL RECREATIONAL VALUE

The total value of kayaking at the Vail Whitewater Park has two components: (1) what people actually pay to kayak (e.g., equipment costs); plus (2) what they would be willing to pay over and above what they currently pay. The first component of value can be represented simply by the expenditures incurred. The second component deserves more explanation. Consumers purchase products in the marketplace because they are better off with the products than they were with the money needed to obtain the products (or whatever else they would have purchased with the money). If that were not true, goods and services would not be exchanged through free will in the marketplace. Similarly, recreational site visits cost money and time, and recreational users would not undertake visits unless the visits yielded net benefits to the participant. Those net benefits are referred to by economists as "consumer surplus," and are measured as WTP.

This case study is divided into several subsections. First, an estimate of kayaking expenditures (for equipment, automobile costs, and travel time) is developed. Second, estimates of WTP "unit values" obtained from the peer reviewed economics literature for recreational whitewater boating are discussed. Third, the use of the kayak course through normal recreational use and through competitive events that attract spectators is described. The following section discusses the economic stimulus to the local community because of the whitewater course. Finally, the value from recreational use of water estimated for the Vail Whitewater Park is compared to the value of uses at other sites.

Kayaking Expenditures

As one component of the value of water recreation in Vail, the research team conducted a preliminary assessment of the costs that might be typically incurred by a kayaker visiting the Vail Whitewater Park. The analysis considers three cost items for all kayakers: kayak equipment, automobile expenses, and travel time. Lodging cost is also included for nonlocal kayakers.

Table I.1 shows the cost calculation for kayak equipment. The estimated cost of purchasing a typical set of new kayak equipment is approximately $2,000. This estimate is based on a review of prices shown at Alpenglow Mountainsport Inc., a retailer of boating gear located in Golden, in 2000. The set of equipment includes a kayak, paddle, helmet, dry top, life jacket, spray skirt, booties, gloves, and throw rope. Equipment costs are amortized over three to five years and it is assumed that the equipment is used on average 15 to 20 days per year. This leads to an average equipment cost ranging between $20 and $44 per user day. In Vail in 2001, the 1-day rental cost for a whitewater kayak is $40 (Mountain Quest Sports). In nearby Breckenridge, Colorado, the 1-day rental cost for a whitewater kayak is $45 (Summit Kayak and Canoe). Hence, the range of $20 to $45 is believed to be a reasonable estimate of the equipment cost per kayaking outing.

Table I.1
Cost of kayaking equipment (2000 USD)

Kayak gear purchase cost	Useful life (years)	Average user days/year	Gear days over useful life	Kayak gear cost per user day
$2,000	5	20	100	$20
$2,000	4	20	80	$25
$2,000	3	20	60	$33
$2,000	5	15	75	$27
$2,000	4	15	60	$33
$2,000	3	15	45	$44

Table I.2
Automobile costs to travel to kayak course (2001 USD)

Average round trip miles to whitewater courses	Auto cost per mile*	Auto cost†
20	$0.345	$6.90
30	$0.345	$10.35
40	$0.345	$13.80

*Equals the federal reimbursement rate as of January 1, 2001.
†Equals round trip miles multiplied by the cost per mile.

The other cost components that are considered for this case study relate to travel costs. Table I.2 shows automobile costs for roundtrip distances of 20, 30, and 40 miles using the federal reimbursement rate of $0.345 per mile. An average 30-mile roundtrip distance is reasonable for the Vail Whitewater Park. User logs available from the park show that approximately 26% of boaters come from outside the area. Hence, assuming a 75%/25% split between local and nonlocal boaters and an average roundtrip travel distance of 20 miles for local boaters and 60 miles for nonlocal boaters, the representative roundtrip travel distance for all users is approximately 30 miles.

A 30 mile roundtrip translates into $10.35 per trip for auto-related expenses alone, as shown in Table I.2.

Studies of recreational expenditures and travel costs also typically include the opportunity cost (value) of travel time in the estimation process (time spent on site also might be included). Assuming an average 50 miles per hour travel rate (including stop signs, etc.), the average travel time given a 30 mile round trip is 0.6 hours. A typical convention in travel cost-based recreational demand studies is to assign a value of one-third of the wage rate to the time spent in transit to and from the site. In this case study, a value of $10 per hour is assumed to reflect this travel time cost, which translates into a $6 travel time cost per visit.

The total cost from the kayak equipment, automobile, and travel time cost components ranges from $36.35 ($20.00 + $10.35 + $6.00) to $61.35 ($45.00 + $10.35 + $6.00) per visit using the range of assumptions regarding kayak equipment cost and assuming an average roundtrip travel distance of 30 miles.

In addition, non local boaters, who make up approximately 25% of park users, are assumed to have additional expenses for lodging. Summer overnight visitors to Vail spend an average of $162 per day on lodging, meals, and other travel expenses (VVTCB, 1999). It is assumed that nonlocal boaters pay for lodging in or near Vail, bring one traveling companion, and stay only one day. Thus out of town boaters are estimated to spend an additional $324 per boater day.

User Values for Kayaking (WTP-based estimates of value)

The method of "benefits transfer" is a standard practice used by resource economists to obtain quick approximations of value when there is no opportunity to undertake primary research by administering a new survey or econometric model. Benefits transfer is conducted by obtaining values per unit of use for similar types of activities from studies conducted in similar locations. Then, those unit values are multiplied by the amount of use. A unit value typically might be the consumer surplus value for an activity such as a fishing day or a hiking trip.

A benefits transfer was conducted using recent, peer-reviewed recreational valuation literature. One set of unit values per day of kayaking was obtained from a 1999 database compiled by John Loomis (Loomis 1999a and 1999b), a professor of economics at Colorado State University and an expert in valuing environmental amenities. Typically, two types of valuation methods are used in the literature, and in the Loomis database: (1) revealed preference (RP) methods such as travel cost models, which use observed recreational behavior to infer values; and (2) stated preference (SP) methods such as contingent valuation, which ask people to state their values or their willingness to trade off different resource commodities. Carson et al. (1996) demonstrated that estimates of use values do not vary substantively whether RP or SP methods are used.

The Loomis database reports values for five regions of the United States. The values used in this report are taken from the values listed for the "Intermountain" region because they apply directly to Colorado. This region had six studies on floatboating, which includes kayaking, rafting, and sailing. The mean value per person per day for the "Intermountain" region is $39.91, in 2001 dollars. For comparison, Loomis found the national average to be $33.06, in 2001 dollars.

The recreation values summarized in the Loomis database are generally consistent with summary values obtained in other analyses, such as Walsh et al. (1980). This study, using the contingent valuation method, found kayaking values on the Crystal, Roaring Fork, and Yampa rivers (all in Colorado) to be $35.62 in 2001 dollars per person per day. In the same study, rafting on these same rivers was valued at $30.81 per person per day. Thus, evidence from the literature indicates that kayaking is a more highly valued activity than rafting. Accordingly, the Loomis value for floatboating may be an underestimate, since it includes kayaking and rafting together. Nonetheless, the Loomis estimate is used as an upper bound in this analysis.

Another study focused on kayaking in the West found that the average user day value for kayaking on the Colorado River is $67.25, in 2001 dollars (Bishop et al., 1989). However, this value is not applied in the analysis because the Colorado River is considered to be a unique resource in the United States and, thus, values for use of this special amenity may be higher than those for similar activities at other sites.

Therefore, for the purposes of this study, a range of $35.62 to $39.91 per person, per activity day, is assumed as the value of kayaking (in 2001 dollars).

Total User Value

Kayaking expenditures per day range from $36.35 to $61.35. The mid point of this range, $48.85, is used in the summary of benefits. WTP per boater day, as described earlier, ranges from $35.48 to $39.91. The mid point of this range, $37.77, is used in the summary. Summing the high value and the low value of these ranges, results in a total user value per day of $71.83 to $101.26.

Nonlocals are also assumed to have an additional cost of $324 per day for lodging, meals, and other travel-related expenses.

LEVELS OF USE

Use of the Vail Whitewater Park can be separated into two categories: nonevent use and event use. Nonevent use includes regular weekend and weekday daily use not organized as specific events. Nonevent use occurs from the late spring through the early fall, with the heaviest use occurring in May, June, and July. Event use consists of boating competitions and exhibitions organized by local boating shops. These events are scheduled in advance and are usually conducted on weekends. The course is one of about 20 man-made top quality whitewater courses in the world. The course attracts competition events for elite as well as nonelite athletes. Every summer there are several whitewater competitions and other events held at the Vail Whitewater Park including the annual Teva Whitewater Festival, a competition that attracts some of the world's best kayakers and large crowds of spectators.

Nonevent Use

At the time of this study, no formal count had been taken of nonevent users at the Vail Whitewater Park. Instead estimates of average and the maximum number of boaters at the park were obtained through from interviews with two local boat shop owners who are also frequent users of the park: Mike Duffy of Alpine Kayak and Canoe and Craig Russom of Mountain Quest Sports. Their estimates of use in the spring and summer are reported in Table I.3.

Multiplying the number of estimated user days by the number of days each month provides us with boater day totals ranging between 1,044 and 2,345 (based on data from Craig Russom and Mike Duffy, respectively, two individuals familiar with the site).

For the purposes of comparison, a simple linear regression model correlating the average boater days with flow was estimated as follows:

$$\text{Boaters Per Day} = -2.97 + 0.076 * \text{cfs} \qquad \text{(I.1)}$$

Flow explained 98% of the total variance (R^2) exhibited in boater use. The model estimates boater use increased by 7.6 boaters per day for every 100 cubic feet per second increase in flow.

A prediction of boater use was developed based on the flows claimed by Vail in their application of water rights and the regression model developed in Equation I.1. The results are summarized by month in Table I.4. The total number of boaters for the season equals 3,474. Because the claimed flows are higher than those observed in 2001, the total boater use estimate is higher than the 2001 use estimates.

Table I.3
Average per day boater use estimates for Vail Whitewater Park*

	Mike Duffy		Craig Russom	
Month	Weekday	Weekend	Weekday	Weekend
April 2001	1	1	2	6
May 2001	30	40	12	24
June 2001	35	45	12	24
July 2001	5	5	0	0

*Mike Duffy and Craig Russom reported that some boaters used the park during other months of 2001, but the use was not significant.

Table I.4
Vail flow and boater use scenario

Month	Claimed flow (cfs)	Boaters per day predicted by model	Predicted boaters per month
March	54	1.1	34
April	227	14.2	426
May	400	27.3	846
June	400	27.3	819
July	400	27.3	846
August	218	13.5	419
September	67	2.1	63
October	48	0.7	20
Total			3,474

As summarized in Table I.5, assuming annual use by 2,345 kayakers and applying the values discussed above, the total use value from nonevent users totals approximately $394,000. This total is composed of $115,000 in equipment, automobile, and travel time costs by all boaters; $190,000 in lodging and other travel-related expenses by nonlocal boaters; and $89,000 in WTP (or consumer surplus) for kayaking above and beyond expenses.

COMPETITIVE EVENT USE

The competitive and other special events held at kayak courses generate other economic benefits besides ones that accrue from the participants.

2001 Teva Whitewater Festival (Vail, Colo.)

Each year for the past 7 years, Vail has hosted a whitewater festival over Memorial Day Weekend that features professional kayaking and rafting competitions. In 2001 the event was sponsored by Teva and Vail's new kayaking park became a major focal point of the event, attracting top kayakers from the United States and around the world. The combination of the high level of competition and the Vail kayak park's convenient location near the center of town attracted the largest crowds in the event's history.

In 2001, approximately 2,300 spectators attended the Saturday competition (Ian Anderson, Vail Valley Tourism and Convention Bureau, personal communication, January 2, 2001), while attendance for the remaining portions of the weekend was "several hundred" (VVTCB, 2001). An event this size provides an important economic stimulus for Vail, especially given the timing of the event during the late spring "shoulder" season (the period after the ski season ends and before the summer recreation season begins), when comparatively few visitors travel to Colorado's mountain resort communities. According to a survey of spectators conducted on behalf of the Vail Valley Tourism and Convention Bureau by RRC and Associates (VVTCB, 2001):

Table I.5

Estimated value of future beneficial uses of waters diverted in the Town of Vail's Whitewater Park (2001 USD)

Beneficial use category	Level of use	Monetary unit value	Beneficial value
Boaters per year*	2,345		
Expenditures		$48.85	$114,553
Additional expenditures nonlocals†		$324.00	$189,945
Consumer surplus		$37.77	$88,559
Spectators at competitive events	>2,300		
Expenditures		$133.00	>$305,900
Consumer surplus		>$0	+‡
Economic stimulus for community			
1.88 times out-of-pocket expenditures§			$1,121,097
Increase in property values			+
Spectators and near-stream recreators (non-event)	>0	>$0	+
Benefits to event sponsors and vendors			+
Beneficial use of water downstream of kayak course (heavy use corridor)			+
Community identity, quality of life			+
Option value			+
Total beneficial use values per year			>$1,820,054
Capitalized value**			>$20.6 million

*Based on Mike Duffy's estimate of 2001 use. Use expressed as number of outings (user days).

†Nonlocal boaters equal 25% of total boaters (586 total) and include one companion. Spending on lodging, meals, and other travel-related expenses in Vail averages $162 per person (VVTCB, 1999).

‡"+" indicates positive benefits that could not be quantified or monetized using readily available data.

§Values shown are net of the expenditures shown above for boaters, and exclude the value of travel time.

**Over a time horizon of 20 years, discounted at 7%.

- 67% came to Vail specifically for the Teva Whitewater Festival.
- 47% came from outside Eagle County, and 26% of these visitors spent at least one night in the area.
- Those staying overnight stayed an average of 3 days, and 30% stayed 4 nights or more.
- The average Saturday spectator spent a total of $133 for food, lodging, and shopping.

The 2001 whitewater festival provided valuable regional and national exposure to Vail. In addition, the Vail Whitewater Park could attract even larger events in the future. In 2001, representatives from Vail entered a bid to host the 2003 World Freestyle Kayak Championships. Despite losing this bid, Vail was considered an attractive location and could present a successful bid to host the biannual event in the future. An event of this size would be a major boost to the Vail economy.

Other Vail Events

In addition to the Teva Whitewater Festival, the Vail kayak park hosts a whitewater rodeo series. In 2001 this series included 11 weekend competitions from May 16 through August 1. Two of these events served as qualifying events for the Teva competition, and the others were a series of competitions leading up to a series final held August 1. These events are cosponsored by two local kayaking shops, Alpine Kayak and Mountain Quest Sports.

Total Value

As summarized in Table I.5, organized events at the Vail Whitewater Park contribute at least $305,900 (2,300 spectators x $133 per spectator) in spending by spectators to the local economy, at the Saturday competition alone. This should be considered a very conservative lower bound because this estimate includes only attendance and spending for a single day (Saturday) at a single event (the Teva Whitewater Festival).

ECONOMIC STIMULUS

Increased expenditures by boaters and spectators generate an economic stimulus for the community. The regional economy is affected through a multiplier effect. The multiplier is a factor that when multiplied by new or increased expenditures (or reductions in expenditures) yields the benefits (or reductions in benefits) to the region. While precise estimates of the multiplier for kayaking in Vail have not been estimated, some related published literature provides guidance.

Cordell et al. (1990) estimated regional economic multipliers of 2.00 and 2.03 for the total economic effects of water-based recreation expenditures on local economies. Norton, Smith, and Strand (1981) estimated a range of multipliers from 2.03 to 2.88 in an analysis of the total economic value of recreational fishing. State of Colorado agencies also use similar or larger multipliers for estimating the total economic impacts of spending activities in one sector (e.g., housing or aeronautics) to the broader economy of the relevant local community or the state as a whole (e.g., estimates of the economic impacts of airport construction-related projects imply a multiplier of slightly over 3.0; Colorado Aeronautics Division, 2001). Literature-based multipliers were, therefore, considered to range from 2.00 (the low end of the range of Cordell et al.) to 2.88 (the high end from Norton, Smith, and Strand), although slightly higher multipliers might also be justified. The multiplier of 2.88 was applied in this study by multiplying 1.88 times boater expenditures (minus the travel time component), the additional expenditures from non-locals, and expenditures by spectators at events, to derive an economic stimulus value of $1,121,000. Consumer's supplies and travel time are left out of this calculation because only actual expenditures feed the regional economic activity directly.

TOTAL VALUE OF BENEFITS

Table I.5 summarizes the total value of the benefits provided by the Vail Whitewater Park in 2001. The table includes use values as described, spectator expenditures, and economic stimulus as described above. The total benefits from these categories are estimated at approximately $1.8 million annually (in 2001 dollars). The total flow claimed by Vail for the whitewater park is approximately 110,000 acre feet from March through October. Thus the lower bound value for this non-consumptive use of water from Gore Creek is $16.36 per acre foot. Over 20 years, assuming a 7% discount rate, the total present value is over $20 million.

Several types of benefits have not been included in this summary. These include, but are not limited to, increases in property values enjoyed by land owners near the park, the consumer surplus of spectators and other non-boating recreators (most notably fishermen who fish in the pools created by the park, improved or renewed community identity and quality of life, and beneficial uses of the waters through the corridors downstream of the park (as all of the water diverted in the Vail course is available for beneficial use downstream). Due to the time limitations of this study these benefits have not been quantified, but are noted.

The total benefits estimated for the Vail Whitewater Park are quite comparable to values estimated for other whitewater parks in Colorado. A 2002 study at a newly constructed whitewater park in Breckenridge found that the lower bound of annual benefits from the park was $1.4 million (in 2001 dollars). A 2000 study of a whitewater park in Golden (which is more easily accessible from the Denver Metropolitan area) found the lower bound of benefits to be in the range of $1.4 million to $2.1 million (in 2001 dollars).

REFERENCES

2002 Recycled Water Task Force. 2002. Economics Work Group Draft White Paper. Department of Water Resources. 18 November. Available: <http://www.owue.water.ca.gov/recycle/docs/WorkgroupWhitePapers.htm>. Accessed March 16, 2004.

Agthe, D.E., and R.B. Billings. 1980. Dynamic Models of Residential Water Demand. *Water Resources Research*, 16(3):476-480.

Agthe, D., R.B. Billings, J.L. Dobra, and K. Raffiee. 1986. A Simultaneous Equation Demand Model for Block Rates. *Water Resources Research*, 22(1):1-4.

American Oceans. 2003. Estuaries on the Edge: The Vital Link Between Land and Sea, Tampa Bay in Florida. Available: <http://www.americanoceans.org/issues/pdf/tampa.pdf>. Accessed October 22, 2003.

Amirafathi, P., R. Narayanan, A. Bishop, and D. Larson. 1985. A Methodology for Estimating In-stream Flow Values for Recreation. Water Resources Planning Series (UWRL/P-85/01). Logan, Utah: Utah State University, Utah Water Research Laboratory.

Anderson, T.L., ed. 1983. *Water Rights: Scarce Resource Allocation, Bureaucracy, and the Environment.* San Francisco, Calif.: Pacific Institute for Public Policy Research.

Anderson, T.L., and D.R. Leal. 1988. Going With the Flow: Expanding the Water Markets. *Policy Analysis* 104 April, 26. Available: <http://www.cato.org/pubs/pas/pa104.html>. Accessed March 16, 2004.

Antle, J. 1983. Incorporating Risk in Production Analysis. *American Journal of Agricultural Economics,* 65:1099-1106.

Aquacraft, Inc. 2003. Commercial/Institutional Water Use Patterns. Boulder, Colo.: Aquacraft, Inc.

Aquacraft, Inc. and Stratus Consulting. 2001. Conservation Plan for Industrial, Commercial, and Irrigation Water Customers in Westminster Colorado. Aquacraft, Inc., Boulder, Colo.

Arbues, F., M.A. Garcia-Valinas, and R. Martinez-Espineira. 2003. Estimation of Residential Water Demand: A State-of-the-Art Review. *Journal of Socio-Economics*, 32(1):81-102.

Ashley, J.S., and Z.A. Smith. 2001. Western Ground Water Wars. *Forum for Applied Research and Public Policy*. Spring 2001:33-39. Available: http://forum.ra.utk.edu/Archives/Spring2001/ashley.pdf. Accessed March 15, 2004.

Australian Government Department of Environment and Heritage. 2003. *Triple Bottom Line Reporting in Australia: A Guide to Reporting Against Environmental Indicators*, p 6. Available: <http://www.deh.gov.au/industry/finance/publications/indicators/>. Accessed February 4, 2004.

AZ Department of Water Resources. 2000. Verde River Watershed Study. Phoenix, Ariz.

Bacher, D. 2003. Alameda Creek Alliance Pushes Steelhead Restoration Plan Forward. Available: <http://www.fishsniffer.com/dbachere/030802alameda.html>. Accessed August 28, 2003.

Bailey, B.. Undated. Minimum In-stream Flow Legislation and the Section 404 Permitting Process: The Impact of State Water Quality Laws. Washington, D.C.: Environmental Defense Fund.

Bay Area Economic Forum. 2002. Hetch Hetchy Water and the Bay Area Economy. San Francisco, Calif.: Bay Area Economic Forum.

Bay Area Monitor. 2003. On the Region's Ballot: Water, Transit. Available: <http://www.bayareamonitor.org/oct02/ballot.html>. Accessed August 21, 2003.

Berrens, R.P., P. Ganderton, and C.L. Silva. 1996. Valuing the Protection of Minimum In-stream Flows in New Mexico. *Journal of Agricultural and Resource Economics,* 21(2):294-309.

Berrens, R.P., A.K. Bohara, C.L. Silva, D. Brookshire, and M. McKee. 2000. Contingent Values for New Mexico In-stream Flows: With Tests of Scope, Group-Size Reminder and Temporal Reliability. *Journal of Environmental Management,* 58(1):73-90.

Billings, R.B. 1990. Demand-Based Benefit-Cost Model by Participation in Water Project. *Journal of Water Resources Planning and Management,* 116(5):593-609.

Billings, R.B., and D.E. Agthe. 1980. Price Elasticities for Water: A Case of Increasing Block Rates. *Land Economics*, 56(1):73-84.

Bingham, T., M.J. Kealy, E. David, M. LeBlanc, T. Graham-Tomassi, and R. Leeworthy (eds.). 1992. Benefits Transfer: Procedures, Problems, and Research Needs. Proceedings from a 1992 Association of Environmental and Resource Economists Workshop, Snowbird, Utah, June 3-5.

Bishop, R.C., and M.P. Welsh. 1992. Existence Values in Benefit-Cost Analysis and Damage Assessment. *Land Economics,* 68(4):405-417.

Bishop, R.C., C.A. Brown, M.P. Welsh, and K.J. Boyle. 1989. Grand Canyon Recreation and Glen Canyon Dam Operations: An Economic Evaluation. In *Western Regional Research Project W-133.* Edited by K.J. Boyle and T. Heekin. Washington, D.C.: USDA.

Black, R., R. McKenney, R. Unsworth, and N. Flores. 1998. Economic Analysis for Hydropower Project Relicensing: Guidance and Alternative Methods. Prepared for the Division of Economics, U.S. Fish and Wildlife Service, and U.S. Department of the Interior. Cambridge, Mass: Industrial Economics Inc.

Boggess, W., R. Lacewell, and D. Zilberman. 1993. Economics of Water Use in Agriculture. Chapter 8 in *Agricultural and Environmental Resource Economics.* Edited by G.A. Carlson, D. Zilberman, and J.A. Miranowski. N.Y.: Oxford University Press.

Booker, J.F., and R.A. Young. 1991. Economic Impacts of Alternative Water Allocations in the Colorado River Basin. Colorado Water Resources Research Institute Completion Report no. 161. Fort Collins, Colo.: Colorado State University.

Boston Globe. 2002. Your Town. Available: <http://yourtown.boston.com>. Accessed September 12, 2003.

Boyle, K.J., and R.C. Bishop. 1987. Valuing Wildlife in Benefit-Cost Analyses: A Case Study Involving Endangered Species. *Water Resources Research,* 23:943-950.

Brookshire, D., M. McKee, and G. Watts. 1994. An Economic Analyses of Critical Habitat Designation in Colorado River Basin for the Razorback Sucker, Humpback Chub, Colorado Squawfish, and Bonytail. Final report to the U.S. Fish and Wildlife Service.

Brown, F., S. Nunn, J. Shomaker, and G. Woodward. 1995. The Value of Water. Draft Report. Prepared for the City of Albuquerque, N.M.

Brown, T.C. 1991. Water for Wilderness Areas: In-stream Flow Needs, Protection, and Economic Value. *Rivers,* 2(4):311-325.

Brown, T.C., and T.C. Daniel. 1991. Landscape Aesthetics of Riparian Environments: Relationship of Flow Quantity to Scenic Quality Along a Wild and Scenic River. *Water Resources Research*, 27(8):1787-1795.

Brown, T.C., and J.W. Duffield. 1995. Testing Part-Whole Valuation Effects in Contingent Valuation of In-stream Flow Protection. *Water Resources Research*, 31(9):2341-2351.

Brown, T.C., J. Taylor, and B. Shelby. 1991. Assessing the Direct Effects of Streamflow on Recreation: A Literature Review. *Water Resources Bulletin*, (6):979-989.

Bryner, G., and E. Purcell. 2003. *The Groundwater Law Source Handbook.* Available: <http://www.colorado.edu/law/centers/nrlc/publications/Groundwater%20Law%20Sourcebook.pdf>. Accessed March 22, 2004.

Butkay, S.M. 1989. The Economic Value of In-stream Flows for Recreation: A Contingent Valuation Approach. Master's thesis. University of Montana, Missoula.

Camp Dresser & McKee. 2003. Town of Reading, Massachusetts. Admission to the Massachusetts Water Resources Authority Waterworks System. Final Environmental Impact Report. September. Cambridge, Mass.

Carson, R.T., and R.C. Mitchell. 1987. Economic Value of Reliable Water Supplies for Residential Water Users in the State Water Project Services Area. SWC Exhibit Number 54. Prepared for The Metropolitan Water District of Southern California.

Carson, R.T., N.E. Flores, K.M. Martin, and J.L. Wright. 1996. Contingent Valuation and Revealed Preference Methodologies: Comparing the Estimates for Quasi-Public Goods. *Land Economics,* 72(1):80-99.

Carver, P.H., and J.J. Boland. 1980. Short- and Long Run Effects of Price on Municipal Water Use. *Water Resources Research*, 16(4):609-616.

Cassuto, A.E., and S. Ryan. 1979. Effect of Price on the Residential Demand for Water Within an Agency. *Water Resources Bulletin*, 345-53.

CBO (Congressional Budget Office). 1997. *Water Use Conflicts in The West: Implications of Reforming the Bureau of Reclamation's Water Supply Policies*. Available: <http://www.cbo.gov/showdoc.cfm?index = 46&sequence = 0>. Accessed March 16, 2004.

Chicoine, D.L., and G. Ramamurthy. 1986. Evidence on the Specification of Price in the Study of Domestic Water Demand. *Land Economics*, 62(1):28-32.

Chicoine, D.L., S.C. Deller, and G. Ramamurthy. 1986. Water Demand Estimation Under Block Rate Pricing: A Simultaneous Equation Approach. *Water Resources Research*, 22(6):859-863.

Christensen, J.D., M.E. Monaco, R.J. Livingston, G. Woodsum, T.A. Battista, C.J. Klein, B. Galeprin, and W. Huang 1998. Potential Impacts of Reduced Freshwater Inflow on Apalachicola Bay, FL Oyster (*Crassostrea virginica*) Populations: Coupling Hydrologic and Biological Models. Silver Spring, Md.: National Oceanic and Atmospheric Administration.

Ciriacy-Wantrup, S.V. 1964. Concepts Used as Economic Criteria for a System of Water Rights. Chapter 15 in *Economics and Public Policy in Water Resource Development.* Edited by S.C. Smith and E.N. Castle. Ames, Iowa: Iowa State University Press, 251-271.

City of Tallahassee. 2003. Public Works, Stormwater Management, Lakes Management Program. Temperature and Water Level Fluctuations. Available: <http://talgov.com/citytlh/stormwater_man/lakes/lakes.htm>. Accessed October 15, 2003.

City West Water Limited. 2004. Available: <http://www.citywestwater.com.au/iis?command = home>. Accessed February 12, 2004.

Colby, B. 1989a. Alternative Approaches to Valuing Water Rights. 1989a. *The Appraisal Journal*, 180-196.

Colby, B. 1989b. Estimating the Value of Water in Alternate Uses. 1989b. *Natural Resources Journal*, 29(2):511-527.
Colby, B. 1993. Benefits, Costs and Water Acquisition Strategies: Economic Considerations in In-stream Flow Protection. In *In-Stream Flow Protection in the West*. Edited by L. MacDonnell and T. Rice. Boulder, Colo.: Natural Resources Law Center, University of Colorado School of Law.
Colby, B.G., J. Leones, C. Mullahy-Koenig, and L. Ryan. 1994. River Recreation and the Economy of Northern New Mexico. Tucson, Ariz.: University of Arizona, Department of Agricultural and Resource Economics.
Cole, R.A., and F.A. Ward. 1994. Optimum Fisheries Management Policy: Angler Opportunity Versus Angler Benefit. *North American Journal of Fisheries Management*, 14(1):22-33.
Colorado Aeronautics Division. 2001. Economic Impacts of Colorado Airports. Available: <http://www.colorado-aeronautics.org> Accessed March 2, 2001.
Conley, B.C. 1967. Price Elasticity of Demand for Water in Southern California. *Annals of Regional Science*, 1(1):180-189.
Cordell, H.K., J.C. Bergstrom, G.A. Ashley, and J. Karish. 1990. Economic Effects of River Recreation on Local Economies. *Water Resources Bulletin,* 26(1):53-60.
Corral, A., A.F. Fisher, and N.W. Hatch. 1998. Price and Non-Price Influences on Water Conservation: An Econometric Model of Aggregate Demand Under Nonlinear Budget Constraint. Unpublished paper.
CPRC (California Policy Research Center). 2001. CPRC Briefing: Water Governance in the San Francisco Bay Area: Challenges and Opportunities. Berkeley, Calif.: California Policy Research Center.
Crandall, K.B., B.G. Colby, and K.A. Rait. 1992. Valuing Riparian Areas: A Southwestern Case Study. *Rivers*, 3(2):88-98, Studies in the Science, Environmental Policy and Law of In-Stream Flow. Tucson, Ariz.: University of Arizona.
Cummings, R.G., P.T. Ganderton, and T. McGuckin. 1994. Substitution Effects in CVM Values. *American Journal of Agricultural Economics*, 76:205-214.
CUWA (California Urban Water Agencies). 1991. Cost of Industrial Water Shortages. Prepared by Spectrum Economics, Inc., San Francisco, Calif. November.
CUWA (California Urban Water Agencies). 1994. *The Value of Water Supply Reliability: Results of a Contingent Valuation Survey of Residential Customers.* Sacramento, Calif.: CUWA. August. Available: <http://www.cuwa.org/library/TheValueofWaterSupplyReliabilityAug94.pdf>. Accessed March 31, 2004.
Dalhuisen, J., R. Florax, H. de Groot, and P. Nijkamp. 2001. Price and Income Elasticies of Residential Water Demand. Tinbergen Institute Discussion Paper TI 2001-057/3. Vrije University, Amsterdam.
Dalhuisen, J.M., R.J.G.M. Florax, H.L.F. de Groot, and P. Nijkamp 2003. Price and Income Elasticities of Residential Water Demand: A Meta-analysis. *Land Economics*, 79(2):292-308.
Danielson, L.E. 1979. An Analysis of Residential Demand for Water Using Micro Time-Series Data. *Water Resources Research*, 15(4):763-767.
Daubert, J.T., and R.A. Young. 1981. Recreational Demand for Maintaining In-stream Flows: A Contingent Valuation Approach. *American Journal of Agricultural Economics*, 63(4):666-676.

Davis, C. 2001. Environmental Ethical Issues in Water Operations and Land Management. Denver, Colo.: AWWA.

Day, L.R. 2001. Testimony of Laura Rose Day, Watershed Project Director, Natural Resources Council of Maine in Support of LD 12488, An Act of Require Public Information Regarding Water Use by Major Water Users, before the Natural Resources Committee (4 April 2001).

Deason, J.P., T.M. Schad, and G.W. Sherk. 2001. Water Policy in the United States: A Perspective. *Water Policy,* 3:175-192.

DeOreo, W.B., P.W. Mayer, D.M. Lewis, A. Dietmann, T. Skeel, and J. Smith. 2001. Retrofit Realities. *Journal AWWA*, 93(3):58-72.

Desvousges, W.H., M.C. Naughton, and G.R. Parsons. 1992. Benefit Transfer: Conceptual Problems in Estimating Water Quality Benefits Using Existing Studies. *Water Resources Research*, 28(3).

Dinar, A. and A. Subramanian (eds.). 1997. Water Pricing Experiences: An International Perspective. World Bank Technical Paper 386. Washington, D.C.: The World Bank.

Douglas, A.J., and R.L. Johnson. 1993. In-stream Flow Assessment and Economic Valuation: A Survey of Non-market Benefits Research. *International Journal of Environmental Studies*, 43:89-104.

Douglas, A.J., and J.G. Taylor. 1998. Riverine Based Eco-tourism: Trinity River Non-market Benefits Analysis. *International Journal of Sustainable Development and World Ecology,* 5(2):136-148.

Driver, B.C. 1997. Western Hydropower: Changing Values/New Visions. Report to the Western Water Policy Review Advisory Commission. August. Springfield, Va.: NTIS.

Duffield, J.W., and D.A. Patterson. 1991. Field Testing Existence Values: An In-stream Flow Trust Fund for Montana Rivers. Paper presented at the American Economics Association Annual Meeting, January 4. New Orleans, La.

Duffield, J.W., T.C. Brown, and S.D. Allen. 1994. Economic Value of In-stream Flow in Montana's Big Hole and Bitterroot Rivers. (No. RM-317). Fort Collins, Colo.: Rocky Mountain Forest and Range Experiment Station, USDA Forest Service.

Duffield, J.W., S. Butkay, and A. Stewart. 1990. Economic Value of Recreation and Preservation Benefits of In-stream Flow. Submitted to U.S. Department of Agriculture, Forest Service, Rocky Mountain Forest and Range Experiment Station, Fort Collins, Colo.

Duffield, J.W., C.J. Neher, and T.C. Brown. 1992. Recreation Benefits of In-stream Flow: Application to Montana's Big Hole and Bitterroot Rivers. *Water Resources Research*, 28(9):2169-2181.

Duffield, J.W., C. Neher, M. Josephson, and R. Josephson. 1991. Market Value of Agricultural Water Leased for In-stream Flows. Report to Montana Department of Fish, Wildlife and Parks, Helena, Mont.

Duffield, J.W., C. Neher, D. Patterson, and S. Allen. 1990. In-stream Flows in the Missouri River Basin: A Recreation Survey and Economic Study. Final report. Helena, Mont.: Montana Department of Natural Resources and Conservation.

Dziegielewski, B., J.C. Kiefer, E.M. Optiz, G.A. Porter, G.L. Lantz, W.B. DeOreo, P.W. Mayer, and J.O. Nelson. 2000. Commercial and Institutional End Uses of Water. Denver, Colo.: AWWA.

EDAW. 2001. Alameda Watershed Management Plan. p.1-2. Alameda, Calif.

Elkington, J. 1997. *Cannibals with Forks: The Triple Bottom Line of 21st Century Business*. Stony Creek, Conn.: New Society Publishers.

Enge, M. 2003a. Agency to Monitor Water Supply. *The Mercury News*. April 10.

Enge, M. 2003b. Water Upgrade Battle Brewing. Available: <http://www.bayarea.com/mld/mercurynews/news/local/4332712.html>. Accessed August 21, 2003.

Epp, D.J., and K.S. Al-Ani. 1979. The Effect of Water Quality on Rural Non-farm Residential Property Values. *American Journal of Agriculture and Economics*, 61(3):529-534.

Espey, M., J. Espey, and W.D. Shaw. 1997. Price Elasticity of Residential Demand for Water. *Water Resources Research*, 33(6):1369-1374.

Farber, S.C., and R. Costanza. 1987. The Economic Value of Wetlands Systems. *Journal of Environmental Management*, 24:41-51.

FERC (Federal Energy Regulatory Commission). 1996. Final Environmental Impact Statement, Upper Penobscot River Basin, Maine, FEIS-0075, September.

Fisher, A., D. Fullerton, N. Hatch, and P. Reinelt. 1995. Alternatives for Managing Drought: A Comparative Analysis. *Journal of Environmental Economic and Management*, 29:304-320.

Flug, M., and R.H. Montgomery. 1988. Modeling In-stream Recreational Benefits. *Water Resources Bulletin*, 24(5):1073-1081.

Flynn, K. 1996. Understanding Wetlands and Endangered Species: Definitions and Relationships. Alabama Cooperative Extension System, ANR-976, May. Alabama A&M University, Huntsville, Ala. and Auburn University, Auburn, Ala.

Foran, P.G., J.A. Beecher, and L.J. Wilson. 1995. Survey of Eastern Water Law. A Report to the Illinois Department of Natural Resources. Available: <http://dnr.state.il.us>. Accessed March 16, 2004.

Foster, H.S., and B.R. Beattie. 1979. Urban Residential Demand for Water in the United States. *Land Economics*, 55(1):43-58.

Foster, H.S., and B.R. Beattie. 1981. On the Specification of Price in Studies of Consumer Demand Under Block Price Scheduling. *Land Economics*, 57(4):624-629.

Franklin, D.R. and R.K. Hageman. 1984. Cost Sharing With Irrigated Agriculture: Promise versus Performance. *Water Resources Research,* 20(8):1047-1051.

Frederick, K.D. 1995. America's Water Supply: Status and Prospects for the Future. *Consequences*, 1(1), Spring. Available: <http://www.gcrio.org/CONSEQUENCES/spring95/Water.html>. Accessed March 17, 2004.

Frederick, K., T. Vandenberg, and J. Hanson. 1997. Economic Values of Freshwater in the United States. Discussion Paper 97-03. Washington, D.C.: Resources for the Future.

Freeman, A.M. III. 1997. On Valuing the Services and Functions of Ecosystems. In *Ecosystem Function and Human Activities: Reconciling Economics and Ecology*. Edited by R.D. Simpson and N.L. Christensen Jr. New York: Chapman and Hall.

Frymier, L.G., and C.H. Mitchell. 1997. A Comparative Analysis of Value Between Users and Non-users of the White River. Proceedings of the 1996 Northeastern Recreation Research Symposium March 31- April 2, 1996, on Lake George in Bolton Landing, New York (pp. 79-81). Randor, Pa.: USDA Forest Service Publications Distribution.

Gaudin, S., R.C. Griffin, and R.C. Sickles. 2001. Demand Specification of Municipal Water Management: Evaluation of the Stone-Geary Form. *Land Economics,* 77(3):399-422.

Gelt, J. 2003. Fountains — Water Wasters or Works of Art? Available: <http://ag.arizona.edu/AZWATER/arroyo/073fount.html>. Accessed December 2, 2003.

Getches, D.H. 1997. *Water Law in a Nutshell*, 3rd edition. St. Paul, Minn.: West Group Publishing.

Gibbons, D.C. 1986. *The Economic Value of Water.* Washington, D.C.: Resources for the Future.

Gibbs, K.C. 1978. Price Variable in Residential Water Demand Models. *Water Resources Research*, 14(1):15-18.

Gillilan, D.M., and T.C. Brown. 1997. *In-stream Flow Protection: Seeking a Balance in Western Water Use*. Washington, D.C.: Island Press.

Giuliano, G., and F. Spaziani. 1985. Water Use Statistics in Industry: Experiences From Regional Surveys and Planning Studies in Italy. *Statistical Journal of the United Nations*, 3:229-235.

Gleick, P.H., ed. 1993. *Water in Crisis: A Guide to the World's Freshwater Resources*. Pacific Institute for Studies in Development, Environment, and Security. N.Y.: Stockholm Environment Institute, Oxford University Press.

Glennon, R. 2002. *Water Follies: Groundwater Pumping and the Fate of America's Fresh Waters.* Washington, D.C.: Island Press.

Goldsmith, R., A. Seidl, and S. Weiler. 2001. Ski-tourism and the Economy of Summit County, Colorado. Agricultural and Resource Policy Report. Colorado State University. Available: <http://dare.agsci.colostate.edu/extension/apr01-04.pdf>. Accessed September 12, 2003.

González-Cabán, A., and J. Loomis. 1999. Measuring the Economic Benefit of Maintaining the Ecological Integrity of the Río Mameyes in Puerto Rico. Research paper PSW-RP-240. Albany, Calif: Pacific Southwest Research Station, U.S. Forest Service, U.S. Department of Agriculture.

Gottlieb, M. 1963. Urban Domestic Demand for Water: A Kansas Case Study. *Land Economics*, 39(May):204-210.

GRI (Global Reporting Initiative). 2003. Available: <http://www.globalreporting.org/guidelines/2002.asp> <http://www.globalreporting.org/divers/economic.asp> <http://www.globalreporting.org/divers/social.asp> <http://www.globalreporting.org/divers/environment.asp> <http://www.globalreporting.org/guidelines/companies.asp>. Accessed August 8, 2003.

Griffin, R., and C. Chang. 1990. Pretest Analysis of Water Demand in Thirty Communities. *Water Resources Research*, 26(10):2251-2255.

Griffin, R.C., and J.W. Mjelde. 2000. Valuing Water Supply Reliability. *American Journal of Agricultural Economics,* 82:414-426.

Grubb, H.W. 1996. The Edwards Aquifer: Conflicts Surrounding Use of a Regional Water Source. *Water Resources Update,* 104:39-45. Available: <http://www.ucowr.siu.edu/updates/pdf/V106_A6.pdf>. Accessed March 24, 2004.

Hanke, S.H., and R.K. Davis. 1973. Potential for Marginal Cost Pricing in Water Resource Management. *Water Resources Research*, 9(4):808-825.

Hansen, R.D., and R. Narayanan. 1981. A Monthly Time Series Model of Municipal Water Demand. *Water Resources Bulletin*, 17(4):578-585.

Harvard University Graduate School of Design. 1999. *Grow Smart North Shore.* Edited by C. Garfield, R. Yaro, M. Racicot, and C. Rogers. June. Cambridge, Mass.

Hazen and Sawyer. 1998. Statement of Estimated Regulatory Costs. Prepared for the Southwest Florida Water Management District, November 25. New York, N.Y.

Heaney, J.P. 2003. Conceptually Sound and Operational Methods for Environmental Valuation. Available: <http://www.ucowr.siu.edu/updates/pdf/V96_A1.pdf.> Accessed October 22, 2003.

Hewitt, J.A. 1993. Watering Households: The Two-Error Discrete-Continuous Choice Model of Residential Water Demand. Ph.D. diss., University of California, Berkeley, Calif.

Hewitt, J.A., and W.M. Hanemann. 1995. A Discrete/Continuous Choice Approach to Residential Water Demand Under Block Rate Pricing. *Land Economics*, 71(2):173-192.

Hogarty, T.F., and R.J. Mackay. 1975. The Impact of Large Temporary Rate Changes on Residential Water Use. *Water Resources Research*, 11(6):791-794.

Hoppe, M.K. 2003. A Fresh Look at Tampa Bay, Complex Science Yields Simple Truth: Fresh Water Vital to Bay's Health, Tampa Bay Soundings. Available: <http://www.baysoundings.com/fall02/freshlook.html>. Accessed October 16, 2003.

Horowitz J.K., and E. Lichtenberg. 1993. Insurance, Moral Hazard, and Chemical Use in Agriculture. *American Journal of Agricultural Economics,* 75(4):926-935.

Horsley & Witten. 2002. Ipswich River Watershed Management Plan. October, p. 6-10. Sandwich, Mass: Horsley & Witten.

Howe, C.W. 1982. The Impact of Price on Residential Water Demand: Some New Insights. *Water Resources Research*, 18(4):713-716.

Howe, C.W., and F.P. Linaweaver. 1967. The Impact of Price on Residential Water Demand and Its Relation to System Design and Price Structure. *Water Resources Research*, 3(1):13-32.

Howe, C.W., and M.G. Smith. 1993. Incorporating Public Preferences in Planning Urban Water Supply Reliability. *Water Resources Research*, 29(10):3363-3369.

Howe, C.W., and M.G. Smith. 1994. The Value of Water Supply Reliability in Urban Water Systems. *Journal of Environmental Economics and Management*, 26:19-30.

Hurd, B.H., J.M. Callaway, J.B. Smith, and P. Kirshen. 1999. Economic Effects of Climate Change on U.S. Water Resources. In *The Impact of Climate Change on the United States Economy*. Edited by R. Mendelsohn and J.E. Neumann. Cambridge, UK: Cambridge University Press, 133-177.

IPCC (Intergovernmental Panel on Climate Change). 2001. *Climate Change 2001: The Scientific Basis.* Edited by J.T. Houghton, Y. Ding, D.J. Griggs, M. Noguer, P.J. van der Linden, D. Xiaosu, and K. Maskell. N.Y.: Cambridge University Press.

IRWA (Ipswich River Watershed Association). 2003a. Ipswich River Watershed Association. Available: <http://www.ipswichriver.org/irwa.html>.

IRWA (Ipswich River Watershed Association). 2003b. Value of Water Survey Results. Ipswich River Watershed Association. Ipswich, Mass.

IRWA (Ipswich River Watershed Association) et al. 2003. IRWA et al. v. Department of Environmental Protection Verified Compliant for Declaratory and Injunctive Relief. April 11.

Jenkins, M.W., J.R. Lund, and R.E. Howitt. 2003. Using Economic Loss Functions to Value Urban Water Scarcity in California. *Journal AWWA,* 95(2):58-70.

Johnson, D.M., and R.G. Walsh. 1987. Economic Benefits and Costs of the Fish Stocking Program at Blue Mesa Reservoir, Colorado. Colorado Water Resources Research Institute Technical Report No. 49. Fort Collins, Colo.: Colorado State University.

Johnson, N.S., and R.M. Adams. 1988. Benefits of Increased Streamflow: The Case of the John Day River Steelhead Fishery. *Water Resources Research,* 24(11):1839-1846.
Jones, C.V., and J.R. Morris. 1984. Instrumental Price Estimates and Residential Water Demand. *Water Resources Research*, 20(2):197-202.
Jones and Stokes Associates. 1993. Draft Environmental Impact Report for the Review of Mono Basin Water Rights of the City of Los Angeles California State Water Resources Control Board. Sacramento, Calif.
Kent, P. 2003. 2003. *Shellfish Officer Town Catch Report, 1996-2001.* Shellfish Officer, Town of Ipswich. Ipswich, Mass.
Koberstein, P. 1997. What's a River Worth? River Values, 8-12. American Rivers.
KQED. 2002. 2002 Election Results. Available: <http://www.kqed.org/w/2002election/local-results.html>. Accessed August 21, 2003.
Kyler, D. 2001. Letter of 9 February 2001 from David Kyler, Executive Director, Center for a Sustainable Coast to the Honorable Roy Barnes, Governor, State of Georgia.
Landry, C. 1998. In-stream Flow Marketing in the Pacific Northwest. Political Economy Research Center. Paper presented at Pacific Northwest Regional Economic Conference: Conflict and Collaboration: How Government, Business, and Academia Shape the Pacific Northwest Economy. May 7-9, 1998. Olympia, Wash.
Lansford, N.H., and L.L. Jones. 1995. Recreational and Aesthetic Values of Water Using Hedonic Price Analysis. *Journal of Agricultural and Resource Economics*, 20(2):341-355.
Leefers, L., and D.M. Jones. 1996. Assessing Changes in Private Property Values Along Designated Natural Rivers in Michigan. Lansing, Mich.: Michigan State University, Department of Forestry.
Leones, J. 1997. Measuring Regional Economic Impacts of Streamflow Depletions. *Water Resources Research*, 33(4):831-838.
Lichtenberg, E. 2001. Harmonizing Agricultural and Environmental Policies. Department of Agricultural and Resource Economics, University of Maryland. College Park, Md. Available: <http://www.fftc.agnet.org/library/article/eb507.html>. Accessed May 28, 2004.
Loomis, J. 1987a. Balancing Public Trust Resources of MonoLake and Los Angeles' Water Right: An Economic Approach. *Water Resources Research,* 23:1449-1456.
Loomis, J. 1987b. The Economic Value of In-stream Flow: Methodology and Benefit Estimates for Optimum Flows. *Journal of Environmental Management*, 24:169-179.
Loomis, J.B. 1998a. Estimating the Public's Values for In-stream Flow: Economic Techniques and Dollar Values. *Journal of the American Water Resources Association*, 34(5):1007-1014.
Loomis, J. 1998b. John Loomis Recreation Meta-Analysis. Ft. Collins, Colo.: Colorado State University.
Loomis, J. 1999a. Meta-Analysis of Non-market Recreation Studies. Colorado State University, Ft. Collins, Colo.
Loomis, J. 1999b. Recreation Meta-Analysis Database. Colorado State University, Ft. Collins, Colo.
Loomis, J., and J. Cooper. 1990. Economic Benefits of In-stream Flow to Fisheries: A Case Study of California's Feather River. *Rivers,* 1(1):23-30.

Loomis, J.B., and M. Creel. 1992. Recreation Benefits of Increased Flows in California's San Joaquin and Stanislaus Rivers. *Rivers*, 3(1):1-13.
Loomis, J., and E. Ekstrand. 1997. Economic Benefits of Critical Owl Habitat for the Mexican Spotted Owl. A Scope Test Using a Multiple-Bounded Contingent Valuation Survey. *Journal of Agricultural and Resource Economics*, 22(2).
Loomis, J., P. Kent, L. Strange, K. Fausch, and A. Covich. 1999. Measuring the Total Economic Value of Restoring Ecosystem Services in an Impaired River Basin: Results From a Contingent Valuation Survey. *Ecological Economics*, 33(2000):103-117.
Lyman, R.A. 1992. Peak and Off-Peak Residential Water Demand. *Water Resources Research*, 28(9):2159-2167.
Lynne, G.D., W. Luppold, and C. Kiker. 1978. Water Price Responsiveness of Commercial Establishments. *Water Resources Bulletin*, June.
Mackin, K., and L. Wagner. 2002. Ipswich River Basin Water Conservation Report Card. Ipswich River Watershed Association, Ipswich, Mass. and Massachusetts Audubon Society, Lincoln, Mass.
MADEP (Massachusetts Department of Environmental Protection). 2003. State Strikes Balance With Water Withdrawal Permits for Ipswich River Basin Communities. May 20. Available: <http://www.mass.gov>.
Mayer, P.W., D. Bennett, W.B. DeOreo, R. Harris, and D. Muir. 2002. Great Expectations-Actual Water Savings With the Latest High-Efficiency Residential Fixtures and Appliances. Proceedings of the Water Sources Conference 2002, Las Vegas, Nev.
Mayer, P.W., W.B. DeOreo, E.M. Opitz, J.C. Kiefer, W.Y. Davis, B. Dziegielewski, and J.O. Nelson. 1999. Residential End Uses of Water. Denver, Colo.: AWWA and AwwaRF.
McBain and Trush. 1999. A Summary of the Habitat Restoration Plan for the Lower Tuolumne River Corridor. March. Prepared for the Tuolumne River Technical Advisory Committee by McBain and Trust, Arcata, Calif.
McGuckin, T. 2000. Value of Water in Municipal Use. Unpublished faculty manuscript. Las Cruces, N.M.: New Mexico State University.
Means, E.G. 2001. *Strategic Assessment of the Future of Water Utilities*. Denver, Colo.: American Water Works Association Research Foundation.
Michelsen, A.M, J.T. McGuckin, and D.M. Stumpf. 1998. Effectiveness of Residential Water Conservation Price and Nonprice Programs. Prepared for American Water Works Association Research Foundation, Denver, Colo.
Minnesota IMPLAN Group. 2003. IMPLAN System (data and software), 1725 Tower Drive West, Suite 140, Stillwater, MN 55082. Available: <www.implan.com; <https://www.implan.com/catalog/product_info.php?cPath=22&products_id=39>.
Mitchell, R.C., and R.T. Carson. 1986. Some Comments on the State of the Arts Assessment of the Contingent Valuation Method Draft Report. In *Valuing Environmental Goods: An Assessment of the Contingent Valuation Method*. Edited by R.C. Cummings, D.S. Brookshire, and W.D. Schulze. Totowa, NJ: Rowman & Allanheld.
Mitchell, R.C., and R.T. Carson. 1989. *Using Surveys to Value Public Goods: The Contingent Valuation Method*. Baltimore, Md.: Johns Hopkins University Press.
Mody, J. 1997. Industrial Demand for Water in Thailand. Ph.D. diss., Boston University, Boston, Mass.
Moncur, J.E.T. 1987. Urban Water Pricing and Drought Management. *Water Resources Research*, 23(3):393-398.

Moore, M. 1999. Estimating Irrigator's Ability to Pay for Reclamation Water. *Land Economics*, 75(4):562-578.

Moore, M., N. Gollehon, and M. Carey. 1994. Multicrop Production Decisions in Western Irrigated Agriculture: The Role of Water Price. *American Journal of Agricultural Economics*, 76:859-874.

Moore, M.R, D.H Negri, and J.A. Miranowski. 1991. Cropland Allocation Decisions: The Role of Agricultural Commodity Programs and the Reclamation Program. In *The Economics and Management of Water and Drainage in Agriculture.* Edited by A. Dinar and D. Zilberman. Dordrecht, The Netherlands: Kluwer, 577-596.

MSU (Michigan State University). 2003. Michigan Tourism Economic Impact Calculator. Available: <http://www.msu.edu/course/prr/840/econimpact/michigan/ecimpest.html>. Accessed September 12, 2003.

Narayanan, R. 1986. Evaluation of Recreational Benefits in In-stream Flows. *Journal of Leisure Research*, 18 (2):116-128.

Narayanan, R., D.T. Larson, A.B. Bishop, and P. Amirfathi. 1983. An Economic Evaluation of Benefits and Costs of Maintaining In-stream Flows. Washington, D.C.: U.S. Department of the Interior.

Neher, C.J. 1989. The Economic Value of In-stream Flows in Montana: A Travel Cost Model Approach. Master's dissertation. University of Montana, Missoula.

Nieswiadomy, M.L. 1992. Estimating Urban Residential Water Demand: Effects of Price Structure, Conservation, and Education. *Water Resources Research*, 28(3):609-615.

Nieswiadomy, M.L., and S.L. Cobb. 1993. Impact of Pricing Structure Selectivity on Urban Water Demand. *Contemporary Policy Issues*, 11:101-113.

Nieswiadomy, M.L., and D.J. Molina. 1989. Comparing Residential Water Demand Estimates Under Decreasing and Increasing Block Rates Using Household Data. *Land Economics*, 65(3):280-289.

Nieswiadomy, M.L., and D.J. Molina. 1991. A Note of Price Perception in Water Demand Models. *Land Economics*, 67(3):352-359.

Norton, V., T. Smith, and I. Strand. 1981. Stripers: The Economic Value of the Atlantic Coast Commercial and Recreational Striped Bass Fisheries. College Park, Md.: University of Maryland.

NPS (National Park Service). 2004. Wild and Scenic Rivers Act (P.L. 90-542, as amended) (16 U.S.C. 1271-1287). Available: <http://www.nps.gov/rivers/wsract.html>. Accessed December 9, 2004.

NRC (National Research Council). 1997. *Valuing Groundwater: Economic Concepts and Approaches.* National Academy Press, Washington, D.C.

NRDC (Natural Resources Defense Council). 2003. The Green Gate. Available: <http://www.nrdc.org/greengate/wildlife/endangeredf.asp>. Accessed August 14, 2003.

Nye, R. 2004. Water Issues Heat up in Texas. *Texas Agricultural Today*. Texas Department of Agriculture. Available: <http://www.ucowr.siu.edu/updates/pdf/V106_A6.pdf>. Accessed March 24, 2004.

Olsen, D., J. Richards, and R.D. Scott. 1991. Existence and Sport Values for Doubling the Size of Columbia River Basin Salmon and Steelhead Runs. *Rivers*, 2(1):44-56.

OMB (Office of Management and Budget). 1992. Guidelines and Discount Rates for Benefit-Cost Analysis of Federal Programs. Circular A-94. Available: <http://www.whitehouse.gov/omb/circulars/a094/a94_appx-c.html>. Accessed May 28, 2004.

Oscher Consulting. 1999. Value of Water Study. Prepared for Tampa Bay Water. July. Tampa, Fla.

OWT (Oregon Water Trust). 1998. Water Rights Acquisition Summary. Portland, Ore.

Paul, D. 2004. Agricultural Economics of the Pacific States: Water Supply Reliability & Pricing. *The Business Forum Journal*. Available: <http://www.bizforum.org/Journal/www_journalDP003.htm>. Accessed May 4, 2004.

Powell Consortium. 1995. Severe Sustained Drought: Managing the Colorado River System in Times of Water Shortage. Arizona.

Public Resources Advisory Group 2001. Groundwater Replenishment System Financial Study. Prepared for Orange County Water District and Orange County Sanitation District, Fountain Valley, Calif. January 26.

Raftelis. 1998. Water and Wastewater Rates Survey. Charlotte, N.C.: Raftelis Environmental Consulting Group, Inc.

Raucher, R.S., J. Agras, D. Mills, M. Harrod, L. Chestnut, D. Burmaster, D. Crawford-Brown, and T. Cox. 2002. *Quantifying Public Health Risk Reduction Benefits*, ISBN 1-58321-192-6. Denver, Colo.: AWWA Research Foundation and American Water Works Association.

RBF Consulting. 2002. San Diego County Water Authority Regional Recycled Water System Study. Irvine, Calif.: RBF Consulting.

Renshaw, E.F. 1982. Conserving Water Through Pricing. *Journal AWWA*. Denver, Colo.: American Water Works Association. January.

Renwick, M.E. 1996. An Econometric Model of Household Water Demand With Endogenous Technological Change Under Demand Side Management Policies. Ph.D. diss., Stanford University, Stanford, Calif.

Renwick, M.E., and S.O. Archibald. 1998. Demand Side Management Policies for Residential Water Use: Who Bears the Conservation Burden? *Land Economics*, 74(3):343-360.

Renwick, M.E., and R.D. Green. 2000. Do Residential Water Demand Side Management Policies Measure Up? An Analysis of Eight California Water Agencies. *Journal of Environmental Economics and Management*, 40:35-55.

Renzetti, S., and D. Dupont. 2002. The Value of Water in Manufacturing. Prepared for the Economic and Social Research Council (ESRC), Centre for Social and Economic Research on the Global Environment (CSERGE). CSERGE Working Paper ECM 03-03. Norwich, UK: CSERGE.

Rigby, R. 1997. Acquiring Water for Flow Augmentation. Special report. Boise, Idaho: US Bureau of Reclamation, Pacific Northwest Region.

RMI (Rocky Mountain Institute). 2003. Commercial, Industrial, and Institutional Water Efficiency. Available: <http://www.rmi.org/sitepages/pid276.php>. Accessed September 16, 2003.

Roger, P., R. Bhatia, and A. Huber. 1998. *Water as a Social and Economic Good: How to Put the Principle Into Practice*. Stockholm, Sweden: Global Water Partnership/Swedish International.

Sanders, L.D., R.G. Walsh, and J.B. Loomis. 1990. Toward Empirical of Total Value of Protecting Rivers. *Water Resources Research*, 26(7):1345-1357.
Sax, J.L, R.H. Abrams, and B.H. Thompson Jr. (eds.) 1991. *Legal Control of Water Resources*. Second edition. St. Paul, Minn.: West Publishing Co.
Sax, J.L., B.H. Thompson Jr., J.D. Leshy, and R.H. Abrams. 2000. *Legal Control of Water Resources: Cases and Materials*. 3rd edition. St. Paul, Minn: West Publishing Co.
Schaefer, J., and J. Liebertz. 2003. Frogs and Toads of Florida. University of Florida, Cooperative Extension Service. Available: <http://edis.ifas.ufl.edu/BODY_UW089>. Accessed October 15, 2003.
Schefter, J.E., and E.L. David. 1985. Estimating Residential Water Demand Under Multi-Part Tariffs Using Aggregate Data. *Land Economics*, 61(3):272-280.
Schneider, M.L., and E.E. Whitlatch. 1991. User-Specific Water Demand Elasticities. *Journal of Water Resources Planning and Management*, 117(1):52-73.
S E A Consultants. 2003. Town of Wilmington Draft Comprehensive Water Resource Management Plan/Environmental Impact Report. October. Cambridge, Mass.: S E A Consultants Inc.
Severn Trent. 2004. The Source. Available: <http://www.stwater.co.uk>. Accessed February 4, 2004.
SFPUC (San Francisco Public Utilities Commission). 2002. Long-Term Strategic Plan for Capital Improvements. May. San Francisco, Calif.
SFPUC (San Francisco Public Utilities Commission). 2003a. About Us: Water. Available: <http://sfwater.org/main.cfm/MSC_ID/16>. Accessed August 8, 2003.
SFPUC (San Francisco Public Utilities Commission). 2003b. Report on Water Rates for Fiscal Year 2004. May. San Francisco, Calif.
SFPUC (San Francisco Public Utilities Commission) and BAWUA (Bay Area Water Users Association). 2000. Water Supply Master Plan. San Francisco, Calif.
Shelby, B., T.C. Brown, and R. Baumgartner. 1992. Effects of Streamflows on River Trips on the Colorado River in Grand Canyon, Arizona. *Rivers*, 3(3):191-201.
Sherk, G.W. 2002. Protecting In-stream Flows: An Economic Benefits Summary. Prepared for the Upper Chattahoochee Riverkeeper.
Shiklomanov, I. 1999. Summary of the Monograph World Water Resources at the Beginning of the 21st Century. Prepared in the Framework of IHP UNESCO. Available: <http://webworld.unesco.org/water/ihp/db/shiklomanov/summary/html/summary.html>.
Shiklomanov, I. 2000. Appraisal and Assessment of World Water Resources. *IWRA*, 25(1):11-32.
Shoreland Connection. 2003. Wildlife. Available: <http://www.theshorelandconnection.org/ng_wildlife.htm>. Accessed October 15, 2003.
Solley, W.B., R.R. Pierce, and H.A. Perlman. 1998. Estimated Water Use in the United States in 1995. U.S. Geological Survey Circular 1200. Denver, Colo: U.S. Geological Survey.
Sommers, C.L. 1997. Water Rights Appraisal. Chino Valley Irrigation District. June. Denver, Colo.
Song, B., and G.A. Doeksen. 2003 Multiplier Analysis for Agriculture and Other Industries. Oklahoma Cooperative Extension Service, Division of Agricultural Sciences and Natural Resources. Available: <http://pearl.agcomm.okstate.edu/agecon/resource/f-821.pdf>. Accessed September 12, 2003.

Southwest Regional Assessment Group. 2000. Preparing for a Changing Climate: The Potential Consequences of Climate Variability and Change. Southwest. Prepared for the U.S. Global Change Research Program. September. Available: <http://www.ispe.arizona.edu/research/swassess/report.html>. Accessed December 2, 2003.

Stevens, T.H., J. Miller, and C. Willis. 1992. Effect of Price Structure on Residential Water Demand. *Water Resources Bulletin*, 28(4):681-685.

Stewart, D. 1998. Splash and Spectacle. *Smithsonian Magazine,* June.

Stratus Consulting. 2000. Preliminary Evaluation of the Beneficial Value of Waters Diverted in the Clear Creek Whitewater Park in the City of Golden. December 7. Boulder, Colo.

Stratus Consulting. 2002. The Beneficial Value of Waters Diverted in the Blue River for the Breckenridge Whitewater Park and in Gore Creek for the Vail Whitewater Park. January 30. Boulder, Colo.

Sydney Water. 2004. Sydney Water Towards Sustainability Report, 2002. Available: <http://www.sydneywater.com.au/html/environment/tsr/pdfs/tsr.pdf>. Accessed February 19, 2004.

Tampa Bay Regional Planning Council. 1999. Economic Impacts of the Proposed Seawater Desalination Facility. Pinellas Park, Fla.

Tampa Bay Regional Planning Council. 2000. Economic Impacts of the Tampa Bay Water Master Water Plan. Prepared for Tampa Bay Water. December. Pinellas Park, Fla.

Tampa Bay Water. 2001. Tampa Bay Water's Master Water Plan. K.R. Herd, Engineering Project Manager. June. Tampa, Fla.

TAMU (Texas A and M University). 2003. Assessing Community Needs. Available: <http://www.rpts.tamu.edu/>. Accessed September 12, 2003.

Taylor, R., and R. Young. 1995. Rural-to-Urban Water Transfers: Measuring Direct Foregone Benefits of Irrigation Water Under Uncertain Water Supplies. *Journal of Agricultural and Resource Economics*, 20(2):247-262.

TCPS (Texas Center for Policy Studies). 2002. Community and Economic Benefits of Texas Rivers, Springs and Bays. Austin, Texas: Texas Center for Policy Studies.

Thomas, B., and D. Rodrigo. 1996. Measuring the Benefits of Non-Traditional Water Resource Investments. In *Proceedings of Conserv 96: Responsible Water Stewardship*, January 4-8, 1996. Orlando, Fla., 195-199.

Tighe & Bond. 2002. Water Rate Survey. Boston, Mass.

UNEP (United Nations Environment Programme). 2002. *Global Environment Outlook 3: Past, Present, and Future Perspectives*. London, UK: Earthscan Publications Ltd.

UNESCO (United Nations Educational, Scientific and Cultural Organization). 2003. World Water Assessment Programme, Water for People, Water for Life, United Nations World Water Development Report. Paris, France: UNESCO.

U.S. Bureau of Reclamation. 2003. *Water 2025: Preventing Crises and Conflict in the West.* Available http://www.doi.gov/water2025. Washington, D.C.: U.S. Department of Interior.

U.S. Census Bureau. 2003. State and County Quickfacts Webpage for Arizona. Available: <http://quickfacts.census.gov/qfd/states/04000.html>. Accessed May 2003.

USDA (U.S. Department of Agriculture). 1991. Economic Value of Recreation and Preservation Benefits of In-stream Flow in Montana.

USDA (U.S. Department of Agriculture). 2002. *Agricultural Resources and Environmental Indicators: Water Use and Pricing in Agriculture*, N. Gollehon, W. Quinby, and M. Aillery, No. (AH722), April. Available at: <http://www.ers.usda.gov/publications/arei/ah722/arei2_1/DBGen.htm>. Accessed May 24, 2004.

USDA (U.S. Department of Agriculture). 2003a. Freshwater Consumption as a Percentage of Local Average Annual Precipitation, Unites States Department of Agriculture, Natural Resources Conservation Service. Available at: <http://www.nrcs.usda.gov/technical/land/meta/m2137.html>. [from chap 3]

USDA (U.S. Department of Agriculture). 2003b. Natural Resources Conservation Service, National Resources Inventory. July. Accessed September 4, 2003.

USDA/NASS (U.S. Department of Agriculture/National Agricultural Statistics Service). 1998. 1998 Farm and Ranch Irrigation Survey.

USDA-RMA (U.S. Department of Agriculture-Risk Management Agency). 2004. Crop Policies. Available: <http://www.rma.usda.gov/policies/>. Accessed May 28, 2004.

U.S. EPA (U.S. Environmental Protection Agency). 1995. Water Trivia Facts. EPA- 810-F-95-001. Washington, D.C.: U.S. EPA, Office of Water.

U.S. EPA (U.S. Environmental Protection Agency). 1996. Liquid Assets: A Summertime Perspective on the Importance of Clean Water to the Nation's Economy. Washington, D.C.: U.S. EPA, Office of Water.

U.S. EPA. (U.S. Environmental Protection Agency). 2000a. Guidelines for Preparing Economic Analyses. EPA 240-R-00-003. September. Washington, D.C.: U.S. EPA.

U.S. EPA (U.S. Environmental Protection Agency). 2000b. Liquid Assets 2000: America's Water Resources at a Turning Point. EPA-840-B-00-001. Washington, D.C.: U.S. EPA, Office of Water. May.

U.S. EPA (U.S. Environmental Protection Agency). 2003. Spokane Tribe of Indians Surface Water Quality Standards. Resolution 2003-259. March 7. Available at: <http://www.epa.gov/waterscience/standards/wqslibrary/tribes/spokane.pdf>. Accessed December 2, 2003.

USFWS (U.S. Fish and Wildlife Service). 1998. 1996 National Survey of Fishing, Hunting, and Wildlife-Associated Recreation: Arizona. Issued March 1998.

USFWS (U.S. Fish and Wildlife Service). 2002. 2001 National Survey of Fishing, Hunting, and Wildlife-Associated Recreation. October. U.S. Department of the Interior, Fish and Wildlife Service and U.S. Department of Commerce, U.S. Census Bureau.

USGS (U.S. Geological Survey). 2003a. Offstream Use Page 9, Table 2, Estimated Use of Water in the United States in 1995. Available: <http://water.usgs.gov/watuse/pdf1995/pdf/summary.pdf>. Accessed September 4, 2003.

USGS (U.S. Geological Survey). 2003b. Public Supply Use, Domestic Use, Commercial Use, Irrigation Use, Livestock Use, Industrial Use, Mining Use, Thermoelectric Use, Estimated Use of Water in the United States in 1995. Available: <http://water.usgs.gov/watuse/pdf1995/html/>. Accessed September 4, 2003.

U.S. NPS (U.S. National Park Service). 2003. Chatahoochee River National Recreation Area. Available: <http://www.sam.usace.army.mil/briefings/ACT-ACF/NPS.pdf>. Accessed September 9, 2003.

U.S. Water News Online. 2003. Arizona Facing Water Crisis With Growth. U.S. Water News Online. July 2000. Available: <http://www.uswaternews.com/archives/arcsupply/tarifac7.html>. Accessed May 2003.

U.S. Water Resources Council. 1979. Procedures for Evaluation of National Economic Development Benefits and Costs in Water Resources Planning. Level C; final rule; *Federal Register* 44, 242, 72892-72976.

U.S. Water Resources Council. 1983. *Economic and Environmental Principles and Guidelines for Water and Related Land Resources Implementation Studies*. Washington, D.C.: Government Printing Office.

UVM (University of Vermont). 2001. The Economic Impact of Tourism in Vermont: The Fall. Prepared by the Department of Community Development & Applied Economics for Vermont Department of Tourism. Available: <http://www.uvm.edu/snr/vtdc/publications/2000_Fall_Impact_Report.pdf>. Accessed September 12, 2003.

Varga, T.G. 1991. The Role of Local Projects to Enhance the Imported Water Supply. Water Supply and Water Reuse: 1991 and Beyond. AWWA, Denver, Colo. June.

Vaux, H.J., and R.E. Howitt. 1984. Managing Water Scarcity: An Evaluation of Interregional Transfers. *Water Resources Research,* 20:785-792.

Verde River Watershed Association. 2000. Confluence: Your Information Source for the Verde River Watershed. January. Available: <http://vwa.southwest-water.org/conjan00.pdf>. Accessed June 2003.

Vogt, W. 1948. *Road to Survival*. New York, N.Y.: William Sloane Associates.

VVTCB (Vail Valley Tourism and Convention Bureau). 1999. Vail Valley Summer Visitor Research. Prepared for the Vail Valley Visitors and Convention Bureau and the Town of Vail. December. Boulder, Colo.: RRC Associates.

VVTCB (Vail Valley Tourism and Convention Bureau). 2001. TEVA Whitewater Festival 2001 Stakeholder Summary. Report on festival held May 26 and 27, 2001. Vail, Colo.: VVTCB.

Wade, W.W., G.M. Mccollister, R.J. McCann, and G.M. Johns. 1988. Estimating Recreation Benefits for In-stream and Diverted Users of Waterflows of the Sacramento-San Joaquin Rivers Watershed. Presented at W-133 Committee Meeting—Outdoor Recreation and the Public Interest. January 22. Monterey, Calif.

Wahl, R.W. 1989. *Markets for Federal Water: Subsidies, Property Rights and the Bureau of Reclamation*. Washington, D.C.: Resources for the Future.

Walsh, R.G., D.M. Johnson, and J.R. McKean. 1988. Review of Outdoor Recreation Economic Demand Studies With Non-market Benefit Estimates, 1968-1988. Fort Collins, Colo.: U.S. Forest Service, Rocky Mountain Forest and Range Experiment Station.

Walsh, R.G., R. Ericson, D. Arosteguy, and M. Hansen. 1980. An Empirical Application of a Model for Estimating the Recreation Value of Instream Flow. Fort Collins, Colo.: Colorado Water Resources Research Institute, Colorado State University.

Ward, F.A. 1987. Economics of Water Allocation to In-stream Uses in a Fully Appropriated River Basin: Evidence From a New Mexico Wild River. *Water Resource Research*, 23(3):381-392.

Ward, F.A. 1989. Efficiently Managing Spatially Competing Water Uses: New Evidence From a Regional Recreation Demand Model. *Journal of Regional Science*, 29(2):229-246.

Ward, F.A., B.A. Roach, and J.E. Henderson. 1996. The Economic Value of Water in Recreation: Evidence From the California Drought. *Water Resources Research*, 32(4):1075-1081.

Water Care. 2004. Water Care: It's in Your Hands. Available: <http://www.watercare.net/>. Accessed January 3, 2004.

Webster's New World Dictionary. 1973. Edited by D. Guralnik. New York: The World Publishing Company.

Williams, M. 1985. Estimating Urban Residential Demand for Water Under Alternative Price Measures. *Journal of Urban Economics*, 18:213-225.

Williams, M., and B. Suh. 1986. The Demand for Urban Water by Customer Class. *Applied Economics* 18:1275-1289.

WIN (Water Infrastructure Network). 2000. *Clean and Safe Water for the 21st Century: A Renewed Commitment to Water and Wastewater Infrastructure*. Washington, D.C.: Water Infrastructure Network. April.

Wong, S.T. 1972. A Model on Municipal Water Demand: A Case Study on Northeastern Illinois. *Land Economics*, 48(1):34-44.

Yaron, D., and A. Dinar. 1982. Optimal Allocation of Water on a Farm During Peak Seasons. *Amer. J. Agr. Econ.,* 64:681-689.

Young, R.A. 1973. Price Elasticity of Demand for Municipal Water: A Case Study of Tucson Arizona. *Water Resources Research*, 9(4):1068-1072.

Young, R. 1996. Measuring Economic Benefits for Water Investments and Policies. Technical Paper No. 338. Washington, D.C.: World Bank.

Young, R., and S. Gray. 1972. Economic Value of Water: Concepts and Empirical Estimates. U.S. National Water Commission Technical Report No. PB210356. March. NTIS: Springfield, Mass.

Young, R.A., and R.H. Haveman. 1985. Economics of Water Resources: A Survey. In *Handbook of Natural Resources and Energy Economics.* Edited by A.V. Kneese and J.L. Sweeney. Vol. II. Amsterdam: Elsevier Science Publishers.

ABBREVIATIONS

AF	acre foot
AGMA	Arizona Groundwater Management Act
AMA	Active Management Area
AWWA	American Water Works Association
AwwaRF	Awwa Research Foundation
BAWSCA	Bay Area Water Supply and Conservation Agency
BAWUA	Bay Area Water Users Association
BT	benefits transfer
CAP	Central Arizona Project
CBO	Congressional Budget Office
$cfsm^2$	cubic feet per second per square mile
CH	critical habitat
CII	commercial, institutional, and industrial
CINI	change in net income
CIP	Capital Improvement Program
CV	contingent valuation
CWA	Clean Water Act
CWS	community water system
DEP	Department of Environmental Protection
EBMUD	East Bay Municipal Utility District
EIR	Environmental Impact Report
EPCWID	El Paso County Water Improvement District
EPWU	El Paso Water Utilities
ESA	Endangered Species Act
FERC	Federal Energy Regulatory Commission
gpcd	gallons per capita per day
gpd	gallons per day
GRI	Global Reporting Initiative
GWRS	Ground Water Replenishment System
ha	hectares
IO	input-output
IRWA	Ipswich River Watershed Association
kgal	thousand gallons

LVWD	Lower Valley Water District
MAF	million acre feet
MAPC	Metropolitan Area Planning Council
MCLs	Maximum Contaminant Levels
mg	million gallons
mgd	million gallons per day
MOU	Memorandum of Understanding
MWD	Metropolitan Water District of Southern California
MWRA	Massachusetts Water Resources Agency
NCWRP	North City Water Reclamation Plant
NGOs	nongovernmental organizations
NPDES	National Pollutant Discharge Elimination System
NPV	net present value
NRC	National Research Council
NTB-WRAP	Northern Tampa Bay Water Resources Assessment Project
O&M	operations and maintenance
OMB	Office of Management and Budget
OPEC	Organization of Petroleum Exporting Countries
ORVs	outstandingly remarkable values
POE	point-of-entry
POU	point-of-use
REUWS	Residential End Uses of Water Study
RO	reverse osmosis
RP	revealed preference
SASE	standard annual shortage event
SDWA	Safe Drinking Water Act
SFPUC	San Francisco Public Utilities Commission
SP	stated preference
SRP	Salt River Project
SWFWMD	Southwest Florida Water Management District
TBL	triple bottom line
TBSD	Tampa Bay Seawater Desalination
TBW	Tampa Bay Water
TDS	total dissolved solids
T&E	threatened and endangered
TECO	Tampa Electric Company
THMs	trihalomethanes
TSI	Trophic State Index

ULFT	ultra-low-flush toilet
USBR	U.S. Department of Interior's Bureau of Reclamation
USD	U.S. dollars
USDA	United States Department of Agriculture
USGS	U.S. Geological Survey
UV	ultraviolet
VWSR	Verde Wild & Scenic River
WIN	Water Infrastructure Network
WTA	willingness to accept
WTP	willingness to pay

IWA Publishing's authorised EU representative for General Product Safety Regulations is Diane D'Arras, 15 rue Duret, 75116 Paris, France, e-mail: safety@iwap.co.uk.

Printed and bound by CPI Group (UK) Ltd, Croydon, CR0 4YY
17/09/2026
02218804-0005